Chlorinated Solvent and DNAPL Remediation

ACS SYMPOSIUM SERIES **837**

Chlorinated Solvent and DNAPL Remediation

Innovative Strategies for Subsurface Cleanup

Susan M. Henry, Editor
URS Corporation

Scott D. Warner, Editor
Geomatrix Consultants, Inc.

American Chemical Society, Washington, DC

Library of Congress Cataloging-in-Publication Data

Chlorinated solvent and DNAPL remediation : innovative strategies for subsurface cleanup / Susan M. Henry, editor, Scott D. Warner, editor.

p. cm.—(ACS symposium series ; 837)

Includes bibliographical references and index.

ISBN 0–8412–3793–X

1. Dense nonaqueous phase liquids—Environmental aspects—Congresses. 2. Organochlorine compounds—Environmental aspects—Congresses. 3. Solvents—Environmental aspects—Congresses. 4. Hazardous waste site remediation—Technological innovations—Congresses.

I. Henry, Susan M., 1954- II. Warner. Scott D., 1961- III. Series.

TD1066.D45 C45 2002
28.5—dc21 2002027741

The paper used in this publication meets the minimum requirements of American National Standard for Information Sciences—Permanence of Paper for Printed Library Materials, ANSI Z39.48–1984.

Distributed by Oxford University Press

PRINTED IN THE UNITED STATES OF AMERICA

Foreword

The ACS Symposium Series was first published in 1974 to provide a mechanism for publishing symposia quickly in book form. The purpose of the series is to publish timely, comprehensive books developed from ACS sponsored symposia based on current scientific research. Occasionally, books are developed from symposia sponsored by other organizations when the topic is of keen interest to the chemistry audience.

Before agreeing to publish a book, the proposed table of contents is reviewed for appropriate and comprehensive coverage and for interest to the audience. Some papers may be excluded to better focus the book; others may be added to provide comprehensiveness. When appropriate, overview or introductory chapters are added. Drafts of chapters are peer-reviewed prior to final acceptance or rejection, and manuscripts are prepared in camera-ready format.

As a rule, only original research papers and original review papers are included in the volumes. Verbatim reproductions of previously published papers are not accepted.

ACS Books Department

Contents

Preface

Chlorinated solvent contamination resulting from the release of chemicals—such as perchloroethylene (PCE), trichloroethylene (TCE), 1,1,1-trichloroethane (1,1,1-TCA), carbon tetrachloride (CTET), and methylene chloride (DCM)—represents a continuing challenge to subsurface remediation. Sites contaminated with chlorinated solvents are particularly difficult to remediate when undissolved separate-phase chlorinated solvents, commonly referred to as DNAPLs (dense non-aqueous phase liquids) are present in the soil and groundwater. Recent technological developments resulting from innovations in the public, academic, and private sectors are providing new tools for cleaning up DNAPL and dissolved–phase chlorinated solvents. Many of these technologies are being pilot tested as a part of cooperative field efforts supported by federally funded projects, regulatory work groups, and industry-based technology development forums; and several have been brought to market in full-scale systems.

This book provides a focus on some of the current technological developments and innovative applications for physical, chemical, and biological remediation of chlorinated solvents present as DNAPL and the dissolved phase in groundwater. These innovations include surfactant flushing to enhance DNAPL removal; in situ chemical destruction by reduction processes involving zero valent iron or related metals; in situ chemical destruction by advanced oxidation processes; and in situ biological destruction by enhanced anearobic bioremediation or natural bioattenuation. Many of the technologies are applicable to other contaminants, and Chapter 12 presents research findings on arsenate and chromate removal by zero valent iron—wholly relevant to chlorinated solvent remediation, because these metals are frequently present as cocontaminants with chlorinated solvents. as well, An emphasis is placed on zero valent iron-based strategies as well in this book, including reaction geochemistry, permeable reaction barrier longevity, rejuvenation of iron walls, and emplacement techniques.

The topics discussed in this book are not without controversy. Some experts in the field argue that sites impacted with chlorinated solvent DNAPLs cannot be completely remediated and have promulgated site management strategies based on *technical impracticality*. Other experts, including those who have authored chapters in this book, believe that, with ongoing innovation and application, remediation solutions are and will become available for effective remediation of sites impacted with DNAPLs. Other controversies revolve around which remediation approaches and site management strategies are *best,* and readers will recognize that the authors of this book have different opinions. Controversy is a good thing in science and engineering: it spurs inquiry and technological advancement.

This book is based on a two-day, 29-speaker symposium titled *Innovative Strategies for Remediation of Chlorinated Solvents and DNAPLS in the Subsurface* presented at the ACS 221st National Meeting held in San Diego, California in April 2001. As such, it is dedicated to the memory of environmental chemist and educator Dr. Charles R. (Chuck) Bennett. During the spring of 2000, Dr. Bennett approached Dr. Susan M. Henry with the idea of co-organizing an ACS Division of Environmental Chemistry symposium, which lead to the symposium on which this book is based. However, Dr. Bennett died of a sudden heart attack on November 12, 2000. Dr. Bennett had been an active member of the ACS Orange County Section (OCS), having held such positions as Chair-Elect, Chair, Past-Chair, Councilor, OCS Representative on the Coordinating Committee of California ACS Sections, and a member of the ACS National Committee on Environmental Improvement. He was Chair of the OCS ACS Public Affairs Committee and the OCS Environmental Committee. He helped establish the Environmental and Industrial Technology (EIT) Committee as well as the Methyl *t*-Butyl Ether (M*t*BE) OCS ad hoc Committee. Dr. Bennett also was an active member of the Executive Committee of the Division of Environmental Chemistry.

We thank the many individuals who have made this book a success, including the creative and hard-working authors; the 27 different individuals who provided peer review of one or more manuscripts; and the ACS Books Department staff, including Stacy VanDerWall, Kelly Dennis, and Margaret Brown. We also thank Ruth A. Hathaway of the ACS Division of Environmental Chemistry who provided assistance with the symposium abstracts and the biography of Dr. Bennett; and the ACS Corporation Associates, who provided a grant for the symposium.

Susan M. Henry
URS Corporation
2020 East First Street, Suite 400
Santa Ana, CA 92705
susan_henry@urscorp.com

Scott D. Warner
Geomatrix Consultants
2101 Webster Street, 12th Floor
Oakland, CA 94612
swarner@geomatrix.com

Chapter 1

Chlorinated Solvent and DNAPL Remediation: An Overview of Physical, Chemical, and Biological Processes

Susan M. Henry[1], Calvin H. Hardcastle[2], and Scott D. Warner[3]

[1]URS Corporation, 2020 East First Street, Suite 400, Santa Ana, CA 92705 (telephone: 714–648–2887; susan_henry@urscorp.com)
[2]Geomatrix Consultants, 330 West Bay Street, Suite 140, Costa Mesa, CA 92627 (telephone: 949–642–0245; chardcastle@geomatrix.com)
[3]Geomatrix Consultants, 2101 Webster Street, 12th Floor, Oakland, CA 94612 (telephone: 510–663–4100; swarner@geomatrix.com)

Remediation of groundwater contaminated by chlorinated solvents is one of the more complex technical challenges faced by environmental engineers and scientists. Early efforts to remediate chlorinated solvent contaminated groundwater involved extraction and treatment of groundwater with the objective of restoring aquifers to pristine conditions. However as these operations continued, it became apparent that restoration of groundwater would not occur within a reasonable period of time. Groundwater scientists attributed the difficulty remediating to the complexity of the aquifer systems and the presence of DNAPLs, the separate phase liquid form of the chlorinated solvents that act as a continuing source of contamination to groundwater. Several leading practitioners have proposed that subsurface soil and groundwater systems contaminated by DNAPLs cannot be completely remediated. Other environmental practitioners have accepted this as an opportunity to advance the development of existing and new, more innovative remediation technologies and solutions. This paper presents an overview of in-situ physical, chemical, and biological processes being used and developed for remediating sites contaminated by chlorinated solvents and DNAPLs. Physical processes involve use and manipulation of the physical properties of chlorinated solvents and DNAPLs to effect their removal from contaminated soil and groundwater. Chemical processes involve the destruction of chlorinated solvents using reductive or oxidative chemical reactions. Biological processes involve the destruction of chlorinated solvents using natural or enhanced biological means. Physical, chemical and biological processes can be and have been used for the in-situ treatment of chlorinated solvents through source zone treatment and reactive treatment zones.

Introduction

The purpose of this chapter is to provide an overview of technologies that are being used or are being developed for addressing chlorinated solvent contamination in subsurface soil and groundwater, and to provide the reader with a guide to some of the additional resources on this topic. As noted in the preface, chlorinated solvents are not the only contaminants discussed in this book. Some technologies that were initially developed for chlorinated solvents are also applicable to other contaminants. For example, zero valent iron permeable reactive barriers (PRBs), originally developed for remediation of dissolved-phase chlorinated solvents, are also effective for heavy metals such as arsenate and chromate, which frequently are co-contaminants with chlorinated solvents. Further, some of the remediation technologies discussed in this chapter and book as applicable to chlorinated solvent dense nonaqueous phase liquids (DNAPLs), are also applicable to other DNAPL compounds.

The general category "chlorinated solvents" refers in a broad sense to any organic compound that contains chlorine substitutions and that has properties of, or was developed for use as, a solvent. In the strict sense of the term, "chlorinated solvents" include both chlorinated aromatic compounds and chlorinated aliphatic compounds, but as a result of common usage, the term "chlorinated solvents" has evolved to represent the chlorinated aliphatic compounds. A comprehensive review of the types of chlorinated solvents, their history of manufacture and use, and their properties is provided in the first chapter of the book *Dense Chlorinated Solvents and other DNAPLS in Groundwater* (1). In summary, the chlorinated solvents that are most commonly detected in groundwater include those that are (or were) most commonly manufactured and used, as well as those that result from the naturally-occurring biological or chemical transformation of "parent compounds." These include tetrachloroethylene (perchloroethylene, PCE), trichloroethylene (TCE), cis-1,2-dichloroethylene (cis-1,2-DCE), trans-1,2-dichloroethylene (trans-1,2-DCE) vinyl chloride (VC), 1,1,1-trichloroethane (1,1,1-TCA), 1,1-dichlorethylene (1,1-DCE), 1,2-dichloroethane (1,2-DCA), carbon tetrachloride (CTET), chloroform (trichloromethane, TCM), and methylene chloride (dichloromethane, DCM)

Chlorinated solvents can be present in the subsurface in several phases: dissolved in groundwater or soil moisture ("dissolved phase"), in soil vapor ("vapor phase"), sorbed to geologic materials in the saturated or unsaturated zone ("sorbed phase"), or as undissolved solvent present as separate phase liquid ("DNAPL"). The presence of chlorinated solvents in the subsurface as DNAPL represents a challenge to detection and remediation (2). Unlike less-dense organic compounds such as petroleum hydrocarbons, chlorinated solvents have a density greater than water. The density and other physical properties of DNAPLs, such as low "wettability" and low solubility, can result in their downward movement through the saturated zone, and their distribution as small globules and ganglia within the geologic matrix and as pools lying on low permeability lenses (1, 3). DNAPLs released to the subsurface are like a well-planned trust fund: slowly paying out an annuity of solvent to the dissolved phase for future generations. It has been estimated that decades or even centuries may be required for natural groundwater processes to deplete DNAPL (1, 4). To successfully remediate sites impacted with DNAPL, the DNAPL itself must be removed.

The understanding, approach, and expectations for remediating sites impacted by chlorinated solvents, particularly those with DNAPL source areas, have evolved greatly over the past two decades. Several leaders in the field have proposed that subsurface systems affected by DNAPLs cannot be completely remediated (1, 4). The concept that remediation of sites affected by DNAPL could be technically impractical was promulgated in guidance produced by the U.S. Environmental Protection Agency (U.S. EPA) in the early 1990s (5). The concept was based both on the technical and economic, or beneficial, viability of restoration as interpreted from the National Contingency Plan (NCP) which stated that. "EPA expects to return usable ground waters to their beneficial uses wherever practicable, within a time frame that is reasonable given the particular circumstance of the site." (6).

Despite the less-than-favorable prognosis, many environmental practitioners maintain that sites impacted with DNAPLs can in fact be remediated (7), and many developed and innovative technologies are currently available for application to DNAPL sites (8). Developed technologies are those that have been used for a long period of time and for which sufficient cost and performance data have been accumulated to support decisions for selecting, designing, and applying the remedial method. Innovative technologies, on the other hand, are those technologies that have been used at fewer sites and have costs and limitations that are not as well understood. Significant progress on testing and assessment of innovative remediation technologies has been made in recent years. During the past decade this progress has been due to the combined efforts of academic, government, industry, and regulatory entities that have teamed to provide the funding, testing, and research on design and use of new remedial methods. Collaborative organizations including the Remediation Technology Development Forum (RTDF) and Interstate Technology Regulatory Council (ITRC), as well as academic-industry consortiums such as the University of Waterloo's (Ontario, Canada) "solvents in groundwater" consortium have championed the development and application of innovative technologies for remediation of chlorinated solvents and DNAPLs.

Remediation Technologies Summary

The concept of remediating contaminated groundwater supplies dates back to the *Laws of Man* proposed by Plato [c.427-347 BC] (9). Plato maintained that if groundwater is contaminated, it needs to be restored. Despite Plato's historic insights, remediation of groundwater to remove chlorinated solvent contamination did not come into practice in the United States until the 1980s. Groundwater contaminated by chlorinated solvents was relatively unknown prior to about 1970 because the analytical methods used to detect these chemicals were not widely used or sufficiently developed (1). Initial remedial efforts focused on isolation and secure storage of the hazardous wastes. By 1980 with the passage of the Comprehensive Environmental Response, Compensation, and Liability Act (CERCLA), remediation began to focus on the detoxification and removal of hazardous wastes from the environment (10).

Early efforts to remediate groundwater typically involved groundwater extraction and treatment (pump-and-treat) to capture groundwater contaminated by dissolved-phase contaminants with the objective of restoring groundwater quality to pristine conditions. However, groundwater scientists soon realized that restoration of aquifers, particularly those impacted with DNAPLs, would require pumping more groundwater than initially thought necessary and that complete restoration may not be possible (11, 12, 13). The difficulty of remediating groundwater impacted by chlorinated solvents is attributable to several causes including (a) the complexity and heterogeneity of groundwater systems; (b) the varied phases in which the chlorinated solvent may occur; (c) the behavior of each phase within the subsurface system; and (d) the ability of DNAPL to remain undetected (11).

However, as experience has been accumulated, it has lead to the realization that pump-and-treat is not the "magic-bullet" solution. This has opened an opportunity for creativity, leading to the development of innovative remediation technologies and solutions. The approaches and technologies described in this overview and in the following chapters of this book are the product of that creativity. In some cases, the approaches have been developed from new applications of old technologies. In other cases, the approaches result from the development of new technologies.

The following sections of this chapter present a brief overview of the major classes of developed and innovative in-situ technologies for remediation of chlorinated solvents and DNAPLS in the subsurface. The focus of this chapter is on in-situ remediation methods. Ex-situ treatment methods for excavated soil or extracted groundwater are not discussed. Physical processes involve use and manipulation of the physical properties of chlorinated solvents to effect their transfer from soil and groundwater to other media where they can be treated or destroyed. Chemical processes involve the destruction of chlorinated solvents using reduction or oxidation chemical reactions. Biological processes involve the natural attenuation and biodegradation of chlorinated solvents. Physical, chemical and biological processes can be and have been used for the in-situ treatment of chlorinated solvents through source zone treatment and reactive treatment zones.

Physical Processes

Technologies involving the use of physical processes for the removal of chlorinated solvents (and for that matter, all chemical contaminants) from the subsurface are applied in consideration of the physical properties of the contaminant. Physical properties that can often be manipulated and exploited for chlorinated solvent removal are volatility, solubility, and mobility. Some chlorinated solvent compounds are relatively volatile (e.g., TCE and PCE) and can be stripped from soil and groundwater using air movement. In cases when the contaminant is not very volatile (e.g., wood treatment compounds), the volatility of the contaminant can be increased by heating the contaminated media to increase the vapor pressure. Likewise, physical properties can be manipulated using surface active agents (i.e., surfactants and solvents) to lower the viscosity and interfacial tension of the contaminant, making the contaminant more mobile or soluble and thereby allowing it to be flushed (14).

Soil vapor extraction (SVE) is perhaps the most widely accepted in-situ technology for chlorinated solvent removal in the unsaturated or vadose zone. SVE is applicable for removing chlorinated solvents that are present in the vadose zone in the vapor, sorbed, and dissolved phases, as well as DNAPL. SVE consists of applying a vacuum to induce air flow through the open interstitial pores in soil, causing the chlorinated solvent to volatilize and partition into the vapor phase. The soil vapor is then extracted under vacuum and typically treated to remove chlorinated solvents prior to atmospheric discharge. SVE is widely used for remediation of chlorinated solvents and other volatile organic compounds (VOCs), and has been designated by the U.S. EPA as a "presumptive remedy" for soils contaminated by VOCs (15).

Variants of SVE technology have been developed to access or expose chlorinated solvents in the saturated zone in order to facilitate removal. Multiphase extraction (MPE) is one such enhancement to SVE whereby a high vacuum (typically 18 to 26 inches of mercury) is applied to the subsurface soils and groundwater to simultaneously extract soil vapor, contaminants dissolved in groundwater, and separate phase liquid (16, 17). MPE, with its application roots traceable back to the 1940s when vacuum was applied to oil wells to degas the wells and to enhance oil recovery (18), describes a general class of in-situ remediation technologies that includes dual phase extraction, two-phase extraction, and bioslurping (16). MPE can also be used to depress the water table or dewater a targeted area to expose chlorinated solvents to removal using SVE. In some instances, MPE can be used to directly remove DNAPLs when such a source can be detected. Application of MPE at chlorinated solvent-impacted sites is common and MPE has also been designated by USEPA as a presumptive remedy for soils and groundwater impacted by chlorinated solvents and other VOCs (17). MPE is best used at sites with low-yielding aquifers, typically less than 5 gallons per minute (16), implying its applicability at sites with finer-grained soils. However, because finer-grained soils are also more difficult to dewater, a well-designed monitoring system should be installed to assess whether the target area is actually being dewatered. Recent surveys of sites using MPE have indicated that the degree of dewatering may actually be less than anticipated or reported (19).

Injection of air beneath the water table, known as in-situ air sparging, can also be used to treat groundwater affected by dissolved-phase chlorinated solvents. Air sparging is used to aerate the groundwater to volatilize chlorinated solvents and other VOCs from the groundwater and transfer the volatilized compounds to the vadose zone. SVE is then used to remove the VOCs from the vadose zone. In some cases, aeration of groundwater also promotes in-situ aerobic biodegradation (20). Use of air sparging at sites where DNAPLs are suspected to be present must be applied cautiously. Where DNAPLs are present, air sparging may exacerbate the extent of contamination by mobilizing DNAPL and causing lateral or vertical migration of DNAPL.

Thermal treatment to enhance removal of dissolved, sorbed, and DNAPL chlorinated solvents and other less-volatile contaminants in the vadose and saturated zones has been used since the mid-1980s (14, 21, 22). As temperature increases, the vapor pressure and volatility of the contaminant also increase, with vapor pressure approximately doubling for every ten degree increase in temperature. Thus relatively non-volatile compounds such as pentachlorophenol, creosote and other polynuclear aromatic hydrocarbons can be treated using SVE when thermal enhancement is used. Forms of thermal treatment include steam stripping, three-phase and six-phase electrical resistive heating, and microwave heating. Limitations of thermal treatment enhancements include high costs associated with steam production and energy required for the heating equipment.

The Southern California Edison Poleyard Superfund Site located in Visalia, California is a demonstration project site for steam stripping and the Dynamic Underground Stripping (DUS) process, which promotes thermally-enhanced removal of DNAPLs. At this site, the DUS process was used to remove creosote DNAPL that was present below the groundwater table. Previous pump-and-treat operations resulted in recovery of less than one gallon per day of creosote. Multi-level steam injection wells were installed at the site in May 1997. After startup of the steam injection system, creosote recovery rates initially increased to more than 1200 pounds per day and then decreased to about 50 pounds per day during 1999. Approximately 50 percent of the creosote removed since 1997 has been recovered as free phase product, while the remainder has been recovered in soil vapor, the aqueous phase, and/or as a thermal combustion product (14). Steam stripping was an economical alternative at this site due to the availability of an on-site source of steam.

Electrical resistance heating (ERH) is a variant of the steam stripping process in which electrical energy is used to thermally heat soil and groundwater in-situ (14). ERH is implemented using three-phase or six-phase electricity (23, 24). Electrodes are installed in a grid pattern which are "out-of-phase" and cause electricity to be conducted throughout the in-situ soil and groundwater. Finer-grained soils, which are more difficult to remediate and where DNAPLs are often trapped, typically provide less electrical resistance and thus are preferentially heated (14). The soil and groundwater are heated to near water-boiling temperatures, which produces steam. Contaminants within the finer-grained soil are then steam stripped from the soil and groundwater. SVE and MPE treatment systems are used to capture the steam and treat vapor and groundwater prior to discharge. Alternative means of heating such as radio frequency heating can also be used to obtain similar results.

As previously mentioned, one of the challenges of remediating chlorinated solvent impacted sites is removal of DNAPL that has become entrained in the aquifer matrix. In-situ flushing of DNAPLs to address this specific challenge has been a focus of research and pilot test efforts for the past ten years. In-situ flushing involves injecting an agent into the subsurface and flushing that agent through the soil and groundwater zones containing DNAPL. The injected agent acts to manipulate the physical properties of the chlorinated solvent to increase the solubility and/or mobility of the DNAPL as the contaminated zone is flushed.

Two types of in-situ flushing approaches are used. One type of flush used in the saturated zone is known as a solubilizing surfactant flood, which is designed to increase the solubility of the chlorinated solvent compound in water. Flushing agents such as surfactants and co-solvents (typically alcohols) are used to increase the solubility of the chlorinated solvent in water and thus increase the mass flux from the DNAPL source to water, which is then extracted and treated. Another type of flush is the mobilizing surfactant flood. The goal of this flushing approach is to increase the mobility of the DNAPL. The flushing agent is injected into the vadose or saturated zones and decreases the interfacial surface tension of the DNAPL until it is mobilized. DNAPL is then displaced by continuous flooding of the subsurface (14). In-situ flushing is described in more detail in Chapters 2, 4, and 10.

Pump-and-treat is recognized as the most widely used remediation technology at chlorinated solvent-contaminated sites and it is the benchmark against which emerging technologies are compared (25). Applications of pump-and-treat are relatively simple: one or more groundwater wells are constructed to extract contaminated groundwater, which is then treated and discharged. Early pump-and-treat efforts focused on capturing the entire volume of the contaminant plume. However, it soon became evident that much more groundwater was being pumped than initially thought necessary to remediate the plume, and in many cases the residual chlorinated solvent concentrations still exceeded remediation objectives (26, 27).

Subsequently, the philosophy of groundwater remediation programs has shifted from "total capture" to containment strategies. These so-called containment strategies involve isolating the source areas using physical structures or hydraulic barriers to control movement of the dissolved-phase plume (in lieu of capture) (1, 26).

Physical structures include slurry barriers, vibrating beam barriers, deep soil mixing, jet grout walls, sheet piling, and geomembrane liners (28). More recently, physical structures such as PRBs have been used to control groundwater movement and to facilitate in-situ treatment. A recent study by the U.S. EPA reported that physical barriers can be used effectively to contain DNAPL source zones. Effective remediation using physical barriers is contingent on minimizing leaks through the barrier, particularly at the "key" area where the barrier is installed into an underlying competent layer such as bedrock or clay (29). Innovations in construction methods are improving the economics of constructing physical barriers and the performance of barriers, and are allowing barriers to be constructed to greater depths.

Hydraulic barriers include using groundwater extraction, injection, or recirculating well systems to contain the extent of dissolved-phase plumes. Extracted groundwater is typically treated prior to surface discharge. In-well treatment systems are also used for in-situ treatment of chlorinated solvents and hydraulic containment of contaminant groundwater plumes. These applications typically involve the use of groundwater circulating wells (GCWs) that recirculate groundwater throughout the portion of the saturated zone that is under the hydraulic control of the GCW (30, 31). The recirculating groundwater pumps, when placed in relative proximity to one another, can establish a "hydraulic barrier" to prevent contaminated groundwater from flowing through the treatment zone. If the treatment system has been designed to strip volatile compounds, groundwater is drawn into a well where the groundwater is aerated to strip volatile constituents. The vapor stream is then recovered under vacuum for aboveground treatment; and the treated water is released back to the formation. In other applications, the treatment system focuses on inducing chemical treatment in the well and/or in the formation. GCWs are often combined with the use of chemical oxidants such as ozone (see Chapter 6) that are circulated throughout the saturated zone and chemically destroy chlorinated solvents in-situ. GCWs may also be modified to circulate groundwater through treatment media such as zero valent iron.

Biophysical removal using wetlands or other vegetation is also being investigated and developed for remediation of shallow groundwater systems. It has been demonstrated that vegetation can promote chlorinated solvent removal by three mechanisms: (a) enhancement of biodegradation by providing nutrients or other elements required by the biodegradation process; (b) destruction of the contaminants; and (c) mass transport to the plant through the root systems and adsorption into plant cells and/or volatilization to the atmosphere (32, 33, 34, 35). Mesocosm studies have been used to demonstrate that uptake of TCE by bald cypress [*Taxodium distichum* (L.) Rich] is related to water use and that TCE can be transported from contaminated groundwater in the plant's root zone to the atmosphere (36). The use of phytoremediation is subject to certain limitations. These include the depths at which a plant's root system can access or influence through water uptake, and other considerations such as the potential toxicity to the root system of high concentrations of chlorinated solvents.

Chemical Processes

The category of "chemical processes" for in-situ remediation of chlorinated solvents and other contaminants in the subsurface includes several technologies that can be applied directly to destroy or, in the case of metals, beneficially modify characteristics of the targeted contaminant. In some cases chemical treatment also may promote the indirect destruction of contaminants through the manipulation of the subsurface geochemical conditions. Although chemical treatment technologies are generally considered innovative, a greater number of both pilot and full-scale applications have occurred in recent years and as a result, the technical aspects and costs for use of this technology are better understood.

Several types of chemical treatment technologies have been developed and applied to the subsurface (see 37, 38, 39, 40, and Chapters 3, 5 and 6). Widely recognized chemical treatment approaches include: (a) direct chemical oxidation, (b) direct chemical reduction, (c) secondary reduction or oxidation, and (d) metal-enhanced dechlorination. Direct chemical oxidation involves the use of one of the following chemical oxidants: hydrogen peroxide (H_2O_2) with a catalyst such as ferrous iron (Fenton's Reagent); sodium (Na) and potassium (K) permanganate (MnO_4^-); perchlorate (ClO_4^-); and ozone (O_3). Direction chemical reduction involves reducing agents such as sodium dithionite ($Na_2S_2O_4$). Secondary reduction or oxidation reactions may occur when reactants influence the oxidation-reduction potential of the aqueous system. Metal-enhanced dechlorination involves the use of reduced, or low valence colloidal or granular iron such as zero valent iron.

Chemically-promoted destruction of chlorinated solvents is the most direct and thus potentially the quickest method to remediate chlorinated solvents in the subsurface environment. However, inherent to any approach using a direct chemical process are the challenges of delivering the chemical reactant to the contaminant, and avoiding competitive reactions that can either lessen the effectiveness of the chemical reagent or render the application completely useless. An additional challenge is avoiding the physical byproducts of both the potentially high-energy delivery methods and/or subsurface reactions that could result in pressure or vapor fronts capable of spreading the contamination. Hydraulically passive systems have been developed that manipulate the subsurface geochemical environment along the migration path of the contaminant with conditions that render the contaminant unstable or removes it from solution (for example, reactive treatment zones such as zero valent iron PRBs discussed in several chapters of this book).

Chemical treatment technologies have been demonstrated to be effective for a wide variety of both organic and inorganic constituents. However, significant limitations on treatment efficacy exist depending on the specific contaminant and the specific chemical reactant to be applied. For example, the chlorinated ethene compounds (including PCE and TCE) are effectively destroyed by chemical oxidation; however, the chlorinated ethane compounds (including TCA) are more resistant to oxidative treatments. Metal-enhanced reductive dechlorination can successfully destroy both chlorinated ethenes and ethanes; however, lower chlorinated methanes (such as methylene chloride) are not readily treated by the corrosion of reduced granular iron in an aqueous system (38).

The application of chemical treatment methods can occur through various methods, including injection of a liquid or colloidal reactant through temporary well points or permanent wells (vertical or horizontal), or the placement of a solid treatment matrix in the subsurface. In all cases, successful delivery and application rely on developing a representative conceptual model of the subsurface through the collection of a comprehensive site characterization data set. Whether or not the chemical treatment approach can chemically destroy a given contaminant often is less important than whether the subsurface conditions will allow for effective application. For example, aqueous systems that consist predominantly of fine-grained sediments (clay, silt, fine sand) are less amenable to remediation by chemical oxidation because of the difficulty in delivering the reactant to a significant volume of affected aquifer. Also, the exothermic and carbon dioxide-producing reactions that can accompany chemical oxidation could negatively force contaminants into tight pore spaces or into otherwise non-contaminated areas.

The concept of the PRB as a mechanism to introduce a chemical reductant has been proven at many sites since the mid 1990s (see Chapter 3). The key to success of the PRB is the ability to move affected groundwater passively through the PRB. This too relies on a sound understanding of the hydrogeologic system as based on comprehensive site characterization data. This technology, unlike the use of direct chemical oxidation, is not suitable for direct treatment of a source area, but may be used to control a source from further affecting downgradient areas.

Biological Processes

After decades of research, the mechanisms of biodegradation of chlorinated solvents are becoming well understood, and biodegradation of chlorinated solvents serves as the basis for several effective remediation technologies (see the bioremediation projects described in 41, 42, 43). Depending on the compound, chlorinated solvents may biodegrade by aerobic and/or anaerobic processes, which has led to the development and application of both aerobic, anaerobic, and sequential anaerobic/aerobic bioremediation applications.

Most bioremediation applications rely on the addition of organic carbon to stimulate biodegradation of chlorinated solvents. Biodegradation of chlorinated solvents is typically a cometabolic process – that is, the bacteria fortuitously transform the chlorinated solvent while growing on another, more palatable source of organic carbon. To date, growth of bacteria on PCE, TCE, the DCE isomers, 1,1,1-TCA, or CTET as a sole source of carbon has not been demonstrated, and field observations support this: at most sites with low concentrations of organic carbon, these chlorinated solvents persist. However, exceptions to this do exist for less-chlorinated compounds such as vinyl chloride, chloroethane, and DCM (41). For example, oxidation of vinyl chloride under low organic carbon conditions has been observed, and recent research has demonstrated that vinyl chloride can serve as a growth substrate for some bacteria (44).

Aerobic bioremediation is applicable to many chlorinated solvents, but not fully-substituted ones such as PCE (C_2Cl_4) and CTET (CCl_4). Numerous aerobic microorganisms, including but not limited to methanotrophs, propane-oxidizers, ammonia-oxidizers, and toluene oxidizers, have been shown to cometabolize certain chlorinated solvents (45). In many cases, inducers such as methane, propane, ammonia, or toluene are needed to stimulate certain bacterial enzymes, monooxygenases and dioxygenases that oxidize chlorinated solvents. Unless naturally present in the groundwater, inducers must be added. In many cases the inducers also provide a source of organic carbon. The addition of a carbon source and an inducer may not be required if the chlorinated solvent plume is commingled with petroleum hydrocarbons, which provide organic carbon and inducers such as toluene. There is, of course, regulatory resistance to the injection into groundwater of compounds such as toluene, which are themselves environmental contaminants. Constitutive mutants of certain toluene oxidizers that express the oxygenase enzyme in absence of an inducer have been developed and are being evaluated for field application (46, 47).

For aerobic cometabolic biodegradation, sufficient dissolved oxygen must be present, both as electron acceptor and to provide molecular oxygen for the oxygenase enzyme to function. Oxygen can be supplied by number of different methods, such as air sparging into the formation; addition of air, oxygen, hydrogen peroxide, or other oxygen sources into recirculated groundwater in an extraction/reinjection system; and injection of commercially-available oxygen-supplying controlled release compounds such as phosphate intercalated magnesium peroxide into the formation. When air sparge/SVE is the primary remedial approach, methane may be added to the sparging system to provide a biological "boost" to the remedial process.

Bioaugmentation for aerobic bioremediation of chlorinated solvents has been evaluated in laboratory and field studies with varying degrees of success (see 41 for summary and references). Toluene oxidizers, methanotrophs, and propane oxidizers have been tested at TCE-impacted sites, and *Pseudomonas stutzeri strain* KC has been tested at a CTET-impacted site. The effectiveness of bioaugmentation has been limited by several factors including (a) the effectiveness with which added bacteria can be transported throughout the target zone without occlusion of porosity, (b) the challenge of maintaining an established population of the added bacteria in the face of competition from native bacteria and other environmental pressures; (c) the costs associated with ongoing bioaugmentation needed to maintain the populations; and (d) regulatory objection to the use of genetically modified bacteria.

Several limitations that are inherent at the biochemical level present a challenge to aerobic bioremediation of chlorinated solvents. Because the process is generally cometabolic, one of the challenges of aerobic bioremediation is competitive inhibition, which results in reduced efficiency. Competitive inhibition is caused by the fact that the growth substrate, the chlorinated solvents, and in some cases the transformation intermediates, compete for the same sites on the oxygenase enzyme (45, 48). Groundwater bacteria may express different oxygenases, some with much slower reaction kinetics, depending on mineral nutrient availability. Because groundwater chemistry controls mineral nutrient availability, transformation rates may be affected by the groundwater geochemistry (45, 49). In the case of chlorinated ethylenes, toxicity of the transformation intermediates reduces the active cell biomass and may reduce the efficiency of aerobic bioremediation (45, 50, 51, 52). The kinetics of aerobic bioremediation may not be as favorable as anaerobic bioremediation for more substituted chlorinated solvents such as TCE; and, as previously discussed, aerobic bioremediation is not effective for PCE. In some applications, anaerobic bioremediation to reductively dechlorinate PCE and TCE may be followed in sequence by aerobic bioremediation that oxidizes the cis- and trans-1,2-DCE and vinyl chloride produced during the anaerobic process (53).

Enhanced anaerobic bioremediation is the most widely applied biological approach for PCE and TCE impacted sites. The basis for anaerobic bioremediation is reductive dechlorination, a bacterially-mediated electron transfer process that results in the sequential removal of chlorine atoms and replacement with hydrogen atoms Complete reductive dechlorination results in production of non-toxic end-products. For example, complete reduction of TCE and PCE results in ethylene and ethane (54, 55).

Since anaerobic biodegradation of chlorinated solvents is a cometabolic process, organic carbon sources are required to stimulate the desired remediation. Addition of organic carbon provides a “food” source for bacteria, and serves as an indirect source of electrons. Hence, organic carbon sources are frequently referred to as “electron donors”. Molecular hydrogen (H_2), the “carrier” of electrons in electron transfer, is currently being evaluated as a direct electron donor (56). Two general approaches have been developed and are being implemented to apply electron donors to the subsurface: (a) injecting soluble organic carbon sources (lactate, acetate, alcohols, sucrose, molasses, potato starch, cheese whey, etc.) into groundwater using an array of groundwater injection and extraction wells; and (b) placing into the formation slow-release electron donors (vegetable oil, soybean oil emulsions, commercially available polylactate esters, etc.) that persist for periods of several months to years and slowly release organic carbon into the groundwater.

An important recent development for anaerobic bioremediation is the discovery of certain bacteria, dehalorespirers, that are capable of direct dehalogenation of chlorinated solvents such as PCE, TCE, DCE, and vinyl chloride, and completely reduce these compounds to ethylene (57, 58, 59). These bacteria are not ubiquitously distributed in nature and have not been found at every chlorinated solvent impacted site (59). Enhanced anaerobic bioremediation can be augmented with the injection of dehalorespiring bacteria, and commercial applications are available (60, 61, 62). One of the challenges of anaerobic bioremediation is that at some sites where PCE and TCE are the target compounds, reductive dechlorination stalls at cis-1,2-DCE. A pilot test conducted at Dover Air Force Base demonstrated that augmentation with dehalorespirers resulted in complete transformation to ethylene (61).

The effectiveness and general applicability of biodegradation for DNAPL source areas has been questioned, in that chlorinated solvents present as separate phase liquid are not readily bioavailable. Groundwater bacteria can only carry out their metabolic activities in the aqueous phase and cannot grow in DNAPL. Therefore, bacteria can only biodegrade the dissolved-phase chlorinated solvents. Because the chlorinated solvent DNAPL itself is not bioavailable, the rate of biodegradation is controlled by dissolution of the chlorinated solvent into the aqueous phase. Further, it was long thought that chlorinated solvent concentrations near DNAPL were high enough to be toxic, and thus inhibitory, to the growth of bacteria capable of biodegrading the chlorinated solvents. However, dehalorespirers have been shown to flourish at the DNAPL/dissolved phase interface, and to cause increased dissolution of the chlorinated solvent into the dissolved phase (63, 64, see Chapter 8). As discussed in Chapter 8, the addition of organic carbon to stimulate anaerobic bioremediation also has been shown to increase the solubility, and hence the bioavailability, of chlorinated solvents. Biodegradation may not be appropriate as the sole remedy for groundwater with significant DNAPL source areas. However, it is certainly applicable to dissolved-phase plumes, as a polish to source removal, and in conjunction with other remedies.

Under some conditions chlorinated solvents in the subsurface will biodegrade naturally. This natural bioattenuation can result in stabilization and shrinkage of a chlorinated solvent plume. Availability of organic carbon plays a key role in natural bioattenuation, and chlorinated solvents typically persist at low organic carbon sites (65, 66). Although certainly not applicable as a "stand-alone" remedy at DNAPL sites or sites where high-concentration sources have not been removed, monitored natural attenuation (MNA) of chlorinated solvents is increasingly becoming a part of the site management strategy for chlorinated solvent sites (e.g. see case studies in 67, 68, 69). It is often used after source removal or in conjunction with other remediation technologies (67, 70). An MNA program includes ongoing groundwater monitoring for indicators of natural attenuation, and numerous protocols for MNA programs have been developed by the U.S. EPA, the Department of Energy, the Air Force, the Navy, various state agencies, corporations, public/private consortiums, and professional industry associations (66, 67).

Conclusions

Remediation of chlorinated solvent and DNAPL contaminated sites is challenging, and the success or lack of success depends upon the interaction of multiple, complicated site-specific factors. Factors that contribute to the challenges of effectively managing chlorinated solvent and DNAPL-impacted sites include, but are not limited to, (a) the complexity and heterogeneity of soil, groundwater and geochemical systems; (b) the varied phases in which the chlorinated solvent may be present; (c) the behavior of each phase within the subsurface system; and (d) the ability of DNAPL, if present, to remain undetected. Selection of a site management strategy, including selection of the remediation technologies, involves consideration of the ability of the selected strategy to eliminate or isolate the source of contamination. This is especially challenging at sites that are impacted with DNAPLs. Returning contaminated sites, particularly those impacted by DNAPLs, to pristine conditions may indeed not be possible. The complexities of site conditions often restrict the end results obtained by the remediation program.

The technologies briefly discussed in this chapter represent the range of technologies currently employed at sites throughout the world – both the developed technologies that have provided a range of results from mediocre to exceptional, and the innovative approaches, represented by the technologies that are evolving. As our ability to understand the interaction of chlorinated solvents and DNAPLs with the environment increases, so does the development of innovative technologies that can more effectively lead to the restoration of our groundwater systems.

References

1. Pankow, J.F.; Feenstra, S.; Cherry, J.A.; Ryan, M.C.; In *Dense Chlorinated Solvents and other DNAPLs in Groundwater*; Pankow J.F.; Cherry J.A., Eds.; Waterloo Press: Portland, OR, 1996, pp. 1-52.

2. Brusseau, M.L.; Gierke, J.S.; Sabatini, D.A. In *Innovative Subsurface Remediation: Field Testing of Physical, Chemical and Characterization Technologies*; Brusseau, M.L; Sabatini, D.A.; Gierke, J.S.; Annable, M.D., Eds.; ACS Symposium Series 725, American Chemical Society: Washington DC, 1999, pp. 2 – 5.
3. Schwille, Freidrich. *Dense Chlorinated Solvents in Porous and Fractured Media*; Lewis Publishers: Chelsea, MI 1988; pp. 1-144.
4. Freeze, R.A. *The Environmental Pendulum, A Quest for the Truth About Toxic Chemicals, Human Health, and Environmental Protection*; University of California Press: Berkeley, CA, 2000; pp. 197-220.
5. U.S. Environmental Protection Agency. *Guidance for Evaluating the Technical Impracticability of Ground-Water Restoration.* Directive No. 9234.2-25. Office of Solid Waste and Emergency Responses, Washington, DC, 1993; pp. 1-28.
6. United States Code of Federal Regulations. *National Contingency Plan*; Title 40 Part 300.430(a)(1)(iii)(F).
7. Jackson, R.E. *Ground Water Monitoring and Remediation* **2001**, 21(3), 54-58.
8. *Treating Dense Nonaqueous-Phase Liquids (DNAPLs): Remediation of Chlorinated and Recalcitrant Compounds*; Wickramanayake, G.B.; Gavaskar, A.R.; Gupta, N., Eds.; Battelle Press: Columbus, OH, 2000; pp.1-256.
9. Saunders, T.J. *Plato, The Laws*; Penguin Books LTD: New York, 1970; p. 348.
10. Reif, G.A. In *Groundwater Hydrology, Contamination, and Remediation*; Khanbilvardi R.M.; Fillos J., Eds.; Scientific Publications Company: Washington, DC, 1986; pp. 419-426.
11. Mackay, D.M.; Cherry, J.A. *Environ. Sci. Technol.* **1989**, 23, 630-636.
12. Travis, C.C.; Doty, C.B. *Environ. Sci. Technol.* **1990**, 24, 1464-1466.
13 National Research Council. *Alternatives for Ground Water Cleanup*; National Academy Press: Washington, DC, 1994; pp. 1-316.
14. Interstate Technology and Regulatory Cooperation Work Group. *Technology Overview Dense Non-Aqueous Phase Liquids (DNAPLs): Review of Emerging Characterization and Remediation Technologies*, 2000; URL http://itrcweb.org/DNAPL-1, pp. 1-59.
15. U.S. Environmental Protection Agency. *Presumptive Remedies: Site Characterization and Technology Selection for CERCLA Sites with Volatile Organic Compounds in Soils*, PB93-963346, 1993; pp. 1-25.
16. U.S. Army Corps of Engineers. *Engineering and Design Multi-Phase Extraction*, EM1110-1-4010, 1999; Chapter 2.

17. U.S. Environmental Protection Agency. *Presumptive Remedy: Supplemental Bulletin Multi-Phase Extraction (MPE) Technology for VOCs in Soil and Groundwater*, PB97-963501, 1997, pp. 1-11.
18. Royston, M.W. *The World of Positive Displacement Blowers, A few facts, memories, and a lot of nostalgia*, 1999; URL http://roystongroup.com /history.htm.
19. Baker, R.; Groher, D; Becker, D. *Ground Water Currents*, **1999**, 33, pp. 1-2.
20. Johnson, R.L; Johnson, P.C.; McWhorter, D.B.; Hinchee, R.E.; Goodman, I. *Groundwater Monitoring Review*, Fall 1993, pp. 127-135.
21. Smith; L.A.; Hinchee, R.E., Eds. *In-Situ Thermal Technologies for Site Remediation*; Lewis Publishers: Boca Raton, FL, pp. 1-209.
22. *Physical and Thermal Technologies: Remediation of Chlorinated and Recalcitrant Compounds*; Wickramanayake, G.B.; Gavaskar, A.R., Eds.; Battelle Press: Columbus, OH, 2000; pp. 1-344.
23. Thermal Remediation Services. *Three-Phase Heating? Six-Phase Heating? Which is Best?* URL http://www.thermalrs.com/TRSpages/WhitePapers. html.
24. Trowbridge, B.E.; Beyke, G.; Heath, W.O.; and Garcia, E. *Environmental Protection*, July **1999**, pp. 14-26.
25. Fountain, J.C. *Technologies for Dense Nonaqueous Phase Liquid Source Zone Remediation*, Ground-Water Remediation Technologies Analysis Center Technical Evaluation Report TE-98-02; URL http://gwrtac.org, 1998; pp. 1-62.
26. U.S. Environmental Protection Agency. *Evaluation of Ground-Water Extraction Remedies: Volumes 1 and 2.* Office of Emergency and Remedial Responses: Washington, D.C., 1989.
27. U.S. Environmental Protection Agency. *Evaluation of Ground-Water Extraction Remedies: Phase II, Volume 1 – Summary Report.* Publication 9355.4-05, Office of Emergency and Remedial Responses: Washington D.C., 1992.
28. Elsevier Science, Inc. *The Hazardous Waste Consultant.* **1999** 17(2), 1.10-1.14.
29. U.S. Environmental Protection Agency. *Evaluation of Subsurface Engineered Barriers at Waste Sites.* EPA 542-R-98-005, URL http://www.clu-in.com, 1998.
30. Gilmore, T.J.; Pinto, M.J.; White, M.D.; Ballard, S.; Gorelick, S.M.; Taban, O.; Spanc, F.A., Jr. *Performance Assessment of the In-Well Vapor-Stripping System.* Pacific Northwest National Laboratory: Richland, WA, 1996.
31. Lakhwala, F.; Langley, W.; Mueller, J.; Stamm, J. *Water Monitoring and Remediation.* **1998**, 2, 97-106.

32. Jordahl, J.L.; Foster, L.; Schnorr, J.L.; Alvarez, P.J.; *J. Environ. Toxicol. Chem.* **1997**, 16, 1318-1321.
33. Schnoor, J.L; Licht, L.A.; McCutcheon, S.C.; Wolfe, N.L.; Carreira, L.H. *Environ. Sci. Technol.* **1995**, 29, 318-323.
34. Schimp, J.F.; Tracy, L.C.; Davis, T.L.; Huang, W.; Erickson, L.E.; Schnoor, J.L. *Crit. Rev. Environ. Sci. Technol.* **1993**, 41-77.
35. Newman, L.A.; Strand, S.E.; Choe, N.; Duffy, D.; Ekuan, G.; Ruszaj, M.; Shurtleff, B.B.; Wilmoth, J.; Heilman, P.; Gordon, M.P. *Environ. Sci. Technol.* **1997**, 31, 1062-1067.
36. Nietch, C. T.; Morris, J.T.; Vroblesky, D.A. *Environ. Sci. Technol.* **1999,** 33, 2899-2904.
37. *Chemical Oxidation and Reactive Barriers: Remediation of Chlorinated and Recalcitrant Compounds;* Wickramanayake, G.B.; Gavaskar, A.R.; Chen, A.S.C., Eds.; Battelle Press: Columbus, OH, 2000; pp. 1-470.
38. Matheson, LJ.; Tratnyek, P.G.; *Environ. Sci. Technol.* **1994,** 28, 2045-2053.
39. U.S. Environmental Protection Agency. *Innovative Remediation Technologies: Field-Scale Demonstration Projects in North America, 2nd Edition, Year 2000 Report.* Office of Solid Waste and Emergency Response, EPA 542-B-00-004, URL http://www.epa.gov/tio/remed.htm, 2000.
40. Yin, Y.; Allen, H.E. Fountain, J.C. *In Situ Chemical Treatment*, Ground-Water Remediation Technologies Analysis Center Technical Evaluation Report TE-99-01; URL http://gwrtac.org, **1998**.
41. *Technical and Regulatory Requirements for Enhanced In Situ Bioremediation of Chlorinated Solvents in Groundwater*; Interstate Technology and Regulatory Cooperation Work Group, December 23, 1998, URL http://itrcweb.org, pp 1-66.
42. *Bioremediation and Phytoremediation of Chlorinated and Recalcitrant Compounds;* Wickramanayake, G.B.; Gavaskar, A.R.; Alleman, B.C.; Magar, V.S., Eds.; Battelle Press: Columbus, OH, 2000; pp. 1-538.
43. *Anaerobic Degradation of Chlorinated Solvents*; Magar, V.S.; Fennell, D.E.; Morse, J.J.; Alleman, B.C.; Leeson, A., Eds.; Battelle Press: Columbus, OH, 2001; Vol. 6(7) pp. 125-313.
44. Verce, M.F.; Gunsch, C.K.; Danko, A.S.; Freedman, D.L. *Environ. Sci. Technol.* **2002**, 36, 2171-2177.
45. Henry, S. M.; Grbic-Galic, D; In *Biodegradation and Bioremediation of Toxic Chemicals*, Chaudhry, R.G., Ed.; Dioscorides Press: Portland, OR 1994, pp. 314-344.
46. Wackett, L.P.; Lange, C.C.; Ornstein, R.L. In *Bioaugmentation for Site Remediation*; Hinchee, R.E.; Fredrickson, J.; Alleman, B.C., Eds.; Battelle Press: Columbus, OH, 1995; Vol. 3(3), pp 105-114.

47. Adams, D.J.; Reardon K.F. In *Bioaugmentation, Biobarriers, and Biogeochemistry*; Leeson, A.; Alleman, B.C.; P.J. Alvarez, P.J.; Magar, V.S., Eds. Battelle Press: Columbus, OH, 2001; pp. 53-39.
48. Henry, S. M.; Grbic-Galic, D. *Applied Environ. Microbiol.* **1991**. 57, 1770-1776.
49. Henry, S. M.; Grbic-Galic, D. *Microb. Ecol.* **1990**, 20, 151-169.
50. Henry, S. M.; Grbic-Galic, D. *Applied Environ. Microbiol.* **1991**. 57, 236-244.
51. Alvarez-Cohen L.; McCarty, P.L. *Applied Environ. Microbiol.* **1991**, 57, 228-235.
52. Oldenhuis, R.; Oedzes, J.Y.; van der Waarde, J.J.; and Janssen, D.B. *Applied Environ. Microbiol* .**1991**, 57, 7-14.
53. Fogel, S.; Lewis, R.; Groher, D.; Findlay, M. In *Bioremediation of Chlorinated Solvents*, Hinchee, R.E.; Leeson, A. Eds.; Battelle Press: Columbus, OH, 1995, pp. 153-160.
54. Freedman, D.L.; Gossett, J. M. *Applied Environ. Microbiol.* **1989**, 55, 2144-2151.
55. Major, D.W.; Hodgind, E.W.; Butler, B.J. In *On-Site Bioreclamation, Processes for Xenobiotic and Hydrocarbon Treatment*, Hinchee, R.E.; Olfenbuttel, R.F. Eds.; Butterworth-Heinmann: Boston, MA, 1991, pp. 147-171.
56. Newel, C.J.; Aziz, C.E.; Haas, P.E.; Hughes, J.B.; Khan, T.A. In *Anaerobic Degradation of Chlorinated Solvents*; Magar, V.S.; Fennell, D.E.; Morse, J.J.; Alleman, B.C.; Leeson, A., Eds.; Battelle Press: Columbus, OH, 2001; pp. 19-26.
57. Maymo-Gatell, X.; Tandoi, V.; Gossett, J.M.; Zinder, S.H. *Applied Environ. Microbiol.* **1995**, 61, 3928-3933.
58. McCarty, P.L. *Science,* **1997**, 276, 1521-1522.
59. Lee, M.D.; Odom, J.M.; Buchanan, Jr. R.J. *Annual Review of Microbiol.* **1998**, 52, 423-452.
60. Harkness, M.R.; Bracco, A.A.; Brennan, Jr., M.A.; Deweerd, K.A.; Spivack, J.L. *Environ. Sci. Technol.* **1999**., 33, 1100-1109.
61. Ellis, D, E.; Lutz, E.J.; Odom, J.M.; Buchanan, Jr. R.J.; Bartlett, C.L.; Lee, M.D.; Harkness M.R.; Deweerd, K.A. *Environ. Sci. Technol.* **2000**, 34, 2254-2260
62. Major, D. W.; McMaster, M. L.; Cox, E. E.; Lee, B. J.; Gentry, E.E.; Hendrickson, E.; Edwards, E.A.; and Dworatzek, S. In *Bioaugmentation, Biobarriers, and Biogeochemistry*; Leeson, A.; Alleman, B.C.; P.J. Alvarez, P.J.; Magar, V.S., Eds. Battelle Press: Columbus, OH, 2001; pp. 27-34.
63. Yang, Y; McCarty, P.L. *Environ. Sci. Technol.* **2000**, 34, 2979-2984.
64. Cope, N.; Hughes, J. B. *Environ. Sci. Technol.* **2001**, 35, 2014-2021.

65. Lapat-Polasko, L.; Aiken, B.S.; Narayanaswamy, K. In *Natural Attenuation of Environmental Contaminants*; Leeson, A.; Kelley, M.E.; Rifai, H.S.; Magar, V.S., Eds. Battelle Press: Columbus, OH, 2001; Vol. 6(2), pp 65-71.
66. Weidemeier, T.H.; Swanson, M.A.; Moutox, D.E.; Gordon, K.; Wilson, J.T.; Wilson, B.H.; Kampbell, D.H.; Haas, P.E.; Miller, R. E.; Hansen, J.E.; Chapelle, F.H. *Technical Protocol for Evaluating Natural Attenuation of Chlorinated Solvents in Groundwater* U.S. EPA, September 1998, 600/R-98/128.
67. *Natural Attenuation for Groundwater Remediation*; National Research Council; National Academy Press: Washington, DC, 2000; pp. 1-274.
68. *Natural Attenuation Considerations and Case Studies: Remediation of Chlorinated and Recalcitrant Compounds;* Wickramanayake, G.B.; Gavaskar, A.R; Kelley, M.E., Eds.; Battelle Press: Columbus, OH, 2000; pp. 1-254.
69. *Natural Attenuation of Environmental Contaminants*; Leeson, A.; Kelley, M.E.; Rifai, H.S.; Magar, V.S., Eds. Battelle Press: Columbus, OH, 2001; Vol. 6(2), pp. 39-145.
70. Wilson, J. T.; Kampbell, Don H.; Ferrey, M. Estuesta, P. *Evaluation of the Protocol for Natural attenuation of Chlorinated Solvents: Case Study at the Twin Cities Army Ammunition Plant*, U.S. EPA, March 2001, EPA/600/R-01/025.

Chapter 2

The Evolution of DNAPL Remediation Practice

Richard E. Jackson

INTERA Inc., 9111A Research Boulevard, Austin, TX 78758

The remediation of dense non-aqueous phase liquids (DNAPLs) and their dissolved components has proven to be an expensive and incomplete chapter in environmental history. In the 1980s, it was assumed by many that ground-water extraction would completely restore aquifers to their precontaminated conditions. By the early 1990s it was apparent that this condition would not be achieved and that pump-and-treat remediation was principally a means of containing dissolved-phase contamination without complete site restoration. This failure prompted experimentation with technologies from different engineering fields, including enhanced oil recovery and waste-water treatment, as potential remediation technologies. Independent performance assessment of all technologies indicates that good site characterization is essential to their successful implementation and that vendor claims of success must be carefully scrutinized. Recent field testing indicates that the benefits derivable from each technology can best be obtained by its use as a unit operation linked to other unit operations in the form of an in-situ "treatment train".

Introduction

The contamination of water-supply wells and suburban neighborhoods by DNAPLs became a national issue in the USA in the late 1970s. Two events caused this sudden awareness of a very much larger problem. The first was the discovery of DNAPL beneath the Love Canal neighborhood of Niagara Falls, New York in 1976 and then in the bedrock beneath nearby chemical plants, causing seepage of contamination into the Niagara River and Lake Ontario. The second was the discovery in 1978-9 of the contamination of municipal wells by dissolved trichloroethylene (TCE), perchloroethylene (PCE) and other chlorinated solvents on Long Island (New York), in the industrial towns around Boston, and in the urbanized alluvial basins of California.

These events produced a number of responses to the crisis posed to public health by the vaporization and dissolution of these DNAPLs. The most profound was the Federal legislation that sought to provide society with the means with which to remediate the problems of contaminated neighbourhoods and wells. This involved the passage of Comprehensive Environmental Response, Compensation, and Liability Act (CERCLA) and the establishment of the Superfund in 1980, its amendment in 1986 (SARA), and of the Hazardous and Solid Waste Amendments to the Resource Conservation and Recovery Act of (RCRA) in 1984.

While Love Canal spawned legislation, the irony of the discovery of TCE contamination of ground water across the USA was that it was accidental. TCE was discovered during the testing of municipal water supplies for the presence of trihalomethanes that was required under the 1978 amendment to the Federal Safe Drinking Water Act (SDWA). With the introduction of the gas chromatograph/mass spectrometer into analytical practice and the adoption of the purge-and-trap device for extracting volatile organic chemicals (VOCs) from ground-water samples in the mid 1970s, the stage was set to reliably detect dissolved TCE and other VOCs (*1*).

The events and the legislation that followed generated the need to enlist and train environmental scientists and engineers who could characterize the hazardous-waste sites contaminated with DNAPLs and could then commission effective remedial operations to protect public health. It is no understatement to say that these demands of site characterization and remediation fell on a technical community that was not only few in numbers but was completely unprepared for the tasks implicit in their execution. They worked without a guiding technical paradigm relating the migration, trapping and dissolution of DNAPLs to their distribution in the subsurface and their detection, quantification and spatial characterization. As a consequence, the engineers and scientists who practiced DNAPL remediation made very limited progress in restoring sites to meet regulatory guidelines without resorting to active and continuing containment. Just as a century before when the germ theory of disease opened the door to vast improvements in public health, so progress in DNAPL remediation was dependent upon the emergence of a paradigm that would explain the migration and fate of DNAPL in the subsurface and would guide research and practice.

This commentary on DNAPL remediation considers the scientific basis of our understanding of DNAPL distribution in the field and the various technologies that have been developed to address the problem. It is in no way a complete survey of the field. Rather it considers the progress made and the failures encountered since Love Canal brought the issue to the public's attention.

The Schwille Paradigm

The term 'paradigm' has become oversimplified in popular use in the years since it was used by T.S. Kuhn in 1962 to describe scientific development (*2*). In Kuhn's own words (*2*, page 10): "By choosing it, I mean to suggest that some accepted examples of scientific practice – examples which include law, theory, application, and instrumentation together – provide models from which spring particular coherent traditions of scientific research." Thus, M.K.Hubbert's analytical theory of ground-water flow in 1940 (*3*) provided a conceptual model based on the principle of conservation of mass and the laws of thermodynamics from which the modern paradigm of ground-water flow developed (*4*).

Friedrich Schwille and his colleagues at the Federal Institute of Hydrology in Koblenz, West Germany developed the conceptual model of the migration and fate of DNAPL as shown in Figure 1. Their 1984 report (*5*) on laboratory experiments of the migration and fate of dense chlorinated solvents established the principal features of the paradigm, including rules and experiments that environmental engineers and scientists would subsequently follow. They showed the patterns of migration and trapping of solvents in sands of various particle size and the retention of this DNAPL by these same sands above and below the water table. Furthermore, they demonstrated the solubilization of the DNAPL to form dissolved-phase plumes, and the importance of the volume of solvent spilled and the geometrical arrangement of zones of differing permeability in determining the spatial distribution of DNAPL in the subsurface. They also demonstrated the importance in the subsurface of the volatilization of chlorinated solvents and the migration of the vapor to depth. A synopsis of this report appeared in English in a 1984 conference proceedings that was little noticed in North America (*6*). Schwille first presented his results in North America in June 1985 at the 2nd Canadian-American Hydrogeology Conference at Banff, Alberta (*7*) and, in 1988, the English translation of the 1984 report (*8*) was published.

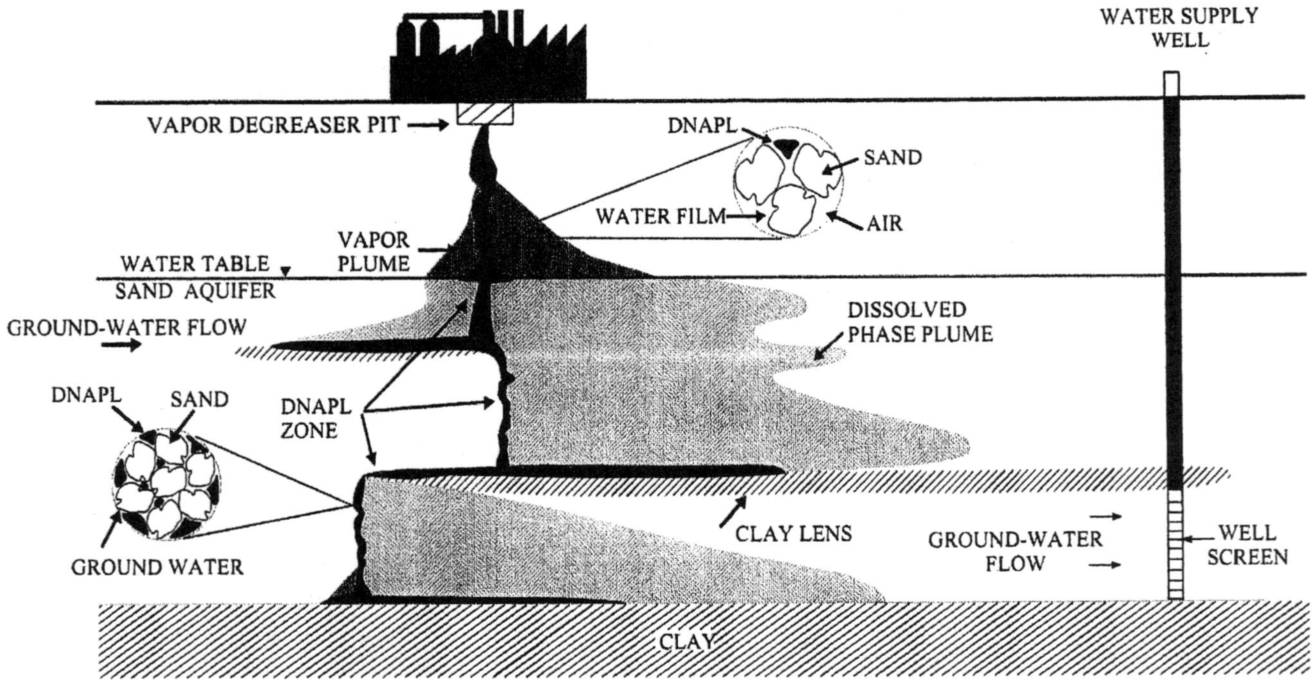

Figure 1. Migration of DNAPL in a sand aquifer and the development of DNAPL and dissolved contamination zones.

Increasingly explicit descriptions of the Schwille paradigm appeared in two reviews published in *Environmental Science and Technology* in 1985 and 1989. The first (*9*) acknowledged Schwille's finding that chlorinated degreasing solvents are retained in significant quantities in the subsurface and act as long-term subsurface sources of contamination. The second review (*10*) described the Schwille paradigm in much greater detail and how it negatively affected attempts to restore contaminated ground water to pristine conditions by pump-and-treat methods. The general acceptance of the paradigm in North America was indicated by presentation of the conceptual model (i.e., Figure 1) in the first (1990) edition of Domenico and Schwartz (*11*). A more complete account of the migration, trapping, volatilization and dissolution of DNAPL was published in 1996 by the Solvents-in-Groundwater Consortium based at the University of Waterloo, Ontario, Canada (*12*).

Thus, Schwille's work provided the rules by which the technical community could begin to develop appropriate subsurface characterization and remediation approaches. He warned (*8*, p. 112) that solvents might remain at the base of aquifers "for time periods ranging between several months to decades." It was this consideration that lead Mackay and Cherry (*10*), in 1989, to state that "the primary challenge in groundwater cleanup is to remove the organics masses that serve, in effect, as subsurface sources and cause the plumes to grow and persist, rather than simply to remove the dissolved contamination that defines the plume."

The Development of Technologies for DNAPL Remediation

Although Mackay and Cherry (*10*) recommended DNAPL removal, the difficulty of that task was apparent from the outset. Wilson and Conrad (*13*) had already shown the impracticability of water-flooding to remove sufficient residual NAPL from alluvium to achieve regulatory goals. As Mackay and Cherry noted prophetically, "very little success has been achieved in even locating the subsurface NAPL sources, let alone removing them." Therefore, increasingly during the 1990s, DNAPL removal came to be seen as an impractical and futile goal by many in industry and the Solvents-in-Groundwater Consortium, which was supported by industrial funding. Alternative approaches were developed, including containment walls borrowed from geotechnical engineering and in-situ chemical oxidation (ISCO) and bioremediation, which were modified forms of waste water treatment methods. Four general classes of DNAPL site remediation technology may be identified (see Table 1): containment, dissolved-phase destruction of VOCs, DNAPL removal from the saturated zone, and soil-vapor extraction to remove DNAPL from the vadose-zone. The first two classes address the dissolved-phase contamination that is generated by ground-water flow through the DNAPL zone; the third addresses the removal of the DNAPL zone itself; and the fourth removes contamination by promoting volatilization.

Table 1: Classes and Examples of DNAPL Remediation Technologies

Class	Example Technoloiges
Containment	"Pump-and-Treat"; Barrier Walls
Dissolved-Phase Destruction	ISCO; bioremediation; PRBs
Saturated-Zone DNAPL Removal	Water, steam, surfactant or alcohol flooding
Vadose-Zone DNAPL Removal	Soil vapor extraction

Containment of Dissolved-Phase Contamination

Mackay and Cherry (*10*) pointed out that the pump-and-treat approach to remediation "is best thought of as a management tool to prevent, by hydraulic manipulation of the aquifer, continuation of contaminant migration". This statement implied that methods of operations research would be useful, in particular those that would derive optimal solutions for containing and efficiently extracting contaminated ground water. Despite the elegance and utility of optimization algorithms, optimal hydraulic capture is seldom attempted at DNAPL sites. In fact, the example DNAPL site used by Gorelick and colleagues in their text on optimal capture and containment (*14*) was commissioned with extraction wells that were placed in a manner that the hydrogeologist responsible felt would ensure maximum ground-water recovery, irrespective of engineering efficiency or the costs of water treatment or the precise axis of the narrow contaminant plume that was unknown to him at the time of well construction. Similarly, the capture zones at other DNAPL sites are most often developed without reference to pumping tests for anisotropic porous media. Such tests are needed to determine minimum extraction rates to achieve capture of the contaminant plume (*15*). In short, although the mechanisms of DNAPL dissolution and plume generation are well understood (*16*) and the tools for efficient hydraulic containment are available (*14,15*), the application of those tools is frequently neglected and containment even less frequently confirmed following several years of pump-and-treat operation.

It is seldom made explicit how important site characterization is to containment system design – whether hydraulic control or barrier walls, which are borrowed from geotechnical engineering practice (*17*). The performance of these walls depends not only on good construction practice but also on good site characterization. At Hill Air Force Base in Utah, the failure of a soil-bentonite slurry wall to envelope the DNAPL zone – leaving about 3000 gallons of DNAPL outside the wall – has been presented publicly (*18*) as part of an effort at Hill to improve remediation engineering performance.

The Hill AFB example shows very clearly the perils of designing and installing a barrier wall without first properly characterizing the DNAPL zone. A similar requirement is necessary with hydraulic control by ground-water extraction in order to ensure that the wellfield installed to contain the aqueous-phase contamination actually fully contains the contaminants, even in the event of failure of one of the extraction wells.

Destruction of Dissolved-Phase Contamination

We may consider three technologies that involve the in-situ destruction of dissolved components of DNAPLs: permeable reactive barriers (PRBs), ISCO, and anaerobic bioremediation. As the name implies, PRB technology combines in-situ destruction with containment to create an entirely new technology, whereas ISCO and bioremediation are borrowed technologies that have involved adapting methods of waste-water treatment engineering.

Warner and Sorel (*19*) provide a superb review of the development of PRB technology and the scientific basis for its function in this volume. They point out the critical dependence of the chemical component of the technology – the substrate or "treatment matrix" that acts on ground water to produce the hydrogen for dechlorination – upon the hydraulic control system that is designed to channel the contamination through the PRB. Given what has been said above about the failure of hydraulic containment systems to contain dissolved-phase contamination, it can be hardly surprising to learn from Warner and Sorel that those PRBs that have failed to function as designed have generally failed for hydraulic control reasons, rather than because the treatment matrix failed. The applicability of PRBs to shallow environments and narrow plumes is well established, although high dissolved concentrations, deep DNAPL contamination, and wide plumes may prove uneconomic for PRB application (*20*).

ISCO relies on electron transfer to oxidize the dissolved contaminant, as with $KMnO_4$(*22*) and other chemical oxidants. Consequently, ISCO requires that the redox reactions occur in an electrolyte such as ground water that has a typical electrical conductivity of 10^{-1}-10^{-2} Siemen-meters (S-m), as opposed to non-electrolytes such as TCE (8×10^{-12} S-m) or PCE (6×10^{-14} S-m). Thus DNAPL dissolution must occur prior to oxidation. As Farquhar and colleagues (*21*) showed in controlled experiments at Borden, Ontario, permanganate oxidation of 8 liters of PCE DNAPL released into Borden sand required some ten months to destroy just 62% of the DNAPL. The slow rate of destruction indicates that these methods should be reserved for systems containing little or no DNAPL, i.e., DNAPL saturations < 0.1% or <~300 mg TCE/kg soil. Therefore, ISCO is suitable as a "polishing" step in aquifers from which DNAPL has been largely removed or in zones of aquifers containing only dissolved-phase contamination.

Cometabolic biotransformation, on the other hand, relies on the reductive dehalogenation of the chlorinated hydrocarbon, typically by a hydrogen-for-chlorine substitution. Recent studies have shown PCE dehalogenation at high aqueous concentrations *(23, 24, 25)* indicating that PCE biotransformation can occur when an electron donor solution may be within the DNAPL zone. As with ISCO, however, the electron transfer is principally occurring in the aqueous phase that constitutes the electrolyte rather than within the DNAPL itself where electron transfer rates are minimal because of its low electrical conductivity.

DNAPL Removal Technologies

DNAPL removal technologies are generally modifications of methods of enhanced oil recovery developed by petroleum engineers prior to the 1990s. Falta (*26*) has provided an extensive review of steam flooding of NAPL-contaminated systems, which developed from the Californian practice of steam-flooding deep heavy-oil reservoirs. The success of steam flooding and other thermal technologies, such as six-phase heating, in removing fuel hydrocarbons is well documented (*27*). However, the inherent mobilization of DNAPL ahead of the steam condensate front raises questions about its suitability for removing DNAPLs such as TCE and PCE (*26*). In this volume, Kaslusky and Udell (*28*) describe an approach involving optimizing the air-to-steam injection ratio to limit the possibility of downward DNAPL migration.

Surfactant and alcohol flooding can also mobilize DNAPL if the viscous forces moving the surfactant and/or alcohol injectate exceed the critical trapping number retaining the DNAPL in place. The efficacy of these two methods of DNAPL removal without this loss of control has been demonstrated in various alluvial aquifers across North America (*29-32*). Surfactant flooding has been shown to remove over 90% of TCE-rich DNAPL (*29, 31*), while alcohol flooding removed approximately 65% of the PCE DNAPL at a dry-cleaners site in Florida (*30*). Therefore, these methods provide a suitable primary means of removing DNAPL prior to the application of the secondary or "polishing" technologies to destroy the dissolved-phase contamination.

Vadose-Zone DNAPL Remediation

A number of other technologies have been developed and applied to DNAPL remediation in the vadose zone. Perhaps the most common is the use of soil-vapor extraction (SVE) for "Multi-Phase Extraction", a technology which was originally developed for the capture and volatilization of fuel hydrocarbons. Many large DNAPL sites are undergoing remediation by SVE. Like pump-and-treat extraction, its efficiency is greatly hindered due to by-passing of DNAPL by the air because of the higher relative permeability to air flow through clean zones. Similarly, air sparging is limited in its effectiveness by channeling of the air through more permeable pathways. Unger and colleagues (*33*) present results of simulations of SVE and air sparging in which the DNAPL is present in the higher permeability zones. It should be noted that the lowering of the water table – to allow SVE to be applied to DNAPL at greater depths – will likely cause DNAPL drainage due to the much lower residual saturations that occur in three-phase systems (air, water, DNAPL) compared with soils of similar texture under two-phase conditions, i.e., water and DNAPL (*34*).

The Futility Argument and its Critics

The front-page Wall Street Journal article (*35*) that appeared in 1991 began the 'futility' argument in earnest. In it, J.A.Cherry of the University of Waterloo maintained that ground-water contaminated by DNAPLs "has the equivalent of terminal cancer". However, despite the evidence of Schwille's experiments and EPA's own study of pump-and-treat sites (*36*), the EPA's deputy director of Superfund maintained in this same article that pump-and-treat remediation would result in achievement of complete aquifer restoration. Consequently, owners of DNAPL sites argued that, if Cherry's position and EPA's survey of sites were correct, then containment of the aqueous phase was the appropriate remedy.

Reluctant to give up on the goal of site restoration to drinking-water guidelines as established by the SDWA – a provision established in the Comprehensive Environmental Response, Compensation, and Liability Act (CERCLA) in 1980 – the EPA modified its approach in 1993 to allow greater use of the CERCLA concept of technical impracticability (TI) at DNAPL sites (*37*). With CERCLA further threatened by the hostility to government regulation that was a hallmark of the 104th Congress, EPA undertook further reforms to CERCLA policies during 1995 that focused future efforts on "hot-spot removal" and containment elsewhere (*38*), which was also a stated goal of the TI policy.

The result of the 1995 Superfund reforms and the TI policy is that remedy selection at Superfund sites has moved away from the stated intent of CERCLA

– according to which preferred remedies are to be implemented causing the volume, toxicity and mobility of the contaminants to be "permanently and significantly" reduced (*39*) – to containment remedies, which are usually considerably cheaper (*38*). Provided human health and the environment are protected, this development was a legitimate policy response by EPA if TI was demonstrated. However, within five years of the 1995 reforms, field demonstrations at Hill AFB in Utah (*29*), Visalia in California (*40*), Jacksonville in Florida (*30*) and elsewhere had shown that it was technologically practicable to remove the majority of DNAPL from a site. Therefore, from EPA's perspective, it was necessary to reconsider the TI policy in light of the technological developments in order to be consistent with CERCLA's of permanent remedies resulting in reductions in the toxicity, volume and mobility of the contaminants. As the critics of the Futility argument were quick to point out, the law (CERCLA) required no less, particularly when the poor performance of containment remedies was leading to a significant number of "ROD re-openers" (*41*) and EPA's Inspector General had himself found that 31% of remedies were not protective of human health and the environment (*38*).

In order to develop a path forward, the EPA formed an expert panel in October 2001, the purpose of which was, from EPA's point-of-view, to determine how to close Superfund sites within 30-50 years and to promote the use of the newly-demonstrated DNAPL removal and in-situ destruction technologies (*42*). Manufacturing industry, the US Department of Energy and the US Air Force were invited to provide the opinions of responsible parties affected by any EPA policy change. Industry's opinion, as presented by DuPont representatives at the Workshop held to inaugurate the Panel's work (*43*), argued against DNAPL source zone cleanup because of its perceived technical impracticability and cost.

Discussion

By the late 1990s the technological capabilities of US environmental remediation practice still had not caught up with the idealistic intent of CERCLA, which had been passed some twenty years before. At least two major issues remained to be resolved.

Firstly, improvements in the practice of DNAPL site characterization are inhibited by outdated technical guidance for assessing the potential of DNAPL occurrence at hazardous-waste sites and, when present, its volume and spatial distribution (*41*). Soil cores and ground-water samples can only be part of DNAPL site assessment and even they provide data of much quantitative uncertainty as the DNAPL remediation tests at Cape Canaveral have shown (*44*). Studies in alluvium at Hill AFB in Utah indicate that the use of ground-penetrating radar, direct-push characterization, continuous soil coring, followed by partitioning interwell tracer testing provide quantitative and reliable estimates of DNAPL volume and spatial distribution (*45*). These field results imply that vendor claims of successful cleanup based solely upon before-and-after soil cores and/or ground-water samples should be considered skeptically.

Secondly, it will be necessary to adopt the "treatment train" concept of using a series of sequential remedial operations – "technology integration" – to achieve the required contaminant fluxes if the benefits of each technology are to be fully realized (*46*). Thus, DNAPL zones would first be characterized in such a manner that volumetric and spatial measurements of DNAPL mass would be obtained. This would allow the remedial design engineer the necessary information to prepare a cost-effective and robust design using a primary DNAPL recovery technology, such as steam or surfactant flooding. [Similar basic information on the subsurface is accepted as essential by design engineers working in petroleum recovery, extractive metal mining, and geotechnical engineering.] That is, the design engineer would provide for a sufficient amount of steam or surfactant to be injected and transported through the entireity of the DNAPL zone to achieve the desired reduction in DNAPL volume. A secondary or polishing technology such as bioremediation or ISCO would follow, this time designed on the basis of predictions of contaminant mass remaining in place following the primary step. Such quantification then allows reliable estimates of effluent concentrations to be generated for waste-water treatment engineers who must deal with the ex-situ engineering at this site, perhaps air stripping followed by carbon adsorption of vapor and aqueous effluents.

Each of these steps – whether conducted in-situ or ex-situ – constitute a unit operation when viewed from the perspective of chemical engineering practice (*47*). This same principle of chemical engineering was advocated by the 1994 Committee on Ground Water Cleanup Alternatives of the National Research Council (*48*) and promises to promote a more quantitative and successful approach to site characterization and remediation in that the inputs must be clearly and accurately defined for each unit operation to function efficiently.

References

1. Jackson, R.E. Recognizing Emerging Environmental Problems: The Case of Chlorinated Solvents in Ground Water. *Technology & Culture*, in press.
2. Kuhn, T.S. *The Structure of Scientific Revolutions*; The University of Chicago Press, Chicago, IL, 1962.
3. Hubbert, M.K. Theory of Ground-Water Motion. *The Journal of Geology* **1940**, *48*, 785-944.

4. Freeze, R.A.; Cherry, J.A. *Groundwater;* Prentice Hall, Inc., Englewood Cliffs, NJ, 1979, chapter 6.
5. Schwille, F.; Bertsch, W.; Linke, R.; Reif, W.; Zauter, S. *Leichtflüchtige Chlorokohlenwasserstoffe in porösen und klüftigen medien: Modellversuche,* Besondere Mitteilungen zum Deutschen Gewässerkundlichen Jahrbuch Nr. 46, Koblenz, West Germany, 1984.
6. Schwille, F. Migration of organic fluids immiscible with water in the unsaturated zone, In: *Pollutants in Porous Media: The Unsaturated Zone between Soil Surface and Groundwater*, Yaron, B.; Dagan, G.; Goldshmid, J., Eds.; Springer-Verlag, Berlin, 1984, pp. 27-48.
7. Schwille, F. Migration of organic fluids immiscible with water in the unsaturated and saturated zones. *Proceedings, Second Canadian/American Conference on Hydrogeology*, Hitchon, B.; Trudell, M.R.; Eds., National Water Well Association, Dublin OH, 1985, pp. 31-53.
8. Schwille, F.; Bertsch, W.; Linke, R.; Reif, W.; Zauter, S. *Dense Chlorinated Solvents in Porous and Fractured Media: Model Experiments,* Translated by James F. Pankow, Lewis Publishers, Chelsea MI, 1988.
9. Mackay, D.M.; Roberts, P.V.; Cherry, J.A. Transport of organic contaminants in groundwater. *Environmental Science and Technology* **1985**, *19*, 384-392.
10. Mackay, D.M.; Cherry, J.A. Groundwater contamination: pump-and-treat remediation: *Environmental Science and Technology* **1989**, *23*, 630-636.
11. Domenico, P.A. and F.W. Schwartz, 1990. *Physical and Chemical Hydrogeology*, 1st edition, John Wiley & Sons, New York.
12. Pankow, J.F.; Cherry, J.A. (Eds.). *Dense Chlorinated Solvents and Other DNAPLs in Groundwater.* Waterloo Press, Portland OR, 1996.
13. Wilson, J.L.; Conrad, S.H. Is physical displacement of residual hydrocarbon a realistic possibility in aquifer restoration? In: *Proceedings of the NWWA/API Conference on Petroleum Hydrocarbons and Organic Chemicals in Ground Water.* National Water Well Association, Dublin OH, 1984, pp. 274-298.
14. Gorelick, S.M.; Freeze, R.A.; Donohue, D.; Keely, J.F. *Groundwater Contamination: Optimal Capture and Containment*; Lewis Publishers, Boca Raton FL, 1993.
15. Bair, E.S.; Lahm, T.D. Variations in capture zone geometry of a partially penetrating pumping well in an unconfined aquifer. *Ground Water 34*, 842-852.
16. Illangasekare, T.H.; Reible, D.D. Pump-and-treat remediation and plume containment: applications, limitations, and relevant processes. In: *Groundwater Contamination by Organic Pollutants: Analysis and Remediation,* American Society of Civil Engineers, Reston VA, 2001, pp.79-119.

17. Mitchell, J.K.; van Court, W.A.N. Barrier design and installation: walls and covers. In: *Subsurface Restoration;* Ward, C.H.; Cherry, J.A.; Scalf, M.R., Eds.; Ann Arbor Press, Inc.: Chelsea MI, 1997; pp. 175-195.
18. Brown, S.G.; Betts, P.C.; Hicken, S.T.; Schneider, J.R. Physical containment of a DNAPL source. In: *Designing and Applying Treatment Technologies: remediation of chlorinated and recalcitrant compounds*, Battelle Press: Columbus OH, 1998, pp. 115-120.
19. Warner, S.D.; Sorel, D. Ten years of permeable reactive barriers: Lessons learned and future expectations. This volume.
20. Cherry, J.A. Comments at the EPA DNAPL Source Remediation Workshop, Dallas TX, October 29-30, 2001.
21. Schnarr, M.; Truax, C.; Farquhar, G.; Hood, E.; Gonullu, T.; Stickney, B. Laboratory and controlled field experiments using potassium permanganate to remediate trichloroethylene and perchloroethylene DNAPLs in porous media. *J.Contaminant Hydrology* **1998**, 29, 205-224.
22. Siegrist, R.L. Overview of source remediation technologies – in situ chemical oxidation. Presentation at EPA DNAPL Source Remediation Workshop, Dallas TX, October 29-30, 2001.
23. Isalou, M.; Sleep, B.E.; Liss, S.N. *Environmental Science & Technology* **1998**, *32*, 3579-3585.
24. Yang, Y.; McCarty, P.L. *Environmental Science & Technology* **2000**, *34*, 2979-2984.
25. Cope, N.; Hughes, J.B. *Environmental Science & Technology* **2001**, *35*,2014-2021.
26. Falta, R.W. Steam flooding for environmental remediation. In: *Groundwater Contamination by Organic Pollutants: Analysis and Remediation,* American Society of Civil Engineers, Reston VA, 2001, pp. 153-192.
27. Udell, K.S. Thermally enhanced removalof liquid hydrocarbon contaminants from soils and groundwater. In: *Subsurface Restoration;* Ward, C.H.; Cherry, J.A.; Scalf, M.R., Eds.; Ann Arbor Press, Inc.: Chelsea MI, 1997; pp. 251-270.
28. Kaslusky, S.F.; Udell, K.S. Optimum air to steam injection ratios to eliminate downward DNAPL migration during steam enhanced extraction operations. This volume.
29. Londergan, J.T.; Meinardus, H.W.; Mariner, P.E.; Jackson, R.E.; Brown, C.L.; Dwarakanath, V.; Pope, G.A.; Ginn, J.; Taffinder, S. DNAPL removal from a heterogeneous alluvial aquifer by surfactant enhanced aquifer remediation. *Ground Water Monitoring & Remediation* **2001**, *21*, 71-81.
30. Jawitz, J.; Sillan, R.K.; Annable, M.D.; Rao, P.S.C.; Warner, K. In situ alcohol flushing of a DNAPL source zone at a dry cleaner site. *Environmental Science & Technology* **2000**, *34*, 3722-3729.

31. Fiorenza, S. et al. (editors). *NAPL Removal: Surfactants, Foams, and Microemulsions.* AATDF monographs, Lewis Publishers, Boca Raton, Florida, 2000.
32. Shiau, B.-J.; Hasegawa, M.A.; Brammer, J.M.; Goodspeed, M.; Carter, T.; Harwell, J.H.; Sabatini, D.A.; Knox, R.C.; Szekeres, E. Field demonstration of surfactant-enhanced DNAPL remediation: case studies. This volume.
33. Unger, A.J.A.; Sudicky, E.A.; Forsyth, P.A. Mechanisms controlling vacuum extraction coupled with air sparging for remediation of heterogeneous formations contaminated by dense nonaqueous phase liquids. *Water Resources Research* **1995,** *31*, 1913-1925.
34. Ewing, J.E., R.E.Jackson and V. Dwarakanath, in review. Remedial implications of air entry into water-saturated alluvium contaminated with NAPLs. Submitted to *Ground Water Monitoring & Remediation.*
35. *Wall Street Journal*, May 15, 1991, vol. LXXXVII, no. 95, p. 1.
36. EPA. *Evaluation of ground-water extraction remedies*, Volume 2. Case Studies 1-19, Interim Final, EPA/540/2-89/054b, Environmental Protection Agency, Washington DC, 1989.
37. EPA. Guidance for Evaluating the Technical Impracticability of Ground-Water Restoration, OSWER, Washington DC, Directive 9234.2-25, 1993, 26p.
38. Nakamura, R. and T. Church, 2000. *Reinventing Superfund: An Assessment of EPA's Administrative Reforms.* National Academy of Public Administration, Washington DC 20005.
39. CERCLA, section 9621, Cleanup Standards. US Code, Title 42, chapter 103, subchapter I.
40. Leif, R.N.; Chiarappa, M.; Aines, R.D.; Newmark, R.L.; Knauss, K.G.; Eaker, C., 1998. In-situ hydrothermal oxidative destruction of DNAPLs in a creosote-contaminated site. In: *Physical, chemical and thermal technologies – Remediation of chlorinated and recalcitrant compounds.* Battelle Press, Columbus, Ohio, pp. 133-138.
41. Jackson, R.E. DNAPL source remediation in alluvial geosystems. Presentation at the EPA DNAPL Source Remediation Workshop, Dallas TX, October 29-30, 2001.
42. Cummings, J. Bad day(s) at black rock: The need for 'traction' on the DNAPL remediation front. Presentation at the EPA DNAPL Source Remediation Workshop, Dallas TX, October 29-30, 2001.
43. Garon, K. Perceived constraints to and incentives for source remediation: some technical, economic and policy issues and challenges. Presentation at the EPA DNAPL Source Remediation Workshop, Dallas TX, October 29-30, 2001.
44. Seventh Interim Report on the IDC Demonstration at Lauch Complex 34, Cape Canaveral Air Station. Battelle Columbus, Ohio, August 15th, 2001.

45. Meinardus, H.W.; Dwarakanath, V.; Ewing, J.E.; Hirasaki, G.J.; Jackson, R.E.; Jin, M.; Ginn, J.S.; Londergan, J.T.; Miller, C.A.; Pope, G.A. Performance assessment of NAPL remediation in heterogeneous alluvium. *J. Contaminant Hydrology* **2002**, *54,* 173-193.
46. Rao, P.S.C.; Jawitz, J.W.; Enfield, C.G.; Falta, R.W.; Annable, M.D.; Wood, A.L. Technology integration for contaminated site remediation: cleanup goals and performance criteria. Presentation at the EPA DNAPL Source Remediation Workshop, Dallas TX, October 29-30, 2001. Also to be published in Ground Water Quality 2001, Third International Conference, University of Sheffield, UK, June 2001, IAHS.
47. Rosenberg, N. Chemical engineering as a general purpose technology. In: *General Purpose Technologies and Economic Growth*, Helpman, E.; Editor. The MIT Press, Cambridge MA, 1998, pp. 167-192.
48. *Alternatives for Ground Water Cleanup*; National Academy Press, Washington, DC, 1994.

Chapter 3

Ten Years of Permeable Reactive Barriers: Lessons Learned and Future Expectations

Scott D. Warner[1] and Dominique Sorel[2]

[3]Geomatrix Consultants, 2101 Webster Street, 12th Floor, Oakland, CA 94612 (telephone: 510–663–4100; swarner@geomatrix.com)
[2]S.S. Papadopulos & Associates, 217 Church Street, San Francisco, CA 94111

The permeable reactive barrier (PRB) has been an innovative *in situ* groundwater remediation concept for more than 15 years, though it has been only over the last 10 years that the PRB has been considered a practical alternative for groundwater treatment. Although close to 50 full-scale and tens more pilot tests have been installed since the first commercial application in 1994, and there have been close to one thousand technical papers and presentations since the early 1990s, the technology is still not considered a developed technology due to a perceived lack of cost and performance data. Over the past ten years, the number of full-scale PRBs installed may have been limited first by the uncertainty in applying a new remedial technology, and second, by the need for the designer and user to collect and understand the comprensive characterization, engineering, and logistical information required for each site specific consideration. However, the use of the PRB will become more commonplace as alternative materials and installation options continue to develop, and more information on successes, failures, and cost information become available.

INTRODUCTION

The permeable reactive barrier (PRB) has been a lexicon of the groundwater remediation industry for nearly a decade. Described as a concept more than 15 years ago by McMurty and Elton (1), this relatively simple concept of passively treating groundwater contaminants using a subsurface zone of reactive material was first subject to practical field demonstrations at the Canadian Base Borden site in 1991 by researchers from the University of Waterloo (2, 3, 4). The first commercial application of the PRB as a final groundwater remedy came three years later with the installation of the PRB at a former manufacturing site in Sunnyvale, California (5, 6). Since that time, PRBs have been installed at close to 50 sites as full-scale groundwater remedies to treat a variety of organic and inorganic chemical constituents. Also, nearly 20 additional pilot tests of PRBs have been completed or are ongoing.

Yet, the PRB concept continues to be characterized as *innovative* (rather than *developed*) due to a perceived lack of data on cost and performance and a consideration that its application remains limited (7). Certainly, the PRB concept first gained acceptance due to the perception that passively treating groundwater *in situ* without the use of pumps or other energy-intensive devises, and without the long-term operations and maintenance demands associated with active treatment methods such as "pump-and-treat," would result in a favorable cost to benefit remedy. For some sites, including the Sunnyvale application, this remains true; for other sites, the applications have either been too recent to accurately assess cost and performance information, or uncertainties in performance may exist.

The purpose of this paper is to recount and assess some aspects of the evolution of the PRB as a remediation concept as it moves into the next decade of development and application. In doing so, we attempt to focus on lessons learned and thus lay the groundwater for assuring the proper and successful use of this remedial concept.

CONCEPT AND RESOURCES

The concept of destroying or immobilizing groundwater contaminants using a subsurface treatment zone, as described by McMurty and Elton (1), is a relatively simple concept by which a material capable of directly (or indirectly through geochemical and biological enhancement of the ambient system) destroying or immobilizing the target chemical constituents is placed in the subsurface to intersect the contaminant plume flowing typically under its own hydraulic gradient. As shown in Figure 1, the PRB could be oriented to capture and mitigate a plume migrating both laterally, as well as vertically downward, though the practical application has been to affect laterally-spreading plumes. The PRB can be placed immediately downgradient of a chemical plume to prevent the plume from migrating further, or immediately downgradient of the contaminant source to prevent a plume from developing. The typical PRB is designed not to impede groundwater flow, although ambient hydraulic conditions can be altered by the PRB system, and such potential hydraulic effects must be considered in the design to avoid unintended performance.

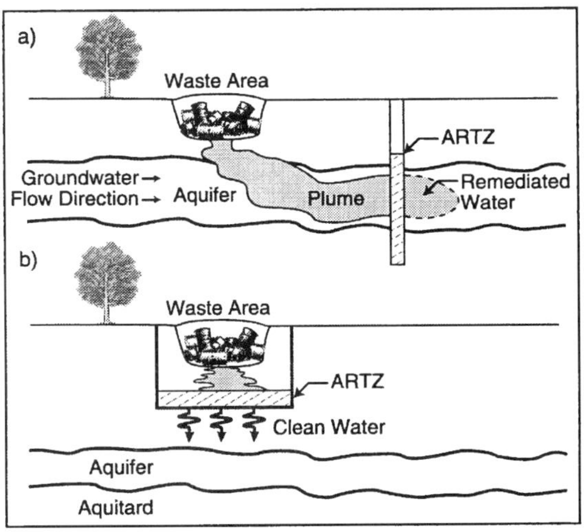

Figure 1. Typical two-dimensional representation of a permeable reactive barrier or applied reactive treatment zone (ARTZ); (a) vertical orientation for lateral capture; (b) horizontal orientation for capture of downward migrating chemicals.

We consider the PRB to be a general category of remediation solutions rather than a specific treatment device because each PRB must be designed for site specific geological, hydraulic, chemical, structural, legal, and economic (e.g., land use) conditions (8). Consequently, consideration of the geometry, design, treatment media, and construction method of each PRB is unique. The consideration of the many parameters necessary to assure a successful deployment of a PRB somewhat explains the difficulty that potential users and regulators have in determining whether a PRB is appropriate for a given site, and in evaluating design criteria prior to implementation.

The site specific uniqueness and consideration of many PRB design variables also partially explains why, since 1991, several hundred technical papers, abstracts, conference proceedings, and more recently, internet web sites, have reported on the development and research of PRBs including field applications, and the chemical and biological reactions of various proposed treatment media. One of the first major conferences to formally devote a session to PRB and related studies was the April 1995 American Chemical Society (ACS) 209th National Meeting held in Anaheim, California. Approximately 40 papers were presented with the focus generally concerning research on the use of the most popular reactive treatment media, zero-valent metals and chiefly zero-valent granular iron, to promote contaminant remediation. Though conferences devoted to *in situ* remediation methods had been held in previous years (9), the 1995 ACS meeting appeared to catalyze industry attention on PRBs as a serious remedial alternative. Since then there have been no less than ten major technical conferences with formal sessions on PRBs and related treatment media (including the 1997 and 2001 ACS national meetings, the 1998 and 2000 Battelle conferences on remediation of chlorinated and recalcitrant compounds, and the 1995, 1997, and 2001 International Containment Technology Conferences).

Table 1 provides a selected list of internet resources available on PRB related topics. These resources are useful for understanding the development of PRBs and treatment media, and are becoming more important for potential users, regulators, and the general public in assessing appropriate PRB design and application. Of note include two internet web sites – the iron treatment media references data base maintained by the Oregon Graduate Institute (OGI), and the Remediation Technology Development Forum (RTDF) – which are devoted to providing technical references and case study information on the use of zero-valent metals as a treatment media, and on PRB deployment, respectively. As of the start of 2001, nearly 500 technical references are listed on the OGI data base and more than 160 references are provided on the RTDF site. The growth in PRB related publication activity is demonstrated in part by the history of publication dates listed on the OGI data base (Figure 3).

PHYSICAL AND CHEMICAL TREATMENT CONSIDERATIONS

The PRB contains two principal components: the treatment matrix, and the hydraulic control system. The treatment matrix is the material or the zone where treatment takes place. Table 2 provides a partial list of treatment materials that have been used or considered for PRB use. The identification and development of alternative treatment materials is becoming a regular occurrence to better

Table 1. Selected Internet Sites on PRBs and Treatment Media

Web Site Name	Web Site Address
Oregon Graduate Institute – Remediation with Iron Searchable database on remediation by zero-valent metals	http://cgr.ese.ogi.edu/iron/
Remediation Technology Development Forum – Permeable Barriers Subgroup Technical literature and case studies on deployment of permeable reactive barriers	http://www.rtdf.org/
Groundwater Remediation Technology Analysis Center Technical literature including technical status reports and case studies	http://www.gwrtac.org/
Interstate Technology Regulatory Cooperation Guidance documents for PRB design prepared by a coalition of state regulatory agencies	http://www.itrcweb.org/
U.S. EPA Technology Innovation Office Technical literature including status reports, case studies, reference lists	http://www.clu-in.org/

remedy complex contaminant mixtures, as well as enhancing the cost effectiveness and longevity of treatment options. For example, Taylor, et al. (10) described laboratory tests evaluating a variety of potential treatment media, including polymer-coated-basalt, enhanced apatite, and crushed pecan shells to mitigate radionuclides. The treatment zone itself may be a trench excavated and subsequently filled with a reactive material, a zone of treatment material that has emplaced via injection, or a zone of geochemically altered native material formed via flushing or injecting a chemical substrate (liquid, solid, or gaseous).

Most PRBs have been implemented to treat groundwater affected by chlorinated hydrocarbon compounds (13). The reactive material of choice for this treatment has been granular zero-valent iron (Figure 2). This treatment material has been the focus of most technical papers on PRBs as demonstrated by comparing paper titles from the 1995 ACS session with this 2001 ACS session which shows that more than 40 of the 50 presentations from the combined sessions are focused on the use of iron as the primary treatment media. What is different today versus 5 to 10 years ago, however, is more a function of practical and economic needs (e.g, lower material costs, assessing methods to increase longevity of reactive materials, developing less intrusive implementation methods, developing deeper implementation methods, etc.) although, there have been some significant developments with respect to understanding the reaction

Table 2. Partial List of Reactive Materials used in PRBs

Treatment Materials	Contaminants Treated	Contaminants not Generally Considered Treatable by Reduced Iron, or with Unknown Treatability
Zero-valent metals (including iron) (may or may not include metal couples)	methanes, ethanes, ethenes, propanes, chlorinated pesticides, freons, nitrobenzene, certain metals (Cr, U, As, Tc, Pb, Cd)	dichloromethane 1,2 dichloroethane aromatic hydrocarbons polychlorinated bi-phenyls, chlorobenzenes, chlorophenols
Ferric Oxides	Mo, U, Hg, As, P, Se	
Zeolites	Sr, Pb, Al, Ba, Cd, Mn, Ni, Hg, certain organics	
Activated Carbon	Mo, U, Tc, cVOCs, BTEX	
Limestone	Cr, Mo, U, acidic water,	
Compost	metals, acidic water	
Alumina	As	
Peat, humate	Mo, U, Cr, As, Pb	
Sawdust, compost	Nitrate	
Oxygen	Aromatic hydrocarbons, MTBE, vinyl chloride	
Phosphates	Mo, U, Tc, Pb, Cd, Zn	

Sources: USEPA (11), USDOE-Grand Junction (12).

Figure 2. Examples of treatment media used in PRB applications: (1) granular iron metal (left), and (2) the zeolite clinoptilolite (right).

mechanisms involved with iron (as well as other reactive materials). For example, although the treatment of chlorinated hydrocarbons was initially thought to rely on the sequential dehalogenation of higher chlorinated compounds (3), the basic premise today is that destruction of the target chlorinated compounds occurs via a two-step beta elimination pathway through acetylene and chloroacetylene as suggested by Sivavec, et al., (14) in 1996 at a meeting of the RTDF in San Francisco. This not-so-subtle change is important in understanding the possible ramifications of incomplete treatment due to not meeting groundwater residence times within the PRB system. We also now have field evidence that these systems, while generally abiotic within their core (15) also promote significant evolution of dissolved hydrogen gas (16) and likely promote reductive and treatment capacity somewhat outside of the rigorous limits of the PRB itself (17, 18). Similar issues in understanding reaction mechanisms also become important in evaluating the longevity of the PRB. As the number of PRBs that have been operating for 3 to 4 years or more increases, the amount of performance data along with the results of special studies designed to assess the potential for porosity loss due to mineralization and particle blinding within the PRB, allows the development of methods to ensure that the treatment reactions are maintained.

Similar to the serendipitous, or rather fortuitous discovery by students and faculty at the University of Waterloo in the mid to late 1980s that reduced met-

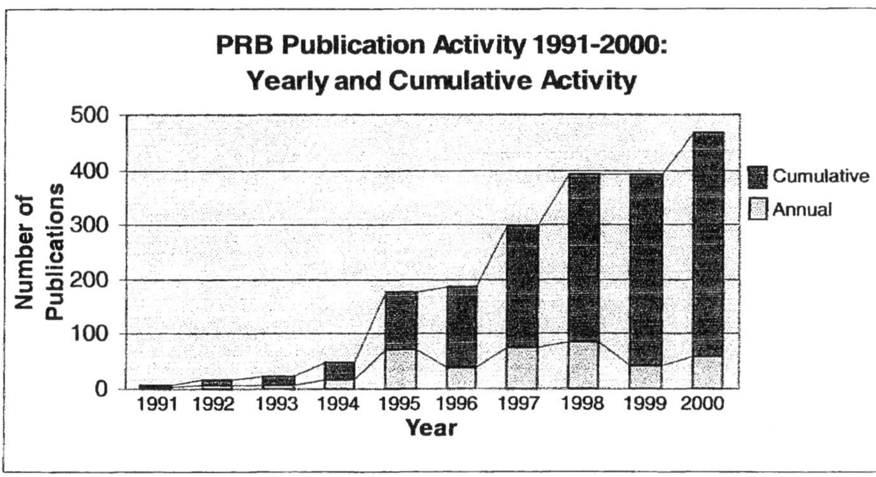

Figure 3. History of PRB Publication Activity 1991 - 2000.

Note: Data compiled from the Oregon Graduate Institute iron references data base. Therefore, numbers indicate publication activity for topics focused on use of zero-valent metals for contaminant treatment.

als, chiefly iron, can degrade chlorinated hydrocarbon compounds in an aqueous system, the approach to developing treatment media for use in PRBs has been less systematic than pragmatic. That is, the treatment media selection process has relied much on an understanding of basic chemical principles that has evolved over the centuries. For example, the basic processes responsible for iron corrosion in an aqueous system, carbonate equilibria response to pH shifts, sorption mechanisms associated with the amount of available organic matter, and ion exchange reactions, form the basis of treatment media selection and application within a PRB. In most cases, the basic chemistry appears to be understood well enough to allow a PRB design to meet treatment objectives. We know that mineralization and corrosion will occur within an iron-based PRB; field data indicate that most precipitation occurs at the PRB interface for many constituents (e.g., iron oxyhydroxides), and is more uniform over the flow through thickness of the PRB for other constituents (19). The results of research also continue to show progress in designing and selecting new mixtures of treatment media that (a) promote the necessary reactions for remediating a greater variety of constituents in a given plume (i.e., the use of the zeolite clinoptilolite [see Figure 2] for removing certain inorganic constituents such as strontium from groundwater), and (b) are less prone to short-term decreases in reaction rates and fouling. For PRBs designed to promote sequential treatment of a chemical plume (e.g., mitigating the migration of inorganic constituents through geochemical reduction, followed by aerobic treatment of aromatic hydrocarbon constituents), the design must consider methods to reduce or eliminate potential competition between the multiple treatment reactions (e.g, high pH conditions from upgradient zero-valent iron zones may limit the ability of certain downgradient oxygen releasing systems to function as intended). Other design considerations also are required to assure hydraulic efficacy between sequential treatment reactive cells or zones. The use of a PRB combined with an approach that relies on natural conditions to provide further degradation of target chemicals downgradient of the PRB (i.e., natural attenuation) will likely be a much more common remedy as the relevant chemical reactions (with respect to both systems) become better understood.

The hydraulic control system is critical to the ultimate performance and effectiveness of the PRB. The hydraulic system functions to (a) route the affected groundwater through the treatment matrix at an appropriate flow rate, and (b) prevent the migration of untreated groundwater around the treatment zone. The hydraulic control system may be simply the ambient groundwater flow system, or it may be augmented by lateral and/or horizontal low permeability barriers constructed to control groundwater pathways and velocity.

The hydraulics of a PRB can be affected by a number of conditions that reflect both the ambient nature of the site and the effects of PRB construction on the inherent hydraulic system. These conditions can include:

- Effect of regional and local precipitation on hydraulic gradient conditions (direction and magnitude).
- Surface water infiltration in vicinity of the PRB.
- Groundwater pumping from nearby areas.
- Smearing or alteration of major groundwater flow zones due to PRB construction.
- Compaction of the treatment material and subsequant loss of permeability within the PRB.

The presence of one more of these conditions can result in unintended hydraulic performance.

It has become apparent that over the past decade of PRB deployment, those PRB systems that have not performed as designed can attribute most of the difficulties to a lack of hydraulic performance rather than to chemical processes. Primary causes for the lack of performance likely are due to incomplete hydraulic characterization of a site, and thus incomplete designs. Most designs to-date would appear to have regular geometries over the entire length of a PRB alignment (e.g., the reactive cell of a long, that is hundreds to thousands of foot-long, PRB system may not be designed to accommodate variability in hydraulic characteristics) which can not effectively respond to changes in ambient groundwater velocities or pressure heads, or a PRB may not have been installed to fully penetrate an underlying low permeability zone along its entire length, thus allowing the potential for underflow to occur. Pilot tests have been one method used to assess the performance of a PRB, and gain valuable design information prior to investing the capital for a full-scale system. However, the pilot test itself can be prone to unintended hydraulic performance, and thus the results can bias (either for or against full-scale installation) the ultimate decision toward the final remedy. If a pilot scale test is applied, it is recommended that the direction and magnitude of the hydraulic gradient be controlled (e.g., potential changes in flow direction are kept to a minimum and the pilot system is oriented to capture the flow as close to perpendicular to the ambient flow direction as possible) to avoid the possibility that groundwater flow would be diverted around the, typically, very short-length of the pilot PRB.

Also, the longevity of the pilot test should be sufficiently long to allow a representative evaluation of the treatment process to be made. The pilot-system installation process itself stresses the subsurface system and causes transient changes to hydraulic and geochemical conditions. The duration of the pilot test should be sufficient to allow these initial effects from construction (e.g., transient water level changes and input of atmospheric oxygen) to dissipate. The reaction process also may require "conditioning" within the subsurface system to allow approximately "steady-state" conditions to develop. A review of the results of numerous pilot and bench-scale tests suggests that more than 20 pore volumes of flow through a PRB composed of zero-valent iron may be required before the reaction-rate can be accurately assessed (J. Vogan, 2001, personal communication). This conditioning process, however, may complicate the assessment of early system performancem, however, as an evaluation of both pilot- and full-scale systems also indicates that treatment media can age (i.e., the reaction rate can decrease) as the number of pore volumes increases (A. Gavaskar, 2001, personal communication).

EMPLACEMENT AND COSTS

The future of PRB use is tied directly to two parameters that each potential application must consider: emplacement approach and cost. Most PRBs to-date have been installed to shallow depths (e.g., less than 50 feet depth) using conventional excavation/fill or trenching techniques. Some of these methods have used sheet piles, trench boxes, or biodegradable slurry solutions to maintain the open trench during infilling of the treatment matrix. Deeper installations (to approximately 120 feet deep) have attempted deep soil mixing, jetting, injection, and fracturing (pneumatic and hydraulic) to emplace the treatment matrix. These developing methods also could be used to place treatment material in depth-discrete chemical migration pathways. Certainly, maintaining the hydraulic control and the groundwater residence time requirements in the deeper installations will be more complicated in the more shallow installations and will require more comprehensive site characterization and design studies to assure a successful application. However, many of the deeper groundwater plumes have few remedies available other that conventional pump-and-treat. Consequently, the development of the more advanced emplacement approaches during the coming years should make the PRB concept more widely acceptable for the plumes in the deeper and more complex aquifer settings.

Cost data on most PRB installations to-date are available only in a general fashion, although general unit costs for the various components of a PRB can be accurately assessed (for example, the average cost of zero-valent granular iron is approximately US$350 to $400 per ton, while hydraulic slurry systems for hydraulic control may have a cost of approximately US$10 per foot). However, the overall costs, which include characterization, design, legal, permitting, and other issues are not well characterized. Generally, the capital cost of a PRB should be in the same range as the capital for many other active groundwater remedies, including groundwater extraction with above-ground treatment ("pump-and-treat"). The savings that a PRB application affords primarily is through lower costs of operation and maintainance activities, compared to the conventional alternatives. As presented during the recently completed U.S. Envionmental Protection Agency training session on PRBs (7), even with then need to replace a PRB on a 10-year interval, the long-term costs for the PRB approach should remain considerably less than the pump-and-treat alternative. As an example, installation of the Sunnyvale PRB (15), which remains the longest running commercial system, reduced annual project expenditures from more than approximately $300,000 to approximately $50,000 following replacement of the former pump-and-treat system at the site. The investment of the capital expenditure for this early PRB, which was on the order of approximately $1million, was thus returned in project cost savings after only a few years of operation (as a note, in 1994 when this PRB was constructed, the unit cost of zero-valent iron was approximately $750 per ton, nearly twice today's price). The annual costs at the site are expected to continue to decrease in the future as well.

LESSONS LEARNED AND FUTURE CONSIDERATIONS

The success of a PRB in meeting the project specific treatment objectives relies on the successful implementation of both the chemical treatment and hydraulic control systems. Significant work continues in the areas of developing enhanced chemical treatment media for a wider variety of target contaminants and under a wider variety of ambient hydrogeological and hydrochemical conditions, and for increasing the longevity of these materials. We anticipate that during the next ten years, research and development of remedial technologies will progress such that most currently known environmental contaminants will be treatable by some reactive media. We also would anticipate that our understanding of the longevity of these treatment measures within a PRB system will also have evolved considerably (20).

Based on a review of case studies, conference proceedings, and technical documents, most of the emphasis over the last 10 years with regard to PRB development, implementation, and monitoring, has been on the chemical treatment portion of the PRB system, with less discussion on how best to design the PRB to incorporate the PRB hydraulics with the chemical processes. Design considerations that more thoroughly address the hydraulic conditions will assure that the chemical treatment processes are as effective as possible in treating affected groundwater.

During the past ten years, the concept of the PRBs as a groundwater remediation alternative has developed from being a hypothetical to a practical solution for many sites. The PRB has been demonstrated to be cost effective for specific sites. We now have more than 5 years of performance data from several sites by which we can better design future PRB installations. From these and more recent installations it is recognized that key aspects to assuring the success of a PRB application include:

- Developing a comprehensive understanding of site characteristics including chemical distribution and hydrogeology.
- Completing a more comprehensive assessment of the relationship between hydraulic design and efficacy of the treatment process.
- Designing the PRB system to optimize longevity.
- Developing monitoring programs that provide representative and accurate performance information.
- Continuing to educate potential users and regulators on the key design aspects of PRB systems.
- Developing a comprehensive cost database for the variety of PRB systems deployed.

Following the April 2001 ACS special symposium on PRBs, a special workshop was convened on April 5, 2001 to discuss the future of PRB technology and to list those considerations that will be paramount to assuring that PRBs are designed effectively and can address sites with greater complexity. The workshop identified the following topics that would appear to be of most interest to the industry at large:

- Treatment approaches –sequential treatment, alternative media
- Emplacement methods – deeper and less expensive methods
- Longevity and methods to rehabilitate PRBs
- Assessing causes for unintended performance of PRBs
- Performance monitoring techniques and strategies
- Developing PRBs for the future

Within the general discussion of the above topics, it was generally agreed to by the participants that the PRB concept remains a valid and potentially powerful tool for effectively mitigating contaminant plumes at a wide variety of sites and protecting potentially threatened downgradient users and resources. However, the group agreed that more work is required to both (a) understand the design criteria necessary to implement a successful remedy, and (b) promote the potential use of the PRB concept at complex sites. As a final note, the group also agreed that collection and interpreting performance information on those PRB sites with unintended results are as important as collecting information on the successful applications so that effective guidance on PRB design and application can be properly formulated and circulated throughout the groundwater remediation industry.

REFERENCES

1. McMurty, D.C.; Elton, R.O. *Environmental Progress*. **1985**, Vol. 4, No. 3, 168-170.
2. Gillham, R.W.; Burris D.R., *Subsurface Restoration Conference*, 3rd International Conference on Ground Water Quality Research, Dallas, **1992,** 66-72.
3. Gillham, R.W.; O'Hannesin, S.F. *Ground Water*, **1994**, Vol., 32, No. 6, 958-967.
4. O'Hannesin, S.F.; Gillham, R.W. *Ground Water*, **1998**, Vol. 36, No. 1, 164-170.
5. Yamane, C.L.; Warner, S.D.; Gallinatti, J.D.; Szerdy, F.S.; Delfino, T.A.; Hankins, D.A.; Vogan, J.L. *209th National Meeting*, American Chemical Society, Anaheim, California, **1995**, Vol. 35, No. 1, 792-795.
6. Szerdy, F.S.; Gallinatti, J.D.; Warner, S.D.; Yamane, C.L.; Hankins, D.A.; Vogan, J.L. *ASCE National Convention*, American Society for Civil Engineers, Washington, D.C., **1996,** 245-256.
7. U.S. Environmental Protection Agency, *Treatment Technologies for Site Cleanup, Annual Status Report (Ninth Edition),* **1999**.
8. Warner, S.D.; Szerdy, F.S.; Yamane, C.L. *Contaminated Soils*, **1998**, 3, 315-327.
9. Gee, G.W.; Wing, N. R., *Thirty-Third Hanford Symposium on Health and the Environment*, **1994**, Vols. 1 and 2.
10. Taylor, T.P.; Sauer, N.N.; Conca, J.L.; Streitelmeier, B.A.; Kaszuba, J.P.; Jones, M.W.; Ware, S.D. *Groundwater Quality 2001*, Third International Conference on Groundwater Quality, University of Sheffield, United Kingdom, **2001**, 110-112.
11. U.S. Environmental Protection Agency, *EPA542/B-00/001*, Office of Solid Waste and Emergency Response, Washington, D.C., **2000**.
12. U.S. Department of Energy, Grand Junction, Research and Application of Permeable Reactive Barriers. April **1998**.

13. U.S. Environmental Protection Agency, *EPA/600-R-98/125*, Office of Solid Waste and Emergency Response, Washington, D.C., 1998.
14. Sivavec, T.M.; Mackenzie, P.D.; Horney, D.P.; Baghel, S.S. *International Containment Technology Conference*, St. Petersburg, Florida, **1997**, 50.
15. Warner, S.D.; Yamane, C.L.; Bice, N.T.; Szerdy, F.S.; Vogan, J.L.; Major, D.W.; Hankins, D.A. *Proceedings of the First International Conference on Remediation of Chlorinated and Recalcitrant Compounds*, Monterey, California, **1998**, 6, 145-150.
16. Sorel, D.; Warner, S.D.; Longino, B.L.; Hamilton, L.A. *Preprints of Papers*, American Chemical Society, Division of Environmental Chemistry, **2001**.
17. Puls, R.W.; Paul, C.J.; Powell, R.M. *Applied Geochemistry*. **1999,** Vol. 14, No. 8, 989-1000.
18. Weathers, L.J.; Parkin, G.F.; Alvarez, P.J. *Environ. Sci. Technol.*, **1997,** 31, 880-885.
19. Phillips, D.H.; Gu, B.; Watson, D.B.; Roh, Y.; Liang, L.; Yee, S.Y. *Environ. Sci. Technol.*, **2000**, 34, 4169-4176.
20. Battelle, *Monitoring plan for the evaluation of permeable reactive barriers at multiple Department of Defense sites*. Prepared for NFESC, Port Hueneme, California, **2000**.

Chapter 4

Field Demonstration of Surfactant-Enhanced DNAPL Remediation: Two Case Studies

Bor-Jier Shiau[1], Mark A. Hasagawa[1], Jeffrey M. Brammer[1], Tracee Carter[1,4], Micah Goodspeed[1,5], Jeffrey H. Harwell[2], David A. Sabatini[3], Robert C. Knox[3], and Erica Szekeres[2]

[1] Surbec-ART Environmental, LLC, 3200 Marshall Avenue, Suite 200, Norman, OK 73072
Schools of [2]Chemical Engineering and Materials Science and [3]Civil Engineering and Environmental Science, University of Oklahoma, Norman, OK 73019
[4]Current address: Tinker Air Force Base, P.O. Box 45663, Tinker Air Force Base, OK 73145–0663
[5]Current address: Parsons Engineering, 4701 Hedgemore Drive, Charlotte, NC 28209

Surfactant flushing field studies have recently been conducted at several DNAPL-contaminated sites by researchers at Surbec-ART Environmental (Norman, OK) and the University of Oklahoma. Results from two demonstrations, Alameda Point, CA (formerly Naval Air Station Alameda), and Spartan Chemical Company Superfund Site, MI, are presented herein. Representative contaminants observed at the sites were trichloroethylene and 1,1,1-trichloroethane (Alameda Point), and a complex mixed waste of methylene chloride, trichloroethylene, ketones and BTEX (Spartan Chemical). These highly successful field demonstrations are all the more impressive given the widely varying site conditions, which supports the technical viability of surfactant enhanced subsurface remediation (SESR). Thus, the use of environmental-friendly surfactants for DNAPL remediation and risk mitigation is very attractive, both economically and technologically, compared to conventional pump-and-treat systems and/or other innovative technologies.

Introduction

Surbec-ART Environmental, Norman, OK and The University of Oklahoma have recently completed several field demonstrations using the surfactant flushing technique. Among these, two case studies will be presented including Alameda Point, CA (formerly Naval Air Station Alameda) and Spartan Chemical Company Superfund Site, MI. Representative contaminants observed at the sites were trichloroethylene (TCE) and 1,1,1-trichloroethane (TCA)(Alameda Point), and a complex mixed waste of methylene chloride ($MeCl_2$), TCE, ketones and BTEX (Spartan Chemical). The major goal for these field tests was to demonstrate that significant NAPL removal (greater than 90 percent) could be achieved by injecting minimal pore volumes of surfactant. Another goal was to reduce the quantity of surfactant used by recycling and reinjecting the surfactant solution treated by an ultrafiltration unit. The performance of these tests was also judged by eliminating the risk of uncontrolled vertical migration of the DNAPL. Since contamination and hydrogeological conditions varied between these test sites, individual site conditions from the two case studies will be briefly introduced, the selection of surfactant system and operations will then be given, and finally significant results and a summary for each demonstration will be presented.

Background

In recent years, surfactant enhanced aquifer remediation has been actively studied at the laboratory and field scales (for example, *1-11*). The major tasks typically associated with surfactant flushing include obtaining regulatory approvals, site investigation, well installation/aquifer testing, modeling, process equipment design, construction and installation, surfactant screening and selection, pre- and post-test tracer testing, system operation, evaluating post-flushing NAPL mass, demobilization and site restoration, residuals management/material storage, and reporting. Some of these tasks are briefly described below.

Well Installation/Aquifer Testing

Preliminary site investigation tasks were conducted to delineate the contaminated zone and to determine the most suitable location for implementing the SESR. After the most suitable location was selected, a few of the study wells, including injection, recovery and monitoring wells, were installed first using common drilling methods, such as hollow stem auger and continuous core sampling, to reach the pre-determined depth. A step-drawdown pump test was conducted at selected newly installed wells and/or at the pre-existing wells. The purpose of the pump test was to determine the hydraulic conductivity (K) and radius of influence (drawdown) at the site. This data was used in the modeling of the site groundwater flow to determine the optimal locations of the system surfactant injection, fresh water injection, recovery, and monitoring well pumping rates. The locations of all the study wells were finalized based on the pumping test data and modeling results (as described below).

Modeling

In general, for proper delivery of the surfactant solution to the target areas, three-dimensional MODFLOW and MT3D analyses are used to, 1) determine potential injection/ extraction well placement locations, 2) evaluate hydraulic capture of the well system, 3) determine potential groundwater production as a result of the recovery wells, and 4) evaluate long-term fate of surfactant not recovered. Prior site investigation reports, hydrogeological information, including hydraulic conductivity, porosity, bulk density, and the storage coefficient, are reviewed and put into the model to simulate aquifer conditions. The resulting hydraulic conductivity from the pumping test was used in the modeling tasks to finalize the installation of well system for the delivery of surfactant.

Process Equipment Design and Installation

A typical surfactant technology design is depicted in Figure 1. Key design parameters for all of the process equipment to be used in a project include flow rates, temperature, influent concentrations, and required effluent standards. The design is purposely versatile and modular in form so that it may be functional regardless of the surfactant system selected for implementation.

The injection/recovery system consists of several pairs of recovery wells, surfactant injection wells, and hydraulic control wells. Each injection and recovery well line is individually controlled with an electric solenoid valve and monitored with an electronic flow meter so the flow to each well is adjustable. The recovery wells are manifolded together. This manifold is connected to an oil-water separator (the initial equipment used for treatment) where any free phase DNAPL is recovered and stored in 55-gallon drums.

The groundwater effluent from the oil-water separator is pumped through pre-filtration units to a VOC separator, such as an air stripper (used at Spartan Chemical), or the Macro Porous Polymer system (MPP-Akzo Nobel; used at Alameda Point) to remove the dissolved NAPL. To reconcentrate and reuse the surfactant recovered in the waste stream, the effluent from the air stripper/MPP system is pumped to the Micellar-Enhanced Ultrafiltration (MEUF) unit. The MEUF system is a spiral-wound cross flow system. The design of the MEUF system is not dependent on the surfactant selected and would not change. However, the filter selected for use in the MEUF is surfactant dependent. In the presented studies, the filters with 10,000 molecular weight (MW) cutoff were used. The permeate from the MEUF is pumped directly into storage tanks. The retentate flows into the surfactant-recycling tank for reinjection.

Surfactant recycling is an important component of the process. When surfactant recycling is not used, surfactant may account for 60 percent to 75 percent of total full-scale project costs, depending on the size of the source area and the surfactant concentration used. To make this technology economically competitive, surfactant recycling and reuse might be necessary, especially when injecting higher surfactant concentrations (> 5 wt%). Even when surfactant recycling is incorporated into the technology, surfactants can account for over 40 percent of total project costs.

Surfactant Screening and Selection

Site soil and groundwater samples are obtained and used to screen for the optimal surfactant system. The laboratory screening activities consist of numerous tests such as contaminant solubilization, surfactant sorption and precipitation, surfactant-NAPL phase properties and contaminant extraction-column studies. Among these, the most important batch tests evaluate surfactant-NAPL phase properties. The objective of these tests is to evaluate the phase behavior of the surfactant (or co-solvent) and NAPL system in an effort to avoid significant reductions in the interfacial tension or unfavorable viscosities or densities that might result in unfavorable flow characteristics. Formation of the middle-phase microemulsion is specific to the contaminant composition and the groundwater system (temperature, surfactant system, ionic strength, etc.) (*1, 7*). These phase behavior studies identify robust surfactant/co-solvent systems capable of achieving high solubilization for the existing groundwater conditions.

Pre- and Post-Tracer Testing

The purpose of a conservative tracer test is to demonstrate hydraulic capture of the injection/recovery system installed. In some cases, a combined conservative/partitioning interwell tracer test (PITT) is conducted prior to the surfactant enhanced subsurface remediation (SESR) demonstration. Partitioning tracer tests are used to evaluate NAPL distribution in the test area, quantify pre-test NAPL concentrations in the flushed zone and target delivery of surfactant during remediation (*12*). Typical tracer concentrations used in the injected solutions range from 500 to 2000 mg/L. Fresh water injection and extraction are typically begun 12-24 hours prior to injection of the tracer to establish capture zones. A 0.15 to 0.3 Pore Volume (PV) tracer solution is mixed and evenly distributed to the injection wells. Groundwater samples collected during the tracer test are analyzed, and the mass balances of each tracer is calculated to make sure that the injected tracers are recovered in the test area by the installed groundwater capture wells.

Results and Discussion

Surfactant system design and results from the two recent case studies are presented in this section. Discussion of test performance and a brief summary will be given respectively for the Alameda Point and Spartan Chemical Sites.

Case 1. Alameda Point (Formerly Navy Air Supply Station), CA

Site Background and Characterization

Data from boring logs, sieve analyses and permeability measurements indicated that two distinct hydrogeological units were present. The upper unit (0 ft – 19 ft) consisted of back-filled sandy material with 1 to 5 ft/d of hydraulic conductivity. The lower section consisted of Bay Mud (19 ft – 30 ft) with a hydraulic conductivity four orders of magnitude less than the fill material. Based on the modeling results, a line drive well configuration, including two surfactant injection wells, four recovery wells, and two hydraulic control wells, was installed. The approach was designed specifically for the hydrogeological conditions at the site. A critical issue was the potential for uncontrolled vertical migration of the DNAPL. Since this project area is underlain by up to 50 ft of low permeability Bay Mud, vertical migration should not be an issue. However, monitoring wells were placed below the test cell in the Bay Mud to confirm that no vertical migration was occurring. In addition, column testing in the laboratory was used to evaluate the potential for mobilization. No vertical migration was observed by the selected surfactant during the 1-D column test (*13*).

Surfactant Selection and Flood

Eight anionic surfactants or mixtures were investigated for their potential use in remediating TCA, TCE, 1,1,-Dichloroethene (DCE), and 1,1,-Dichloroethane (DCA) mixtures. Based on the screening batch and one-dimensional column tests, Surbec concluded that the following surfactant system was the best candidate for this test: disodium hexadecyldiphenyloxide disulfonate (Dowfax 8390) (5wt%), sodium dihexyl sulfosuccinate (AMA-80I) (2%), NaCl (3%), and $CaCl_2$ (1%). Column elution of TCA by a mixture of Dowfax/AMA/NaCl/$CaCl_2$ is depicted in Figure 2. Significant TCA was dissolved in the effluent without mobilizing it. Five tracers were used, including conservative--bromide and 2-propanol (IPA), and partitioning--hexanol, 2,4-dimethyl-3-pentanol (DMP), heptanol. Injection of the surfactant solution was conducted from August to September of 1999. Surfactant samples were analyzed with an HPLC system, and the VOC and alcohol samples were analyzed onsite using the GC/FID method.

Contaminant Breakthrough and Recovery

Key contaminants detected at the site in declining order are TCA, TCE, DCA and DCE. TCA and TCE are the main constituents of the DNAPL collected (over 80%). As a result, the data presented illustrates the combined TCA and TCE recovered. Example of the contaminant mass recovered from the recovery wells is shown in Figure 3.

Based on the flow rates and dissolved VOC concentrations in the recovery wells, it is estimated that the total mass of TCA and TCE recovered was 320 kg (65 gallons; refer to Figure 3), while the total NAPL recovery was approximately 80 gallons (including DCE and DCA). The mass recovery trend is a steep linear relationship through day ten (four pore volumes), followed by a significant reduction in slope. This is consistent with the depletion of DNAPL as the project proceeds. The graph also illustrates the rate of mass recovered from each well. The mass recovery and depletion of DNAPL trends are similar despite the differences in absolute values between wells. After 10-days of surfactant injection, an intermittent injection/recovery approach was used to evaluate the possible rebound of contaminant concentrations during the surfactant flood. The system was shut down for 12 hours and resumed the surfactant injection. No concentration rebound was observed as a result of this intermittent injection (see Figure 3).

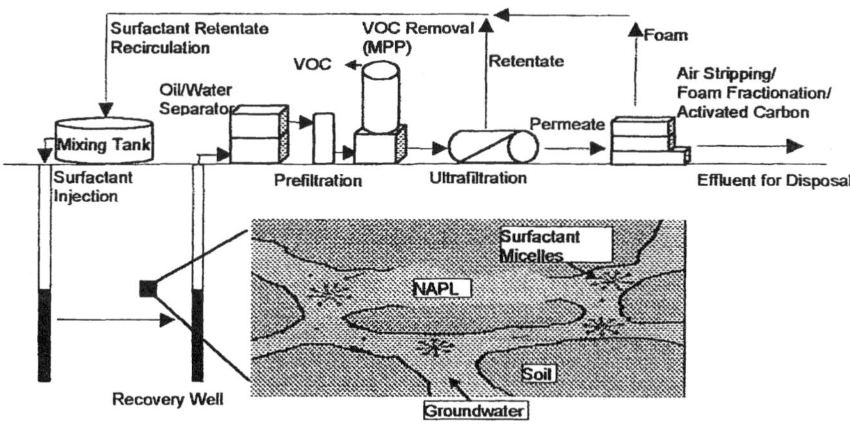

Figure 1: Process Flow Schematic

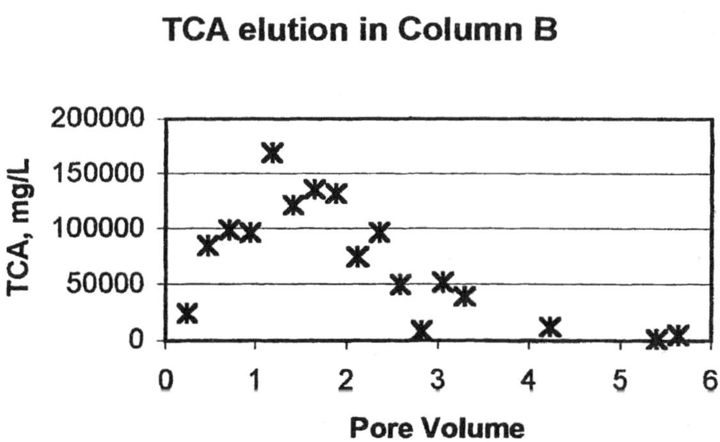

Figure 2: TCA elution by 5%Dowfax/2%AMA/3%NaCl/1%$CaCl_2$ in 1-D column

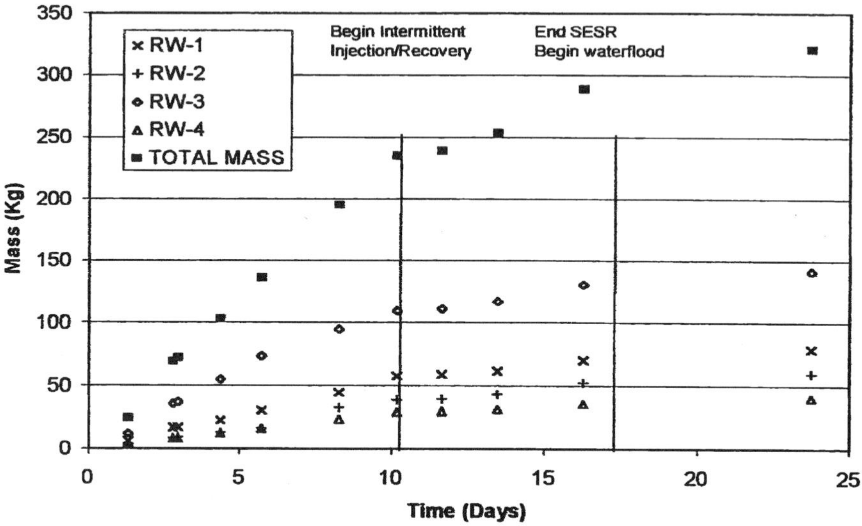

Figure 3: Mass TCA+TCE Recovery from Wells

Comparison of PITT and Soil Coring Results

The mass of DNAPL present within the demonstration cell was estimated using two techniques, PITTs and soil coring data. Based upon the pre-PITT calculations, the estimated DNAPL volume in the cell ranged from 100 to 169 gallons as summarized in Table 1. This data was estimated by calculating the first moment of each tracer breakthrough curve. Results of post-PITT indicated that the NAPL volume left in the cell after the surfactant flood ranged from 0 to 3 gallons (Table 1). The swept pore volumes measured before and after surfactant flood were similar (8499 gallons vs. 8677 gallons). Based on the PITTs, it is estimated that more than 99% of the DNAPL present in the demonstration cell was removed during the surfactant injection and recovery.

The second technique used to estimate total NAPL mass was collecting soil core data. Eight soil corings were completed, and samples collected before and after the surfactant flood from the soil cores were evenly distributed throughout the quarter sections. Results of the pre- and post-test TCA and TCE concentrations in soils are listed in Table 2. The total mass was converted to a volume of 34 gallons (pre-) and less than one gallon (post-flushing) using an assumed density of the mixed DNAPL. A 97% reduction of NAPL mass

Table 1. Results from PITT for the Alameda Point Site

	RW-1	*RW-2*	*RW-3*	*RW-4*	*Total*
Pre-Test NAPL Volume (gal)					
Hexanol*	30	53	29	57	**169**
DMP*	16	32	21	31	**100**
Post-PITT NAPL Volume (gal)					
DMP*	0	0	0	0	**0**
Heptanol*	3	0	0	0	**3**

*These alcohols were used as the partitioning tracers during the PITT.

Table 2: Pre- and Post-Test TCA and TCE Soil Concentrations

Sample Locations	*Depth Ft. bgs.*	*Contaminant Concentrations (mg/L)*			
		TCA-Pre	*TCA-Post*	*TCE-Pre*	*TCE-Post*
IW-1	15.0	20.0	ND[1]	2.60	ND
	16.0	1700.0	NS[1]	190.0	NS
	17.0	63.0	ND	66.0	ND
IW-2	15.0	31.0	ND	7.0	ND
	16.0	36.0	NS	8.0	NS
	17.0	5800.0	0.57	3500.0	0.18
MRW-1/ RW-1[2]	15.0	2.10	0.19	0.82	ND
	17.0	32000.0	300.0[3]	5400.0	120.0
MRW-2/ RW-2	15.0	5.4	ND	3.0	ND
	17.0	42.0	1600.0	23.0	1100.0
MRW-3/ RW-3	15.0	54.0	ND	5.5	ND
	17.0	36.0	630.0	58.0	100.0
MRW-4/ RW-4	15.0	30.0	ND	19.0	ND
	17.0	4.9	25.0	29.0	24.0
MLS-1A	15.0	8.2	ND	2.9	24.0
MLS-1B	16.0	3100.0	ND	770.0	ND
MLS-1C	17.0	32000.0	15000.0	8300.0	730.0
MLS-2A	15.0	39.0	0.14	6.3	0.08
MLS-2B	16.0	270.0	ND	130.0	ND
MLS-2C	17.0	210.0	1.2	180.0	1.5

Note: 1. ND = non-detected; NS = not sampled.

2. Data were combined for these close-distance wells, including the monitoring wells (MRW) and the corresponding recovery wells (RW).

3. The shaded areas indicate samples collected in the clay aquitard beneath the aquifer which will yield anomalous data.

occurred in the demonstration cell after the surfactant flood based on the soil borings data.

Groundwater Contaminant Concentrations

Groundwater samples were taken from seventeen locations including various depths in the treatability study cell boundary prior to SESR implementation. These samples set a baseline for groundwater quality. Samples were also collected from eight borings completed during the site investigation phase.

Table 3 summarizes the pre- and post-test TCA and TCE concentrations in groundwater. The shaded areas showed locations with slight increases in the groundwater concentrations after the SESR. MRW-1, MRW-2, MRW-3, MRW-4 are located near the edge of the treatment zone. The slight increase of post-TCA and TCE concentrations at MRW-2 and -3 are probably due to NAPL outside the cell that has been pulled toward these two wells.

The final groundwater concentrations within the cell boundary showed an 80% decrease in mass of the TCA and a 56% decrease in mass of the TCE. This study demonstrated that proper design of SESR could significantly reduce the contaminant concentrations of soil and groundwater.

Surfactant Recovery

A total of 1540 gallons of Dowfax and 1045 gallons of AMA were used to prepare the injection solutions. Since the actual mass of surfactant injected into the subsurface included surfactant recovered using the MEUF system, the total mass of surfactant injected was actually higher. The total mass recovery of Dowfax and AMA from the subsurface from the subsurface was 117% and 90%, respectively, as compared to the surfactant mass injected.

Volatile Organic Compounds (VOC) Removed by the MPP System

The removal of TCA and TCE by the MPP was less effective while surfactant concentrations were highest, during the surfactant injection stage. On the average, the MPP removed 80% (during peak surfactant concentrations) to 95% (lower surfactant concentrations) of the contaminant from the waste stream.

Table 3: Pre- and Post-Test TCA and TCE Groundwater Concentrations

Sample Locations	*Depth Ft. bgs.*	*Contaminant Concentrations (mg/L)*			
		TCA-Pre	*TCA-Post*	*TCE-Pre*	*TCE-Post*
IW-1	13.0-17.0	210.0	15.0	37.0	5.00
IW-2	13.0-17.0	180.0	1.20	58.0	1.70
FWI-1	13.0-17.0	91.0	1.80	54.0	2.90
FWI-2	13.0-17.0	65.0	1.30	45.0	2.10
MRW-1/ RW-1[1]	13.0-17.0	120.0 (NS)	30.0 (28.0)	37.0 (NS)	9.20 (11.0)
MRW-2/ RW-2	13.0-17.0	49.0 (NS)	61.0[2] (9.7)	18.0 (NS)	68.0 (6.60)
MRW-3/ RW-3	13.0-17.0	190.0 (NS)	120.0 (80.0)	35.0 (NS)	86.0 (32.0)
MRW-4/ RW-4	13.0-17.0	110.0 (NS)	24.0 (110.0)	47.0 (NS)	47.0 (57.0)
2A-AE	6.0-16.0	5.0	1.10	2.1	2.70
MLS-1A	12.75-13.0	480.0	28.0	120.0	9.70
MLS-1B	14.75-15.0	370.0	18.0	88.0	5.10
MLS-1C	16.25-16.5	580.0	5.60	150.0	2.10
MLS-1D	21.75-22.0	12.0	0.12	3.5	0.082
MLS-2A	12.75-13.0	150.0	0.11	54.0	0.13
MLS-2B	14.75-15.0	400.0	97.0	120.0	59.0
MLS-2C	16.25-16.5	440.0	310.0	120.0	170.0
MLS-2D	21.75-22.0	190.0	3.80	68.0	3.70

Note: 1. Data were combined for these close-distance wells, including the monitoring wells (MRW) and the corresponding recovery wells (RW); RW data were indicated in parentheses; NS = not sampled.

2. The shaded areas showed locations with slight increases in the groundwater concentrations after the SESR.

Overall, the MPP system provided an effective, trouble free performance and is a viable technology for contaminant separation from solution.

Summary

The variation in estimates between soil coring and the PITT method (34 gallons NAPL vs. 100 to 160 gallons NAPL) reflects a number of difficulties in quantifying the DNAPL volumes within the subsurface. The differences may be due to insufficient soil boring density or DNAPL pooling or preferential flow paths affecting the sensitivity of PITT results. The pre-PITT results showed fairly substantial variations, and the precision to assign to this analysis is still not clearly defined (*14*). The post PITT and post soil coring results consistently showed little to no DNAPL remaining within the test cell.

The results of this project indicate that SESR is effective for DNAPL mass removal at the subject site. Over 320 kg of TCA and TCE were recovered using four recovery wells. Soil and groundwater concentrations within the demonstration cell decreased significantly (see Tables 2 & 3). Using the most conservative estimates, both the PITT and soil coring results showed that DNAPL removal efficiencies exceeded the 95% target removal. The surfactant system (Dowfax, AMA, NaCl, and CaC12) significantly enhanced the DNAPL solubility and extraction efficiency. The post-remedial ground water concentrations decreased 80% in mass of the TCA and 56% in mass of the TCE. The MPP efficiently decontaminated the surfactant solution for reinjection during the demonstration. This demonstration project thus illustrated the effectiveness and robustness of SESR for remediating subsurface DNAPL contamination at the subject site.

Case 2. Spartan Chemical Site, Michigan

Site Background and Characterization

The Spartan Chemical Company Superfund Site (Spartan) is located in the City of Wyoming, Kent County, Michigan. Surfactant Enhanced Subsurface Remediation (SESR) was the option chosen by the Michigan Department of Environmental Quality (MDEQ) and the Snell Environmental Group (main contractor) for treating the DNAPL contamination at the base of the sand aquifer, approximately 76 ft to 92 ft below ground surface (bgs). Contaminants found at the site included $MeCl_2$, TCE, acetone, methyl ethyl ketone (MEK), methyl isobutyl ketone (MIBK), toluene, ethylbenzene, m/p-xylene, o-xylene, chloroform, and 1,2-dichloroethane. The primary objective of the study was to determine the effectiveness of the SESR treatment system to remove contaminants that cannot be effectively removed using conventional water-based pumping techniques.

System Well Installation

The two recovery wells, RW-1 and RW-2, were completed with 10 ft of screen from approximately 82 ft to 92 ft just above the clay interface at 92.5 ft. The three injection wells, including IW-1, WAT-1, and WAT-2, were screened across the bottom 5 ft of the aquifer. The two monitoring wells, MW-1D and MW-4D, were screened at various depths across the aquifer to monitor for the presence of surfactant and/or enhancement/movement of volatile organics in these zones. The locations of these wells are depicted in Figure 4.

Bench-Scale Study

Laboratory screening tests, as described previously, were conducted to select the optimal surfactant system for application at the site. Since a site-specific NAPL sample was not available, two synthetic NAPL mixtures were used to perform NAPL/surfactant phase behavior studies. The selected surfactant system consisted of 4wt% AMA, 4.5% alkylated naphthalene sulfonate (SMDNS), and 2% NaCl. An example phase diagram for synthetic NAPL2 ($MeCl_2$, 44%; acetone, 25%; TCE, 17%; toluene, 14%) and surfactant mixture is depicted in Figure 5. The optimal Winsor Type III (middle-phase) microemulsion was realized at 2% NaCl (see Figure 5).

Conservative Tracer Test

A Conservative Tracer Test (CTT) was first conducted using bromide (Br^-) and IPA but with little success, mainly due to the interference of IPA present in the site groundwater and low Br^- recovery (30.3% of injected mass). The reason for this low Br- recovery was unclear. A second CTT was conducted using fluorescein, a common fluorescent dye tracer, at a concentration of 19.1 mg/L. The capture results (measured by a fluorometer) indicated that 89.5% of the fluorescein had been recovered by the start of the surfactant injection. Additional recovery occurred during the surfactant injection. The measured contaminant concentrations during the first CTT are shown in Table 4. The VOC concentrations were not measured during the second CTT.

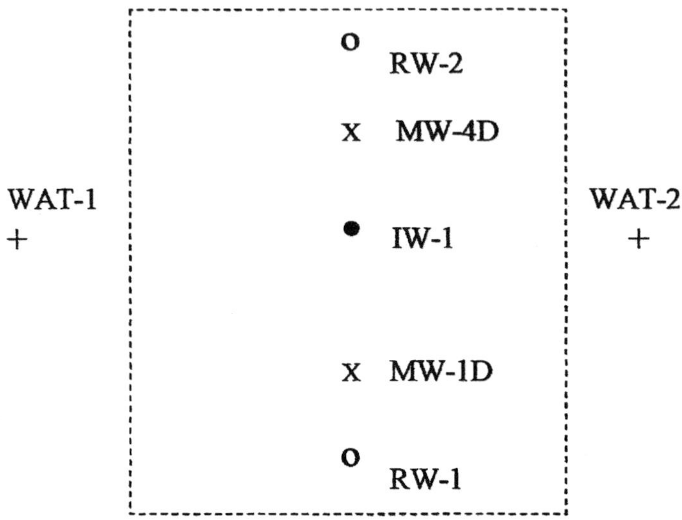

Figure 4: Well Locations at Spartan Chemical Site

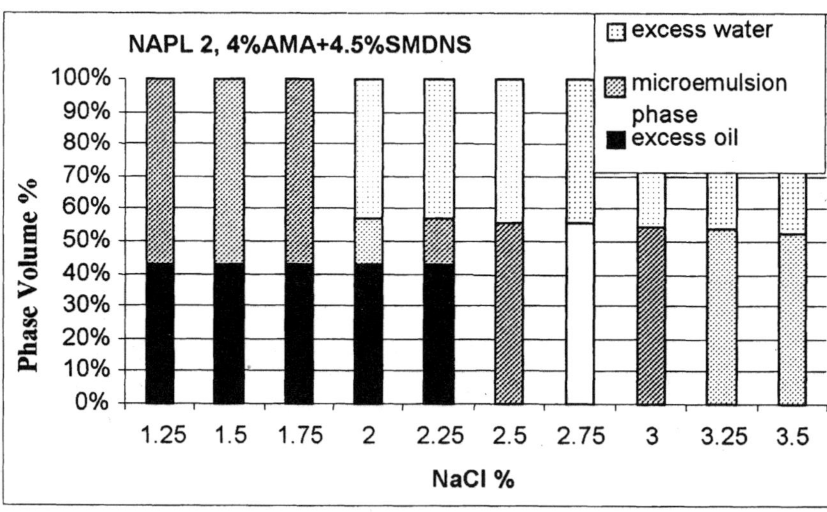

Figure 5: Phase Behavior of Spartan NAPL and 4%AMA/4.5%SMDNS

Table 4: VOC Analyses Conducted by MDEQ during the CTT, July 11th through 12th, 2000

Sample Event	*Time (hr)*	*Contaminant (ug/L)*					
		Acetone	*$MeCl_2$*	*MEK*	*TCE*	*MIBK*	*Toluene*
RW-1							
1	0.0	560,000	32,000	200,000	7,300	120,000	58,000
3	1.0	430,000	30,000	150,000	6,200	100,000	53,000
7	3.0	440,000	25,000	110,000	5,300	91,000	47,000
31	12.0	320,000	30,000	83,000	7,000	95,000	60,000
40	22.0	110,000	17,000	37,000	4,200	42,000	37,000
RW-2							
1	0.0	23,000	28,000	8,400	5,300	6,300	49,000
3	1.0	28,000	31,000	11,000	5,400	11,000	55,000
7	3.0	34,000	25,000	14,000	4,800	16,000	48,000
31	12.0	80,000	20,000	25,000	4,300	26,000	45,000
40	22.0	75,000	12,000	23,000	2,600	18,000	29,000
MW-1D							
3	1.0	110,000	24,000	33,000	4,300	37,000	47,000
MW-4D							
3	1.0	36,000	35,000	19,000	6,700	55,000	65,000
7	3.0	40,000	37,000	18,000	7,500	73,000	69,000

Surfactant/Freshwater Injection

In July of 2000, surfactant injection was initiated into IW-1 (at 10 gpm flow rate). A total 3.9 PV of the surfactant solution was flushed through the treatment area. After surfactant injection, 2.5 PV of fresh water was flushed through the treatment area to remove the surfactant solution from the injection area.

VOC Enhancement

It should be noted that these contaminants are highly soluble and a significant amount of contaminant mass was recovered from RW-1 and RW-2 during pre flooding activities (first CTT), as indicated in Table 4. The contaminant mass removal for acetone, $MeCl_2$, MEK, TCE, MIBK, and toluene was calculated for the time period (22 hrs) and the pumping rates (10 gpm per recovery well) during the first tracer test. The total mass recovered from RW-1 and RW-2 was 35.7 kgs (78.6 lbs). The mass removal started very high as pumping began and reduced steadily as the tracer test continued. By the time surfactant was injected, mass removal had stabilized at a relatively low level. Figures 6a & 6b show how the concentration of methylene chloride varied in monitoring wells, MW-1D and MW-4D, over time during the surfactant/freshwater injection. Additional data for individual contaminants are summarized in Table 5. Monitoring wells, MW-1D and MW-4D, were located in the cell between IW-1 and two recovery wells, RW-1 and RW-2 (see Figure 4). These monitoring wells were indicative of true enhancement within the cell as the analytical results would not be effected by dilution from outside the cell as was true in the recovery wells.

During the test, the contaminant concentrations at the recovery wells (RW-1 and RW-2) remained fairly consistent throughout the surfactant/freshwater injection since much of the water in these two recovery wells comes from outside the cell. The contaminant concentrations from MW-1D during the surfactant/fresh water injection showed two distinct patterns. All of the contaminants in MW-1D showed a similar trend. The contaminant concentration initially is relatively low, then enhancement (i.e. contaminant concentration increases) was observed during the surfactant injection. The concentration then trends downward until a second peak concentration was observed during the freshwater injection. The initial enhancement peak was due to the introduction of the surfactant. The second peak was potentially due to freshwater pushing surfactant that may have been tied up in the soils through the cell to RW-1. Another factor of influence could be that flow from the injection well to MW-1D and then to RW-1 is upgradient. Another encouraging result shown during the surfactant injection was that the highly miscible contaminants, acetone, MEK, and MIBK were also observed to show an

Table 5: Summary of Post-Surfactant Flood Contaminant Concentrations in Ground Water

Sample Event	*Contaminant (ug/L)*					
	Acetone	*$MeCl_2$*	*MEK*	*TCE*	*MIBK*	*Toluene*
RW-1						
1	36000	8300	13000	2300	17000	19000
10	42000	7400	14000	2100	16000	17000
20	34000	6000	12000	1600	13000	14000
30	29000	4400	8900	1200	9500	10000
33	17000	3600	5600	1000	6600	8900
RW-2						
1	31000	5600	10000	1300	8700	15000
10	35000	5100	11000	1200	8900	14000
20	37000	4500	12000	1100	9800	12000
30	27000	2900	7300	700	5800	8500
33	18000	2400	5700	670	4500	7400
MW-1D						
1	560	180	190	130	290	1500
11	1200	230	250	300	330	3100
20	13000	2200	3200	510	6400	7800
29	14000	2200	3100	530	3700	6700
33	8600	1400	1900	550	2300	7500
MW-4D						
1	620	390	210	110	520	750
11	ND	1300	ND	ND	1900	950
20	34000	1900	9000	ND	7600	3400
29	66000	1600	10000	ND	7900	3200
33	5200	480	1000	120	1300	960

enhancement upon injection of the surfactant. Typically, those compounds respond less favorably to the surfactant flushing technique.

Contaminant Mass Removal during the Surfactant Injection

Contaminant mass removal was calculated for acetone, $MeCl_2$, MEK, TCE, MIBK, and toluene for the time period (108 hrs) during the surfactant injection. The total mass recovered from RW-1 and RW-2 was 42.5 kgs (93.6 lbs). Table 6 summarizes the enhancement observed in MW-1D and MW-4D throughout the surfactant injection. The average enhancement factors of the individual compounds ranged from 4 (TCE) to 62.8 (acetone) between the high concentrations and background levels (Table 6). Based on the monitoring well

a

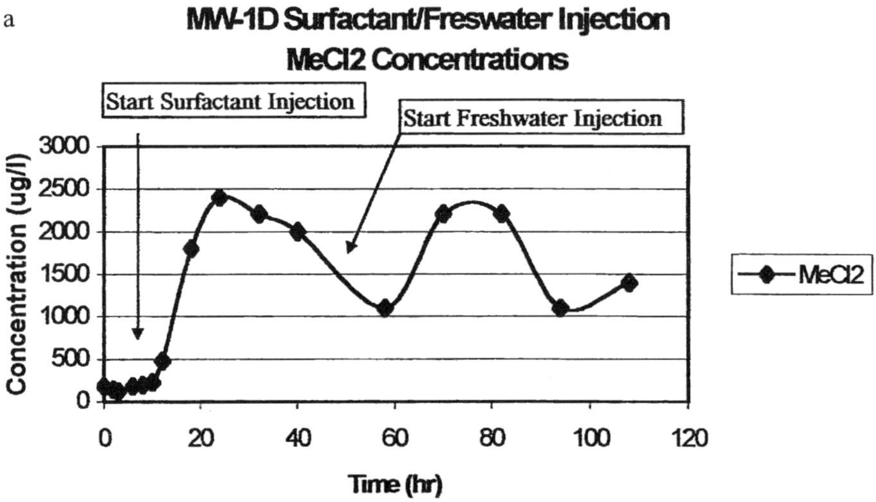

b

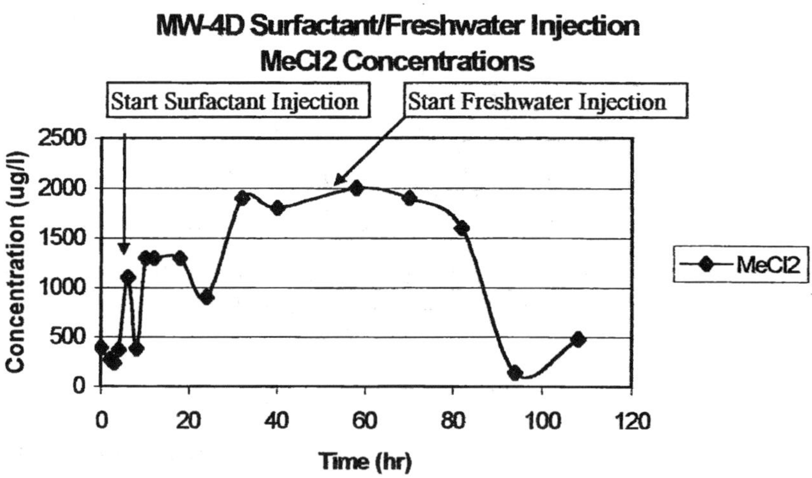

Figure 6: $MeCl_2$ Concentration in MW-1D (a) and in MW-4D (b)

Table 6: Summary of Contaminant Enhancement for the Spartan Chemical Site

Well Number	*Contaminant (ug/L)*								
	$MeCl_2$			*Toluene*			*TCE*		
	Initial[1]	High	Factor[2]	Initial	High	Factor	Initial	High	Factor
MW-1D	160	2400	+15.0	1300	7800	+6.0	114.5	550	+4.8
MW-4D	330	2000	+7.3	590	3700	+6.3	110	350	+3.2
Average			**+11.1**			**+6.2**			**+4**
	Acetone			*MEK*			*MIBK*		
	Initial	High	Factor	Initial	High	Factor	Initial	High	Factor
MW-1D	560	13000	+23.2	190	3200	+16.8	290	6400	+22.0
MW-4D	645	66000	+102.3	205	11000	+53.7	435	10000	+23.0
Average			**+62.8**			**+35.3**			**+22.5**

Note: 1. Average of two background samples, Events 1 and 3; Event 2 was not analyzed by MDEQ.

2. Factor = the enhancement factor observed from the background to the high concentration detected during the surfactant flushing.

data, it is anticipated that the predicted mass removal enhancement should increase at least several fold by surfactant injection, as compared to the water-only flushing. Note that the high pumping rates used at RW-1 and RW-2 (a total of 30 gpm) probably mitigated the impact of the total enhancement due to the dilution effect. The contaminant mass removal was provided to show the potential for mass removal by surfactant injection and groundwater recovery. From a total mass basis more contaminant was removed per hour during pre-flood operations, but the mass removal had dropped significantly prior to the injection of surfactants. Based on the mass recovery in the CTT (35.7 kg) and surfactant flood (42.5 kg), 46% of NAPL mass was recovered during the pre-flood operations in this study.

Surfactant Recovery and Waste Stream Treatment Efficiency

Surfactant breakthrough curves indicated that the final concentrations of AMA and SMDNS surfactant detected in the recovery wells ranged from 8 mg/L to nondetected. AMA was detected as high as 9,721 mg/L in RW-1 and 10,990 mg/L in RW-2. The SMDNS high concentration for RW-1 and RW-2 was 9,883 and 2,495 mg/L, respectively. Hence, it appears that the surfactant was removed from the test cell area based on the mass balance.

For treatment efficiency assessment, samples were collected from the recovered water prior to and after treatment. In this test, most of the VOC mass in the waste stream was removed by a 35-foot packed column air-stripper installed at the site. After air-stripping, the waste stream was then subjected to an ultrafiltration unit for surfactant removal from solution. As shown in Table 7, the treatment efficiency was very high. Typically, acetone is miscible with water, and so the high level of acetone found in the recovery wells is quite unusual. This indicated that the NAPL trapped at Spartan site was quite complicated as compared to a pure acetone system.

Summary

Fluorescein is a good candidate tracer (detection and capture calculation) for the Spartan site. Alcohols are not suitable tracers for this site since numerous alcohols were stored at the site in the past and leaked into the groundwater aquifer. It is not clear if inorganic tracers such as bromide are suitable as tracers for the site. Hydraulic capture was observed during the second conservative tracer test.

Initially, water flooding was effective in removing a large amount of the

Table 7: Influent/Effluent Sample Results

Contaminant	*Influent[1] Concentration Range (ug/L)*	*Effluent[2] Concentration Range (ug/L)*	*Efficiency %*
Acetone	42,000 to 170,000	30 to 730	>98.3
Methylene Chloride	1,400 to 32,000	ND to 21 (1)[3]	>98.5
MEK	11,000 to 72,000	ND to 20 (10)	>99.8
TCE	1,400 to 6,200	ND (1)	100
MIBK	9,900 to 58,000	ND to 38 (1)	>99.6
Toluene	3,800 to 59,000	ND to 13 (1)	>99.7
Ethylbenzene	ND to 3,600 (1)	ND (1)	100
m/p-xylene	ND to 9,700 (2)	ND to 2.2 (2)	>99.9[4]
o-xylene	ND to 4,100 (1)	ND to 1.5	>99.9[4]
Chloroform	ND (1)	3.3 to 5.0	CC[5]
1,2-dichloroethane	ND to 1,100 (1)	ND to 1.2 (1)	>99.9[4]
Bromodichloromethane	ND (1)	1.3 to 1.7	CC
Dibromochloromethane	ND (1)	ND to 1.1 (1)	CC
1,3,5-trimethylbenzene	ND to 28,000 (1)	ND to 1.4 (1)	>99.9[4]
1,4-dichlorobenzene	ND (1)	1.7 to 3.2	CC

Notes: 1. Influent sample collected in process trailer prior to sediment filtration unit, after oil/water separator.

2. Effluent sample collected at manhole on Wrenwood Street.

3. ND = not detected at or above the laboratory detection limit that is indicated in parentheses.

4. % is skewed as highest influent concentrations was used in calculation.

5. CC = can not calculate since influent range starts at non-detect.

more soluble contaminant mass. However, after several days of pumping the rate of mass removal dropped substantially. After surfactant injection, enhanced solubilization (from 4 to 62.8 folds) was observed for various contaminants throughout the test cell during the surfactant injection phase. Concentrations of the highly water-miscible contaminants, acetone, MEK, and MIBK, also showed significant increased during the surfactant injection phase. Greater enhancement was observed in the downgradient direction of the flow in the cell. Hence, a line drive in the direction of flow would probably be the optimal injection recovery configuration. The freshwater injection was successful in removing remaining surfactant from the test cell. The reason for the increase of different ketones in waste stream was unclear, but it might be due to the release of the trapped NAPL from low permeable zones as a result of reducing interfacial tension after surfactant was injected.

References

1. Dwarakanath, V.; Pope, G. A. *Environ. Sci. Technol.* **2000**, *34* (22), 4842-4848.
2. Fountain, J. C.; Starr, R. C.; Middleton, T.; Beikirch, M.; Taylor, C.; Hodge, D., *Ground Water* **1996**, *34* (5), 910-916.
3. Jawitz, J. W.; Annable, M. D.; Rao, P. S. C.; Rhue, R. D. *Environ. Sci. Technol.* **1998**, *32* (4), 523-530.
4. Knox, R. C.; Shiau, B. J.; Sabatini, D. A.; Harwell, J. H. In *Innovative Subsurface Remediation: Field Testing of Physical, Chemical, and Characterization Technologies*; Brusseau, M. L.; Sabatini, D. A.; Gierke, J. S.; Annable, M. D. Eds.; ACS Symposium Series 725, American Chemical Society: Washington, DC, 1999; pp. 49-63.
5. Martel, R.; Gelinas, P. J. *J. Contam. Hydrol.* **1998**, *29* (4), 319-346.
6. Martel, R.; Rene, L.; Gelinas, P. J. *J. Contam. Hydrol.* **1998**, *30* (1-2), 1-31.
7. Sabatini, D. A.; Harwell, J. H.; Knox, R. C. *Progr. Colloid Polym. Sci.* **1998**, *111*, 168-173.
8. Sabatini, D. A.; Knox, R. C.; Harwell, J. H.; Wu, B. *J. Contam. Hydrol.* **2000**, *45*, 99.
9. Shiau, B. J.; Sabatini, D. A.; Harwell, J. H., *J. Environ. Eng.* **2000**, *126* (7), 611-621.
10. Shiau, B. J.; Sabatini, D. A.; Harwell, J.H.; *Environ. Sci. Technol.* **1996**, *30* (1), 97-103.
11. Weerasooriya, V.; Yeh, S. L.; Pope, G. A. In *Surfactant-Based Separations: Recent Advances*; Scamehorn, J. F.; Harwell, J. H. Eds.; ACS Symposium Series 740, American Chemical Society: Washington, DC, 1999; pp. 23-34.
12. Jin, M.; Delshad, M.; Dwarakanth, V.; McKinney, D. C.; Pope, G. A.; Sepehrnoori, K.; Tilburg, C.; Jackson, R. E.; *Water Resour. Res.* **1995**, *31* (5), 1201-1211.
13. Surbec Environmental, In *Technical Memorandum for Surfactant Enhanced DNAPL Removal Treatability Study at Alameda Point (Site 5)*, Section 5, Department of the NAVY Engineering. Field Activity West, San Bruno, California 94066-2402, 1999.
14. Knox, R. C.; Goodspeed, M.; Sabatini, D. A. University of Oklahoma; 2001. unpublished.

Chapter 5

Permanganate Oxidation Schemes for the Remediation of Source Zone DNAPLs and Dissolved Contaminant Plumes

X. David Li and F. W. Schwartz

Department of Geological Sciences, The Ohio State University, Columbus, OH 43210

This study utilized a series of laboratory experiments to explore the feasibility of applying oxidation techniques to clean up ground water and soil contaminated by chlorinated ethylenes. We investigated problems that could potentially impact the oxidation of TCE with permanganate, such as MnO_2 precipitation around zones of high DNAPL saturation, and permeability-related flow bypassing. 1-D column and 2-D flow tank experiments provided data on mass removal rates and related flushing efficiencies. The results indicate that mass removal rates are greatly influenced by MnO_2 precipitation and flow bypassing. It is anticipated that in actual field settings, the issue of flushing efficiency needs to be considered in the design. A permanganate reactive barrier system (PRBS) was designed and a proof-of-concept experiment was carried out in a 2-D flow tank. Results show that the PRBS was able to delivery permanganate at a stable, constant, and controllable rate. The design works well in reducing the concentration of the TCE contaminant plume. This idea could be developed as a long-term reactive barrier for the remediation of plumes of chlorinated solvents. Installation would be relatively quick, and cost efficient. Once installed, its long-term management cost would be minimal, and it would require only sporadic recharging with permanganate.

Introduction

Subsurface contamination by chlorinated organic solvents presents a serious threat to the nation's soil and ground-water quality. The most common contaminants are chlorinated ethylenes, like tetrachloroethene (PCE) and trichloroethene (TCE), which are listed at the top of the national priority list for environment clean up (*1*). Contamination by chlorinated ethlylenes usually occurs in source areas as dense non-aqueous phase liquid (DNAPL), along with a plume down gradient. Often, site remediation must address both types of contamination.

The *in-situ* destruction of chlorinated ethylenes by potassium permanganate has been studied by many researchers (*2, 3, 4, 5, 6*). Yan and Schwartz (*4*) elucidated the reaction pathway and the products of TCE oxidation by MnO_4^-. Using isotopically labeled materials, they found that MnO_4^- oxidized TCE to CO_2 with carboxylic acids as intermediate products. Another key product in this reaction is manganese oxide (MnO_2), which precipitates as a solid.

Based on the mass balance from the laboratory experiment, Schnarr et al. (*3*) proposed the following general equation for the overall reaction of permanganate with chlorinated TCE

$$C_2HCl_3 + 2MnO_4^- \rightarrow 2MnO_2 + 2CO_2 + 3Cl^- + H^+ \quad (1)$$

From the storichiometric relationships in Equation (1), the destruction of TCE can be monitored by tracking the accumulation of the chloride over time.

Previous studies have shown that permanganate is efficient in oxidizing chlorinated ethylenes in an aqueous phase, with half-lives on the order of tens of minutes. Ultimately, all of the chlorinated compounds and intermediates are mineralized to produce harmless products. With these obvious advantages, this technology has been rushed into field applications concerned with the removal of source-zone DNAPLs at various sites around the nation (*7, 8*). At these field sites, investigators reported a dramatic decrease in the contaminant concentrations when permanganate flushing was applied. However, in some cases, concentrations rebounded after the treatment was stopped. Without detailed and careful observations, it is difficult to determine whether this rebound was caused by the remaining DNAPL or simply by the re-establishment of the contaminant plume from upstream.

Clearly, there is a need to better understand the *in-situ* oxidation process, especially with respect to the behavior of the solid reaction products and their effects on the porous medium. We used visualization-type experiments to evaluate the *in-situ* oxidation process. Experience has shown visualization approaches to be useful in understanding processes in the subsurface (6). Thus, an objective of the study is to assess the extent to which the precipitation of MnO_2 impacts the oxidant flood.

Oxidation of chlorinated ethylenes by permanganate has been developed primarily as a remediation method for source-zone DNAPL cleanups. To date, there has been limited work in applying permanganate for the treatment of dissolved organic contaminants in plumes. The most important problem in this respect is how to deliver the chemical to the target area in an efficient and economical fashion. Two difficulties in permanganate delivery are its high solubility and strong tendency to react with other materials, including the aquifer solids. The high solubility makes it difficult to use solid $KMnO_4$ in a conventional reactive barrier wall. Its strong reactivity requires that the delivery be on target before it reacts during transport. In order to solve these problems, a permanganate reactive barrier system (PRBS) was designed (Figure 1). The design includes a series of vertical wells to deliver solid potassium permanganate into the subsurface, through diffusion out of the well casing. The idea is to form a zone of localized reaction through dispersive mixing with dissolved TCE (Figure 1). Thus, as the contaminant plume passes the PRBS, *in-situ* oxidation will destroy the contaminant and prevent further spreading. This semi-passive reactive barrier system is similar to concept proposed by Devlin and Barker (*9*), in which a permeable nutrient injection wall was used to deliver potassium acetate to stimulate bioremediation.

A second objective of this study is to use 1-D and 2-D column and flow tank experiments to investigate issues such as contaminant removal rate, and the effect of reaction products on flow. Digital visualization techniques form the basis for this proof-of-concept experiment.

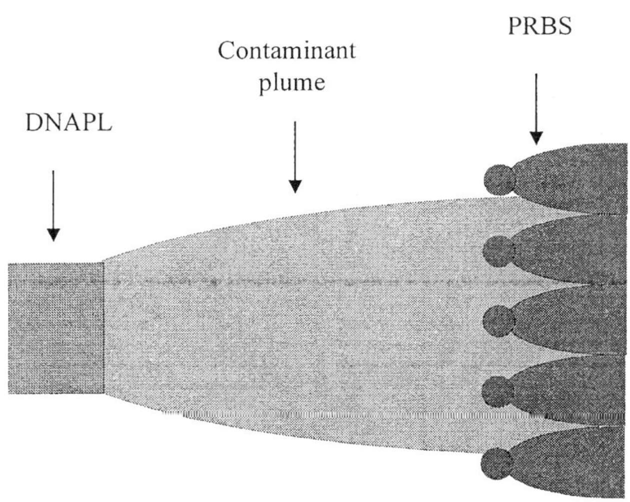

Figure 1. Design of the permanganate reactive barrier system (map view)

Methods

Experiment setup

1-D Column Experiment. The 1-D column experiment was conducted using a glass column with Teflon end fittings and packed with medium silica sand (45 micron) (US Silica, Ottawa, IL) (Figure 2). Details of the experimental set up are summarized in Table 1. Once the column was packed, it was positioned upside down. One mL of TCE was added to the column evenly across the opening to create a zone of residual DNAPL saturation. The column was returned to an upright position and fluid was pumped upward through the column using an Ismatec tubing bed pump (Cole Parmer Instrument Co. Vernon Hills, IL). Effluent samples were collected periodically for chemical analysis. After the experiment finished, several porous medium samples were taken along the column. The samples were treated with thiosulfate to dissolve the MnO_2. The resulting solution was analyzed by ICP-MS to quantify the distribution of MnO_2 along the column.

Table 1. Experimental design for the 1-D column and 2-D flow tank experiments

Experiment	*Dimension (mm)*	*Medium*	*Q (mL/min)*	*$KMnO_4$*	*Porosity*
Column	605(L)x50 (ID)	Silica sand	1.2	1 g/L	0.385
Flow tank	305(L)x50(H)x3 (T)	Glass beads	0.5	0.2 g/L	0.42
PRBS	305(L)x50(H)x3 (T)	Silica sand	0.05	Solid	0.385

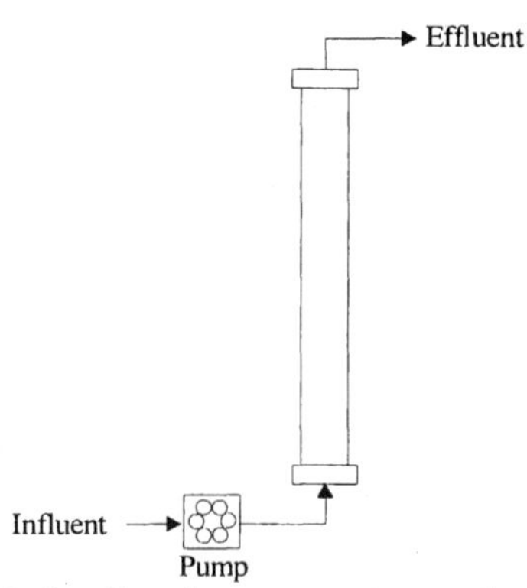

Figure 2. 1 – D column experimental setup

2-D Flow Tank Experiments. A thin, small 2-D glass flow tank with Teflon end fittings was constructed for these experiments (Table 1). The flow tank was filled with transparent borosilicate glass beads with a mean grain size of 1 mm. Two PTFE (polytetrafluoroethylene) tubes with inside diameter of 1.32 mm were installed in the two ends of the column to function as recharge and discharge wells. Both tubes had ends open at a depth of 0.5 cm from the bottom of column. Once the column was packed, it was saturated with Milli-Q water. A Spectroflow 400 solvent delivery system (Kratos analytic, NJ) provided flow into the tank. An Ismatec pump removed fluid at the outlet (Figure 3).

With both pumps running at the same steady rate, the ambient flow was horizontal along the length of the tank. One mL of the TCE was added to the tank from the top to form a zone of residual DNAPL saturation across the vertical depth of the tank and a small DNAPL pool at the bottom. The tank was flushed for about two weeks with $KMnO_4$. Effluent samples were taken three times a day for chemical analysis. Images of the column were taken with a digital camera to monitor the development of MnO_2 precipitation.

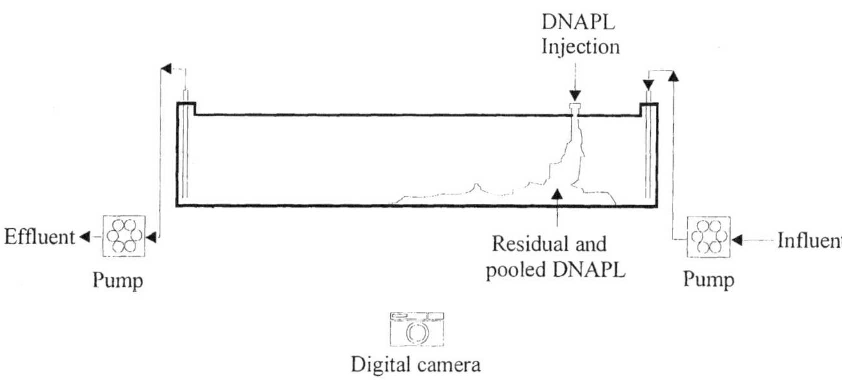

Figure 3. 2–D flow tank experimental setup

Test of the PRBS Design. The 2-D flow tank was modified slightly to test the efficacy of PRBS. A 1 mm (inside diameter) Teflon tube was installed in the flow tank to represent an inlet well in the PRBS. On one side of the Teflon tubing wall, micro holes 1 mm in diameter was drilled to release permanganate into the porous medium. The MnO_4^- was released as small quantities of water entered the tube to slowly dissolve the solid $KMnO_4$ particles inside the tube.

An ambient horizontal flow was established using pumps on the inlet and outlet. The experiment began by flowing de-ionized water through the tank for two days, at that time, the inlet pump was switched to deliver dissolved TCE at a concentration of TCE of 25,000 μg/L. This unusually large concentration was used in part to facilitate the visual monitoring of the experiment. At the outlet, effluent samples were collected at regular time intervals. The narrow tank permitted back lighting and visual monitoring of the MnO_2, which formed an observable brown precipitate.

Visualization

During the 2-D flow tank and PRBS experiment, digital photography was used to monitor the flooding and plugging with MnO_2. With either the glass beads or the silica sand, the flow tank is semi-transparent with light from the back of the tank. The formation of the reaction product could be monitored by the digital images as a function of time using a Nikon 950 digital camera. The visual monitoring was carried out inside a dark room to avoid interference from ambient light. After the experiment was completed, the images were downloaded to a PC, and were processed with Photoshop to correct the slight size differences among the images during the extended experiment.

Chemical analysis

The effluent from the column and flow tank was collected periodically for chemical analysis to monitor TCE degradation. Concentrations of TCE, Cl^- and MnO_4^- were monitored during the entire experimental period. We utilized the Cl^- concentration to estimate the quantity of TCE that was oxidized, and the overall efficiency of the cleanup scheme.

TCE concentration was measured with a Fisons Instruments 8060 gas chromatograph equipped with a Ni^{63} electron capture detector and a DB-5 capillary column (J&W Scientific, Rancho Cordova, CA). Pentane was used for liquid-liquid extract of TCE. Cl^- ion concentration was measured with a Buchler Digital Chlorodometer. Permanganate concentrations were measured with a Varian Cary 1 UV–visible spectrophotometer at a wavelength 525 nm. The samples of MnO_2 were dissolved and analyzed by a Perkin-Elmer Sciex ELAN 6000 Inductively Coupled Plasma Mass Spectrometer (ICP-MS).

Results

The 1-D column experiment confirms the ability of MnO_4^- to oxidize the TCE present in the column. Mass balance calculations, based on the measured TCE and Cl^- concentration in the effluent from the column, indicated the nearly complete removal of the pure TCE in a relatively short time (Figure 4). Integrating the TCE removal curve shows that after 365 hours of permanganate flushing, 96.9% of the initial TCE was removed from the column. The rate of TCE removal was highest at early time when MnO_4^- first entered the column, and decreased significantly with time. Ninety percent of the TCE was removed in the first 115 hours of the experiment, while another 250 hours was required to remove the last 7%. Flushing by MnO_4^- was halted when Cl^- measurements suggested that TCE had been removed from the column. Flushing continued with Milli-Q water pumped through the column. Interestingly, TCE concentrations rebounded to about 130 μg/L. Apparently, the Cl^- measurements were not sensitive at the lower level of detection to the small quantities of Cl^- produced by the oxidation of TCE that remained in the column. It is estimated that several hundred additional hours would be needed to bring TCE concentration below 5 μg/L, the regulatory limit.

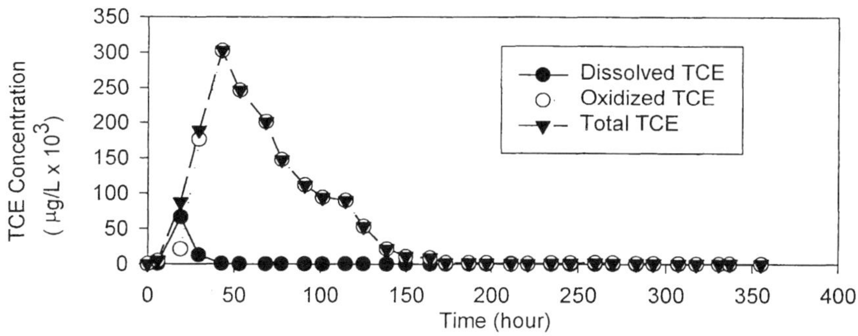

Figure 4. Variation in TCE concentration with time in the effluent of the 1 – D column and the calculated removal based on Cl^- stoichiometry.

The distribution of MnO_2(s) in the column at the end of the experiment is presented in Figure 5. The majority of the MnO_2 was located at or closely adjacent to the original DNAPL zone. These precipitates tended to plug the column toward the end of the experiment, making flushing increasingly difficult.

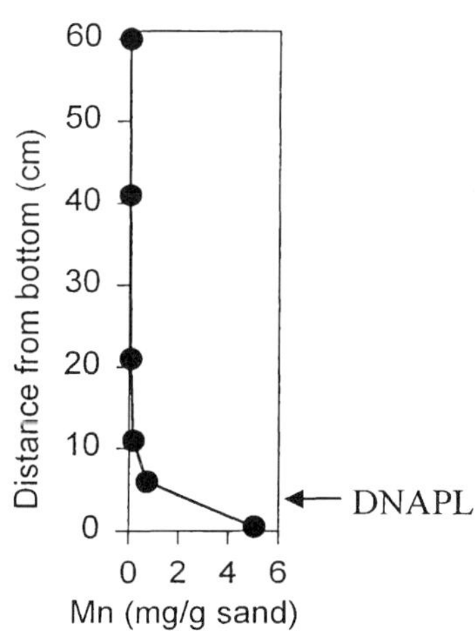

Figure 5. Distribution of Mn along the column after in-situ oxidation of DNAPL.

The results from the 2-D flow tank experiment (Figure 6) were similar to the 1-D experiment, except that the removal efficiency was noticeably less. After 313 hours of flushing with MnO_4^-, only 34.9% of the initial TCE was removed from the tank. Half of the TCE was removed in the first 25% of the elapsed time. Note that in Figure 6 at the end of the experiment the TCE concentration again rebounded, once the injection of MnO_4^- stopped. With increased TCE concentration, the residual permanganate in the flow tank reacted and produced an increase in Cl^- concentration at the end of the experiment.

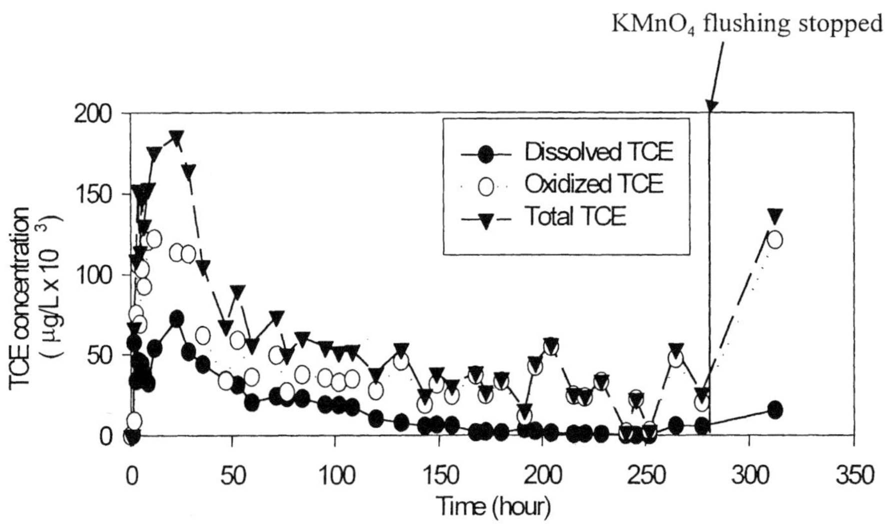

Figure 6. Effluent TCE concentration change with time in the 2 – D column experiment.

The digital imaging system provided a useful way to monitor the growth in the zone of MnO_2 precipitation (Plate 1). Precipitation started once MnO_4^- came in contact with dissolved TCE that occurred in close proximity to the zone with pure TCE DNAPL. The zone of MnO_2 precipitation was marked by a light brown color down gradient of where the residual DNAPL occurred (Plate 1(b)). There was a tendency for the MnO_4^- flood to bypass the zone with the highest DNAPL saturation, moving instead through a much less saturated zone in the upper portion of the tank (Plate 1 (c)). With time, the precipitation of MnO_2 reduced the permeability across most of the tank. MnO_2 rapidly formed a precipitation rind above the small DNAPL pool located near the bottom of the tank (Plate 1(d), (e), (f)). A greater injection pressure was required to maintain the flow near the end of the experiment. The experiment was halted when MnO_2 plugged the tank nearly completely and MnO_4^- could no longer be injected. Carefully examination of the tank after the experiment indicated the presence of tiny CO_2 bubbles, produced from the oxidation reaction. The gas bubbles likely played a role in reducing the permeability and the flow in the system. Flushing with MnO_4^- appears effective in removing residual DNAPL. However, much of the original volume of the pool of DNAPL at the bottom of the tank was evident at the end of the experiment.

Figure 7 shows how the concentration of MnO_4^- changed with time at the outlet of the flow tank during the test of PRBS. The concentration of MnO_4^- stayed at around 0.5 g/L after a decrease at early time, which was caused by solid $KMnO_4$ particles dropping out of the tubing during and following the initial loading. The results indicate that the system is capable of delivering the permanganate at a relative constant rate to the plume. At the outlet, the TCE concentration was below 20 μg/L most of the time (Figure 8), which is a three to four orders of magnitude decrease as compared with the input concentration. The design works well in reducing the concentration of dissolved TCE. The measured chloride concentration at the outlet (Figure 9) also indicates the break down of TCE with time. The digital images show the progressive accumulation of MnO_2 on the downstream side of the PRBS (Plate 2). This plugging may have reduced the hydraulic conductivity causing lower MnO_4^- concentrations in the latter part of the experiment. Also, the MnO_4^- concentration distribution in the vertical direction is affected by density-driven flow. Density effects need to be taken into consideration in the future designs.

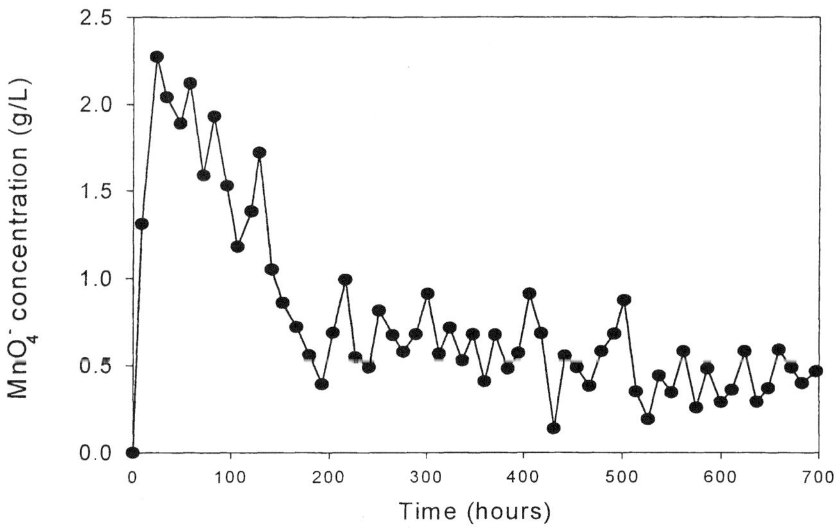

Figure 7. KMnO4 concentration at the outlet, downstream of the PRBS

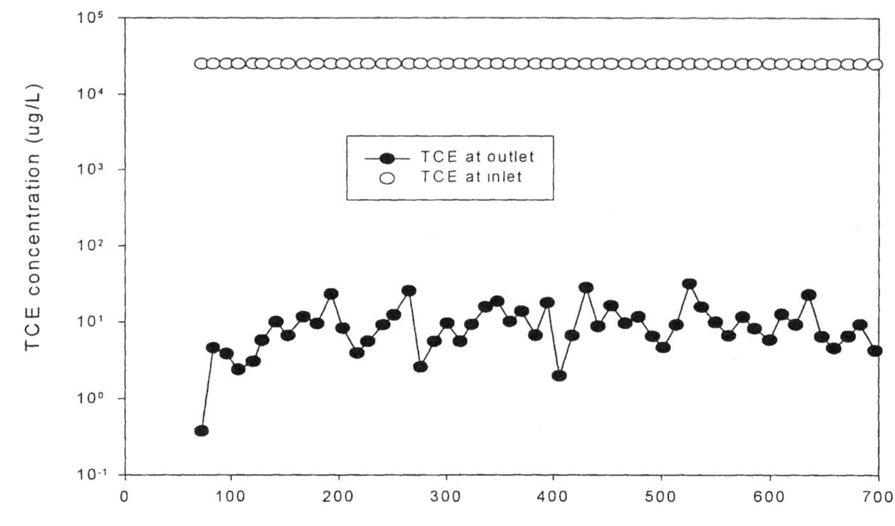

Figure 8. TCE concentration at both inlet and outlet of the flow tank containing the PRBS

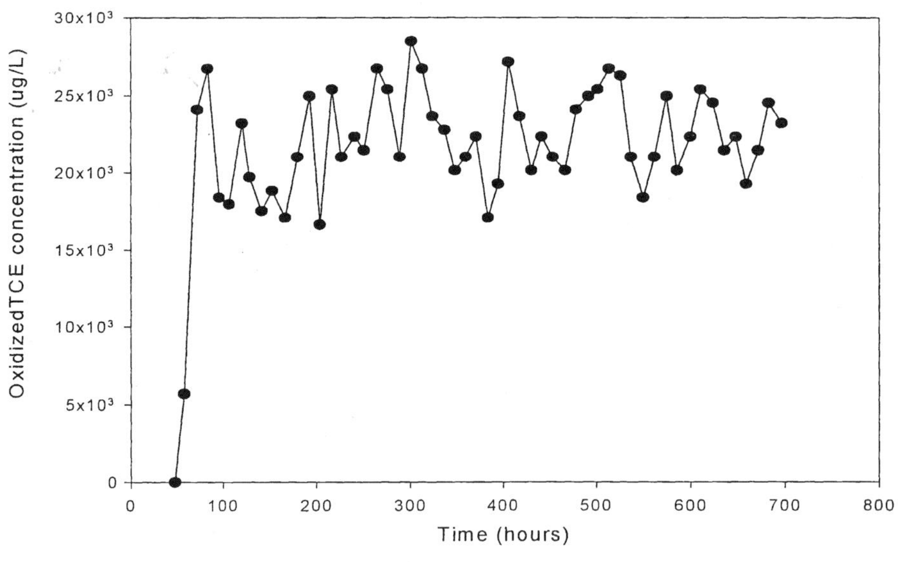

Figure 9. Oxidized TCE concentration at outlet, downstream of the PRBS.

(a) Time: 0. Two-dimensional sand tank at the begainning of the experiment. The arrow shows the flow direction.

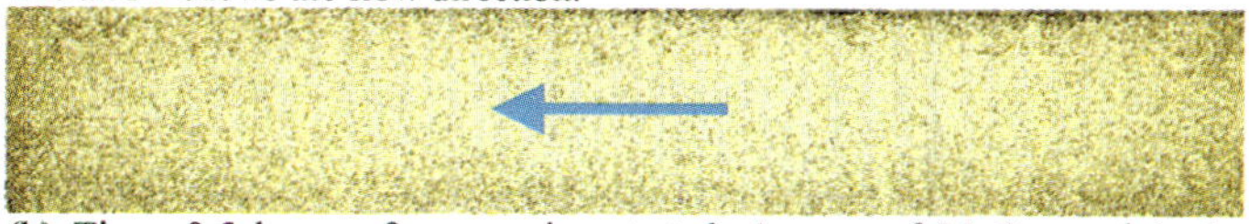

(b) Time: 3.5 hours after pumping started. A zone of MnO_2 precipitation is formed immediately downstream of where the residually saturated TCE is present.

(c) Time: 12 hours after pumping started. Much more extensive MnO_2 precipitation is observed. A small pool of TCE has became evident at the bottom of the tank.

(d) Time: 48 hours after pumping started. The more intensive precipitates indicated that TCE is being oxidized. However, the DNAPL pool is relatively unaffected.

(e) Time: 7 days after pumping started. Flow of MnO_4^- is now mainly occurring at the top of the tank.

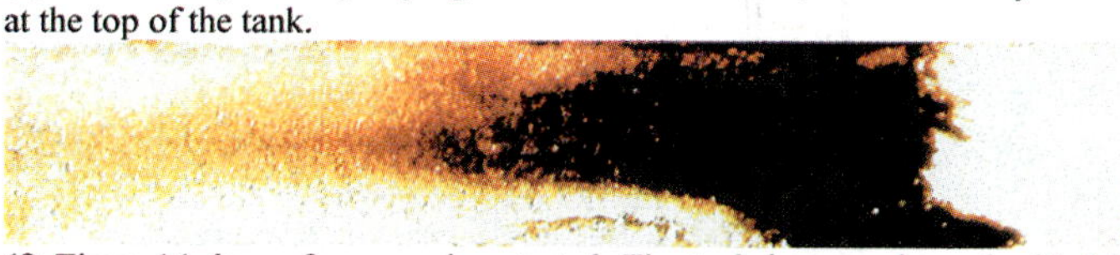

(f) Time: 14 days after pumping started. The tank is now plugged with MnO_2. Note how the TCE pool has been sequestered.

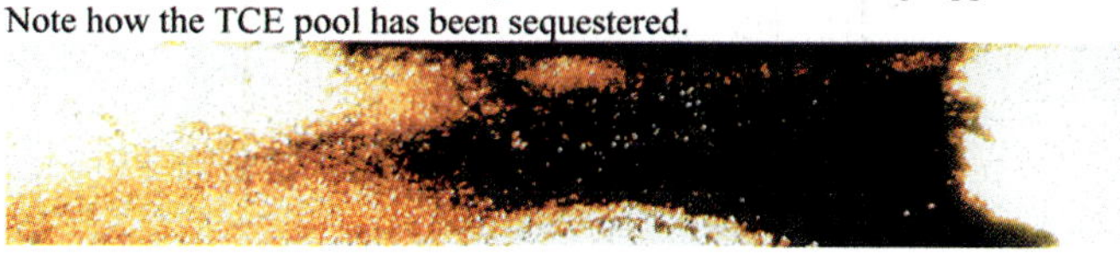

Plate 1. DNAPL removal in the 2-D flow tank experiment. The dark brown color indicates the presence of $MnO_2(s)$, and the light color at the mid-bottom of the image (f) is the remaining pool of DNAPL after flushing had stopped

(a) Time: 1 hour after the start of pumping Milli-Q water. MnO_4^- is diffusing from the tube.

(b) Time: 4 hours after the start of pumping Milli-Q water. The dissolved MnO_4^- created downward density-driven flow that helps to facilitate mixing.

(c) Time: 3 days after the start of pumping TCE solution. Considerable MnO_2(s) is noticeable downstream of the PRBS.

(d) Time: two weeks after the start of pumping TCE solution.

(e) Time: four weeks after the start of pumping TCE solution. MnO_2(s) continue to accumulate.

Plate 2. Experiment involving the proof-of-concept of the PRBS. Early images capture the diffusion and the sinking of the $KMnO_4$ solution in (e.g., b). MnO_2 precipitates downstream and accumulates through the experiment. (e.g., d, e).

Discussion

Results from both the 1-D and 2-D experiments indicate that early in the flushing process, the DNAPL mass removal rate was greatest. As the treatment proceeded, the rate decreased dramatically. Conducting the experiments to achieve complete DNAPL removal was frustrated by the tailing effect in the TCE concentration toward the end of the treatment. This tailing effect has been noted by researchers studying DNAPL dissolution (*10, 11*). Towards the end of their experiments, when a large proportion of the initial DNAPL was dissolved, small droplets remained in hydraulically stagnant zones in the medium. The stagnant zones are not flushed because preferential flow paths develop even within the homogeneous media. With MnO_4^- oxidation schemes, the tendency for preferential flow paths to develop will be further promoted as MnO_2 precipitates in the zones of higher DNAPL saturation or CO_2 bubbles are trapped in the porous medium. The rapid oxidation and relatively slow mass transfer rate from the DNAPL to the aqueous phase means that MnO_2 would tend to precipitate at or immediately adjacent to the DNAPL, which results in a zone of more concentrated precipitates around the zones of greatest saturation. Flow would tend to bypass these zones and follow a more permeable flow path, causing the DNAPL oxidation process to become diffusion controlled. Although CO_2 bubbles in the subsurface can dissolve into the aqueous phase according to Equation 2, the rapid accumulation of CO_2 early in the flushing led to the production of the bubbles. Flow bypass also reduced the possibility for water in contact with the bubbles, to dissolve them. Thus, the effect of the bubbles in reducing the permeability is persistent.

$$CO_2 + H_2O \Leftrightarrow HCO_3^- + H^+ \qquad (2)$$

These effects are most obvious in the 2-D tank experiment where DNAPL is present at high residual saturation as a small pool. The pooled DNAPL persisted because the interfacial area was relatively small and MnO_4^- moved around but not through these zones. In zones of lower residual saturation, TCE oxidation was more efficient because MnO_4^- could move through the zones. With the opportunities for MnO_4^- to move around zones of DNAPL saturation in the two-dimensional system, overall destruction rate of TCE was significantly less than that in the one-dimensional system. Similar observation has been reported by other researchers (5). These results suggest that with 3-D flow conditions, evident at actual contaminated sites, the efficiency of removal would probably be smaller. It appears that MnO_4^- oxidation is more effective in cleaning up residually saturated DNAPLs rather than pools of DNAPL. Clearly, in actual field settings, the issue of flushing efficiency will be of concern in the design. Experiments are underway to find an effective additive that minimizes the precipitation effects.

The proof-of-concept experiment demonstrated how the PRBS could deliver MnO_4^- to a dissolved plume at a stable, constant, and controllable rate. The experiment also shows that the density-driven flow and clogging due to MnO_2 precipitates are factors to be considered in the design. This idea of PRBS could be developed as a long-term reactive barrier for the remediation of chlorinated solvents in plumes. Once installed, the long-term management cost would probably be low compared with other alternatives, and the maintenance may be limited to sporadic recharging with permanganate.

Conclusions

The column and flow tank experiments continue to indicate that the oxidation of chlorinated ethylenes by permanganate as a useful remediation technology. The kinetics of the oxidation reaction are generally fast and capable of reducing the mass of pure-phase product, especially early in the treatment. However, MnO_2 precipitates and CO_2 gas bubbles present significant delivery problems. The rapid buildup of these reaction products in zones with high DNAPL saturation leads to pore plugging, which lowers the permeability, and causes flow bypassing. The flushing efficiency would be reduced without control of this problem. For this technology to be practical, there is a critical need to find a solution that would control or remove the MnO_2. The test of the permanganate reactive barrier system in this study has shown positive results. With its potential benefit of cost and time efficiency, PRBS can be developed as a field technology for the remediation of the contaminant plume.

Acknowledgements

This paper is based on research support from the Department of Energy under grant no. DE-FG07-00ER15115.

References

1. *Alternatives for Groundwater Cleanup;* USNRC (U.S. National Research Council), National Academy Press, Washington, DC, 1994; p 26.
2. West, O. R., S. R. Cline, W. L. Holden, F. G. Gardner, B. M. Schlosser, J. E. Thate, D. A. Pickering, T. C. Houk, 1997, A full-scale demonstration of in situ chemical oxidation through recirculation at the X-701B site, Oak Ridge National Laboratory, Oak Ridge, TN, ORNL/TM-13556, 101p.

3. Schnarr, M.; Truax, C.; Farguhar, G.; Hood, E.; Gonullu, T.; Stickney, B. *Journal of Contaminant Hydrology*. 1998, 29(3): 205-224.
4. Yan, Y.E.; Schwartz, F. W. *Journal Of Contaminant Hydrology*. 1999, 37(3-4): 343-365.
5. Thomson, N. R., Hood, E. D., MacKinnon, L. K. In *Chemical Oxidation and Reactive Barriers: Remediation of Chlorinated and Recalcitrant Compounds;* Wickramanayake, G. B.; Gavaskar, A. R.; Chen, A. S. C., Ed.; Battelle Press, Columbus, OH, 2000, pp 9-16.
6. Reitsma, S., Marshall, M. In *Chemical Oxidation and Reactive Barriers: Remediation of Chlorinated and Recalcitrant Compounds;* Wickramanayake, G. B.; Gavaskar, A. R.; Chen, A. S. C., Ed.; Battelle Press, Columbus, OH, 2000, pp 125-134.
7. Mott-Smith, E.; Leonard, W. C.; Lewis, R.; Clayton, W. S.; Ramirez, J. In *Chemical Oxidation and Reactive Barriers: Remediation of Chlorinated and Recalcitrant Compounds;* Wickramanayake, G. B.; Gavaskar, A. R.; Chen, A. S. C., Ed.; Battelle Press, Columbus, OH, 2000, pp 125-134.
8. McKay, D. J.; Stark, J. A.; Young, B. L.; Govoni, J. W.; Berini, C. M.; Cronan, T. J.; Hewitt, A. D. In *Chemical Oxidation and Reactive Barriers: Remediation of Chlorinated and Recalcitrant Compounds;* Wickramanayake, G. B.; Gavaskar, A. R.; Chen, A. S. C., Ed.; Battelle Press, Columbus, OH, 2000, pp 109-1116.
9. Devlin, J. F. and Barker, J. F., 1999. Field demonstration of permeable wall flushing for biostimulation of a shallow sandy aquifer: *Ground Water Monitoring & Remediation* 19(1): 75-83.
10. Powers, S.E., Abriola, L.M. and Weber, Jr., W.J., 1992. An experimental investigation of nonaqueous phase liquid dissolution in saturated subsurface system: Steady state mass transfer rates. Water Resour. Res., 28(10): 2691-2705.
11. Imhoff, P.T., Jaffé, P.R. and Pinder, G.F., 1994. An experimental study of complete dissolution of a nonaqueous phase liquid in saturated porous media. Water Resour. Res., 30(2): 307-320.

Chapter 6

Microbubble Ozone Sparging for Chlorinated Ethene Spill Remediation

William B. Kerfoot

K-V Assocoiates, Inc., 766 Falmouth Road, Unit B, Mashpee, MA 02649

The use of engineered microbubble systems with ozone has proved a powerful means of targeting and effectively decomposing chlorinated ethenes, like tetrachloroethene (PCE), in-situ to harmless byproducts of dilute hydrochloric acid (HCl) and carbon dioxide (CO_2)(1,2). The reaction involves a low molar ratio of ozone to contaminant molecule to attached the central band of the ethene molecule, resulting in the decomposition of the chlorinated fragments, cleaving into a carbonyl compound and a hydroxy hydroperoxide, which rapidly lose HCl in the presence of water (3,4). The stoichiometry of PCE, trichloroethene (TCE), 1,2-dichloroethene (1,2-DCE), and chloroethene (vinyl chloride) is presented. For field applications, a site example is given with mass oxidation demand and equations for chloroethene attenuation.

Ozone treatment of groundwater and soil has become one of three generally recognized oxidation systems for chloroethene treatment. From 1999 to 2001 the number of states using the microbubble systems for on site treatment increased from 11 to over 30. Over 70% of the applications involved **chlorinated solvent spill remediation.**

Field use over the past five years has frequently demonstrated economical operation, effectiveness, and cleanliness which may qualify the process for use of removing chlorinated ethenes in drinking water aquifers or discharges to water streams. Despite this, effectiveness could be improved further by a better understanding of the chemical process. The treatment has been shown in various groundwater conditions, over a wide breadth of pH, ranging from acidic sands to basic carbonate limestones. The use of microbubble injection by microporous materials (C-Sparging™) has repeatedly shown the capacity to lower chlorinated solvent spill aqueous plume regions below drinking water MCLs (1,5). Normally, a pH change of only .5 units towards acidity is observed when treating water concentrations in the 1 to 10 ppm (mg/L) range (2). An increase in dissolved oxygen occurs when ozone is supplied in quantities greater than the decomposition reaction, particularly during treatment of PCE where oxygen is an end product. Third party economic analysis has shown operating costs below standard air sparging (6).

The purpose of this paper is to review the state of science in ozone microbubble remediation technology for chloroethenes and to provide guidance to engineers for application. A case study is presented with stoichiometry and simple approaches to answering the critical questions of how much and for how long.

Treatment Technology Overview - Criegee-Like Oxidation and Microbubble Treatment

The KVA C-Sparge™ process combines the unit operations of air stripping and oxidative decomposition into a single process. This patented process involves injecting sparged air and ozone directly into the groundwater and soil column (7,8). The wastewater treatment industry typically uses sparged air to remove VOCs from the liquid phase (e.g., contaminated groundwater and wastewater). When removed from the liquid phase, the VOCs must be treated in the recovered air stream by one or more processes prior to the air being discharged.

When air is bubbled through wastewater, dissolved VOCs transfer from the liquid to gas phase in accordance with Henry's Law. The spargers used in the pilot test process (C-Sparger®) provide extremely small "microbubbles" .5 to 200 micron size with a very high surface area to volume ratio. This high surface area to volume ratio maximizes VOC transfer from the liquid phase to the gas phase. If the air bubbles contain sufficient ozone for decomposition, the VOCs react with the ozone and are destroyed while still in the water column. This in-situ combined VOC recovery and destruction not only obviates the need for an additional process step but also enhances the physical and chemical kinetics of the process.

Some scientists believe that bubbles propagate only in limited situations and that injected gases travel along explicit pathways as a continuous gas phase rather than bubbles. In my experience, this occurs when bubble size is not controlled with suitable pressure, gas volume is too great for pore entry, and when pulsing is not used to generate liquid wavefronts to assist fine bubble movement through soil pores. Fine bubble collectors, combined with downwell video cameras can check the propagation through the soil during pilot testing if necessary.

For practical field applications, the process engineer needs to define the oxidant requirement for the site. This normally depends upon:

1. The stoichiometric oxidation requirement.
2. The soil oxidant demand.
3. The aqueous oxidant needs for metals, carbonates, and sulfides.
4. The in-situ decomposition rate of the ozone.

The theory section attempts to define the stoichiometric oxidant demand. A pilot test is often desirable in the field, particularly with mixtures of target organics. Later, a field test is given as an example, oxidant demand computed, and field attenuation rates analyzed.

Theory

When ozone is bubbled into an aqueous solution containing dissolved VOCs, ozonation may occur in either the aqueous phase or the gas phase. That is, the ozone gas may dissolve in to the aqueous solution and then react with the dissolved VOCs. Dissolved VOCs also may transfer from the aqueous phase into the ozone bubble and react in the gas phase. Those VOCs not destroyed by reaction are liberated into the chamber headspace according to Henry's Law.

Whether the VOC transfers into the ozone-containing bubble and is destroyed in the gas phase or the ozone dissolves in the water and destroys the VOC in the aqueous phase is primarily dependent upon the rate of reaction of each VOC with ozone. Table I shows the reaction rates for the VOCs of concern with ozone in both the aqueous and gas phases.

Table I. Reaction Rates for VOCs of Concern in Aqueous and Gas

Contaminant	*Aqueous Phase(10,11)* *(L/mol/sec)*	*Gas Phase(9)* *(sec-1)*
Tetrachloroethene (PCE)	<0.1	332
Trichloroethene (TCE)	17	[250]
cis-1,2-Dichloroethene (1,2-DCE)	540	22
trans-1,2-Dichloroethene (1,2-DCE)	6500	150
Chloroethene (vinyl chloride)	14000	3.9

Sources: Reproduced from source 9. Copyright 1998, American Chemical Society
Hoigne, J. and Bader (10,11)
[ ] Kerfoot (12), assuming proportionality

For instance, the reaction kinetics shown in Table I indicate that chloroethene and cis- and trans-1,2-dichloroethene (1,2-DCE) react much faster in the aqueous phase than in the gas phase (10,11). This means that the ozone that dissolves into the solution will react with the chloroethene and cis- and trans-1,2-DCE more quickly than any chloroethene and DCE that transfer into the gas will react with the gaseous ozone. Thus, the aqueous phase reaction predominates over the gas phase reaction. On the other hand, tetrachloroethene (PCE) and trichloroethene (TCE) react much faster in the gas phase and transfer into the ozone bubble and reacts with the gaseous ozone more quickly than dissolved ozone reacts with the aqueous phase.

In accordance with Henry's Law, the dissolved VOCs will be driven into the gas phase, and the gaseous ozone will be driven into the aqueous phase. This will result in the various reactions occurring at the bubble-liquid interface, whether in the gas-film or liquid-film of the bubble (Figure 1). Whether the primary decomposition reaction is occurring in the gaseous or liquid phase, the chloroethenes are driven by partitioning into the bubble environment. The smaller the bubble, the greater the surface-to-volume ratio and ability to "strip" volatile organics (1).

Kerfoot developed a procedure involving pressure and vacuum treatments to separate reactive phases (12). Reaction data for trichloroethene (TCE) were not available in the gas phase in current scientific literature (9). Through a comparison of the reaction rates in paired comparisons in groundwater, it appears that TCE reacts faster in the gaseous phase than in the aqueous phase, when microbubbles are suspended under pressure.

Greater detail on the reaction kinetics of the chloroethenes can be found in Army Environmental Center reports (13,14).

Gas Phase Chemical Mechanisms

The gas phase ozonation of PCE has not been studied widely. A proposed mechanism is provided (Figure 2). As the reaction proceeds, the dichloroformaldehyde formed (Reactions 1,2) decomposes to carbon dioxide and hydrochloric acid (Reaction 3).

Carbon dioxide and hydrochloric acid are assumed to be the only reaction products, following the stoichiometry in Reaction 4.

$$C_2Cl_4 + 2O_3 + 2H_2O \rightarrow 4HCl + 2CO_2 + 2O_2 \qquad (4)$$

In laboratory tests performed under pressure with iron silicate slurries (16), and in field tests with iron silicates added (13), the reaction is stabilized and occurs more readily.

Aqueous Phase Chemical Mechanisms

The aqueous phase reaction of ozone with ethene compounds results in cleavage of the carbon-carbon double bond through the formation of dipolar ionic complexes or zwitterionic (German for double ion) complexes that rapidly collapse into ozonides. The mechanism is described and verified through experimental testing by Criegee and others (9). Criegee's pioneering studies were first cited in English by Bailey (20). This is briefly described for chlorinated ethene compounds and is shown in Figure 3.

In the aqueous phase, ozone first attaches to the ethene, forming the primary zwitterionic complex (Reaction 5) and then attaches across the carbon-carbon double bond, forming an unstable cyclic trioxide (Reaction 6). This compound rapidly decomposes to form a carbonyl compound and the tertiary zwitterions (Reaction 7); the secondary zwitterion complex is not shown. In water, the tertiary zwitterions further decomposes into a similar carbonyl compound and hydrogen peroxide (Reaction 8). For ethene compounds containing chlorine, both carboxylic acid and hydrochloric acid also will be produced.

In the case of the aqueous phase ozonation of TCE (Figure 4), ozone cleaves the double bond, forming three compounds: chloro-hydroxymethyl hydroperoxide (Reaction 9), dichloroacetic acid (Reaction 10), and chloroformic peracid (Reaction 11). The chloro-hydroxymethyl hydroperoxide rapidly decomposes into formic peracid (Reaction 12), which further decomposes into either formic acid and hydrogen peroxide (Reaction 13) or into carbon dioxide, hydrochloric acid, and water (Reaction 14). The chloroformic peracid decomposes into carbon monoxide and hydrochloric acid (Reaction 15). The percentage of each set of reaction products that might occur is shown by A, B,

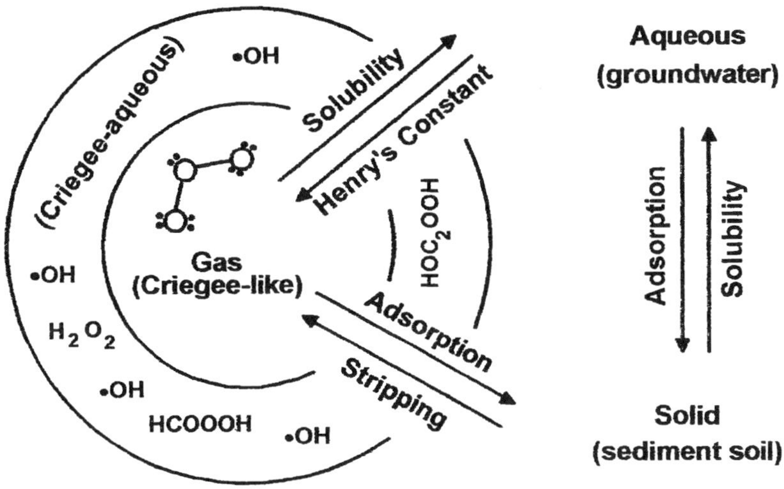

Figure 1. Microbubble organic oxidation reactions and partitioning environment for ozone reactions.

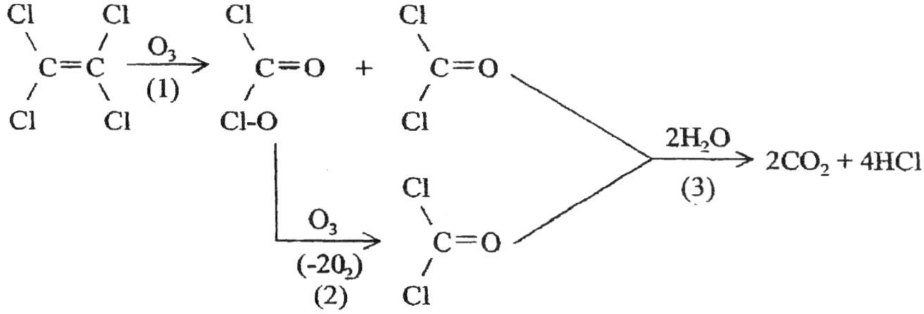

Figure 2. Gas phase ozonation of tetrachloroethene (PCE)

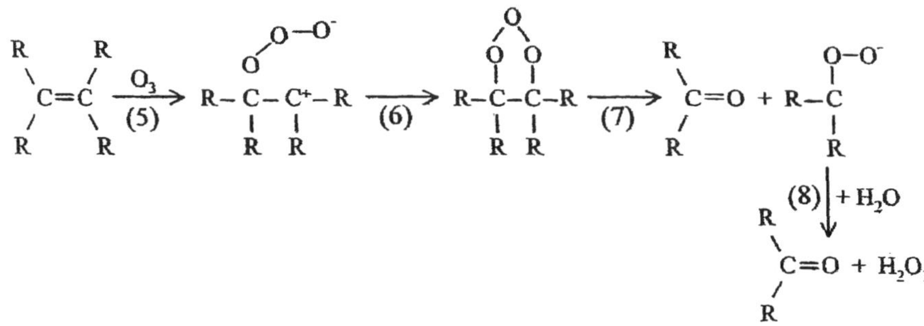

Figure 3. Aqueous phase ozonation of ethane

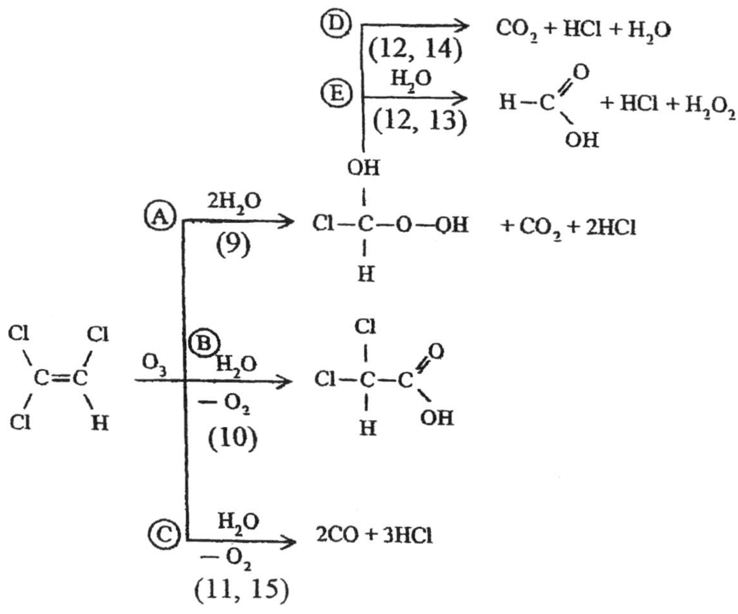

Figure 4. Aqueous phase ozonation of trichloroethene (TCE)

C, D, and E for Reactions 9,10,11,14, and 13, respectively. These values must be determined experimentally.

Experiments indicate that A = 97%, B = 4%, C = 3%, D = 11%, and E = 82% (calculated from Dowideit and von Sonntag [9] data), resulting in the stoichiometry for the aqueous phase ozonation of TCE in Reaction 16:

$$100C_2HCl_3 + 100O_3 + 264H_2O \rightarrow 82HC(0)OH + 4Cl_2HC_2(O)OH + 292HCl + 6CO + 104CO_2 + 82H_2O_2 + 7O_2 \qquad (16)$$

In the case of the aqueous phase ozonation of cis- and trans-1,2,-DCE (Figure 5), ozone cleaves the double bond (Reaction 17), forming the same two compounds: hydroxymethyl hydroperoxide and formyl chloride. The hydromethyl hydroperoxide rapidly decomposes into formic peracid (Reaction 18), which further decomposes into either formic acid and hydrogen peroxide (Reaction 19) or into carbon dioxide and water (Reaction 20). The formyl chloride originally produced decomposes into either carbon monoxide and hydrochloric acid (Reaction 21) or into formic acid and hydrochloric acid Reaction 22). Because the two initial reaction products (Reaction 17) can each follow two different reactions, four sets of final reaction products are possible. The percentage of each set of reaction products that might occur is shown by A, B, C, and D for Reactions 21,22,19, and 20 respectively. These values must be determined experimentally.

Experiments indicate that A = 97%, B = 3%, C = 98%, and D = 2% (calculated from Dowideit and von Sonntag [9] data), resulting in the stoichiometry for the aqueous phase ozonation of cis- and trans-1,2-DCE of Reaction 23:

$$100C_2H_2Cl_2 + 100O_3 + 99H_2O \rightarrow 101HC(O)OH + 200HCl + 97CO + 2CO_2 + 98H_2O_2 \qquad (23)$$

For the aqueous phase ozonation of chloroethene (vinyl chloride) (Figure 6), ozone cleaves the double bond, forming either chloroformaldehyde and hydroxymethyl hydroperoxide (Reaction 24) or formaldehyde and chloro-hydroxymethyl hydroperoxide (Reaction 25). The chloroformaldehyde decomposes to form carbon monoxide and water (Reaction 26) or hydrolizes to formic acid and hydrochloric acid (Reaction 27). The chloro-hydroxymethyl hydroperoxide decomposes to acetic acid, hydrochloric acid, and hydrogen peroxide (Reactions 28,29), and the peroxide combines with any formaldehyde present to form additional hydroxymethyl hydroperoxide (Reaction 30). The percentage of each set of reaction products that might occur is shown by C & G for Reactions 25,28,29, and 30, and D for Reactions 24,26, and 27. These values must be determined experimentally.

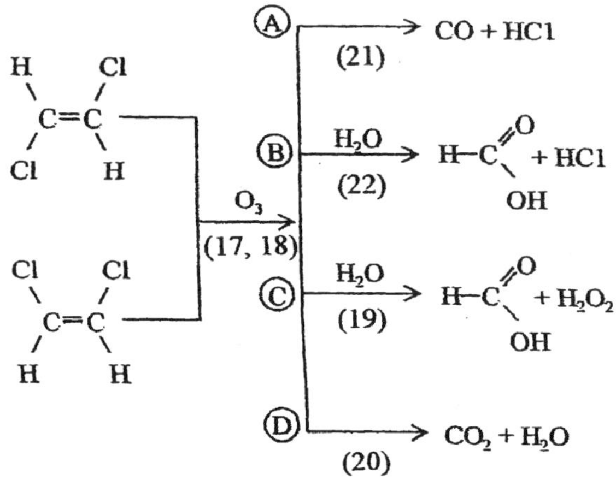

Figure 5. Aqueous phase ozonation of cis- and trans-1,2-dichloroethene (DCE)

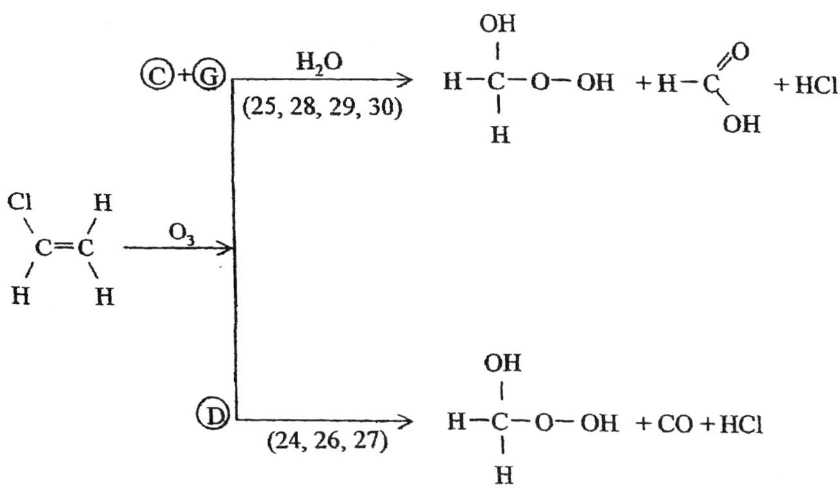

Figure 6. Aqueous phase ozonation of chloroethene (vinyl chloride)

Experiments indicate that C + G = 6% and D = 94%. The complete reaction mechanism and percentage for each reaction product are shown in Figure (calculated from Dowideit and von Sonntag [9] data). This results in the stoichiometry for the aqueous phase ozonation of chloroethene in Reaction 31:

$$100C_2H_3Cl + 100O_3 + 6H_2O \rightarrow 6HC(O)OH + 100H_2OHCOOH + 100HCl + 94CO \quad (31)$$

Table II. Ozonation Reactants and Products for Selected VOCs of Concern

Contaminant	*PCE*	*TCE*	*1,2-DCE*	*Vinyl Chloride*
Reactants (Moles per mole of contaminant)				
Ozone	2.00	1.00	1.00	1.00
Water	2.00	2.64	0.99	0.06
Products (Moles per mole of contaminant)				
Formic Acid ($HC(O)OH$)	0.00	0.82	1.01	0.06
Dichloroacetic Acid ($HCl_2C_2(O)OH$)	0.00	0.04	0.00	0.00
Hydroxymethyl-hydroperoxide ($H_2OHCOOH$)	0.00	0.00	0.00	1.00
Methanol (CH_3OH)	0.00	0.00	0.00	0.00
Hydrochloric Acid (HCl)	4.00	2.92	2.00	1.00
Carbon Dioxide (CO_2)	2.00	1.04	0.02	0.00
Carbon Monoxide (CO)	0.00	0.06	0.97	0.94
Hydrogen Peroxide (H_2O_2)	0.00	0.82	0.98	0.00
Oxygen (O_2)	2.00	0.07	0.00	0.00

Microbubble Ozonation Kinetics

Thin-film reactions described by Henley and Seader (17) have relevance to the fine bubble systems.

For a microbubble ozonation system, the reactions are described by the following irreversible, reaction-in-series equation:

$$A \rightarrow B \rightarrow S \tag{32}$$

In a gas-phase microbubble-ozonation reaction, A is the VOC dissolved in the aqueous phase, B is the VOC absorbed into the gas phase in accordance with Henry's Law, and S is the reaction product from the VOC reacting with gaseous ozone.

For a liquid-phase microbubble-ozonation reaction, A is gaseous ozone in the bubble, R is ozone dissolved into the aqueous phase, and S is the reaction product from the VOC reacting with gaseous ozone.

The rate equations for the three components, A, R, and S are:

$$r_A = dC_A/dt = -k_1C_A \tag{33}$$
$$r_B = dC_R/dt = k_1C_A - k_2C_R \tag{34}$$
$$r_S = dC_S/dt = k_2C_R \tag{35}$$

Rearranging and integrating each equation yields Equations 36,37,38. These equations describe the concentrations of components A, B, and S over time (18):

$$C_A = C_{A0}e^{-k_1t} \tag{36}$$

$$C_B = C_{A0}k_1 \left(\frac{e^{-k_1t}}{k_2 - k_1} + \frac{e^{-k_2t}}{k_1 - k_2} \right) \tag{37}$$

$$C_S = C_{A0} \quad 1 + \frac{k_2}{k_1 - k_2} e^{-k_1t} + \frac{k_1}{k_2 - k_1} \quad e^{-k_2t} \tag{38}$$

For a gas phase reaction, where the reaction of VOC with ozone is much faster than the mass transfer of VOC into the ozone bubble ($k_2 >> k_1$), Equation 38 becomes:

$$C_S = C_{A0}\,(1 - e^{-k_1t}) \tag{39}$$

For a gas phase reaction, where the mass transfer of VOC into the ozone bubble is much faster than the reaction of VOC with ozone ($k_1 >> k_2$), Equation 38 becomes:

$$C_S = C_{A0} (1 - e^{-k_2 t}) \quad (40)$$

These equations describe a reaction in series where there is no interaction between the two phases except for the n necessary step-by-step continuation. However, a microbubble-ozonation system is much more dynamic and interdependent, both chemically and physically. In a thin film microbubble ozonation reaction, the two reactants are physically driven toward one another under the influence of Henry's Law, overwhelming the occurrence of random physical collisions. In this case, the speed, or rate of chemical reaction, is enhanced by the physical mechanism of Henry's Law.

Under standard air stripping conditions, the reaction rate will decrease as the VOC concentration in the gas phase approaches equilibrium with the VOC dissolved in the e aqueous phase. However, in a microbubble ozonation system, the VOC in the gas phase reacts with the ozone, leaving no VOC in the gas phase to slow the VOC transfer. Therefore, superior phase transfer under Henry's Law, or the removal efficiency from the liquid phase, is obtained by the ozonation in the bubble or at the bubble interface.

Thin film theory describes the mass transfer of a reactant across a liquid and a gas film before it contacts the other reactant. However, in a microbubble ozonation system, not only is the VOC driven toward the ozone bubble, but the gaseous ozone also is driven toward the liquid phase. Physically, the intense agitation provided by the microbubbles decreases the thickness of the liquid film, and Henry's Law drives mass transfer of ozone through the gas film, bringing the reactants together at the interface. Therefore , it is likely that the chemical reaction occurs at the interface of the gas and aqueous phases. If overvolume of gas is supplied resulting in channelizing of gas, the ozone gas would have less contact with the aqueous phase, thus reducing the aforementioned rates of reaction.

Analytical Radial Transport Model

Clayton (19) developed a simple mode of ozone transport for a chemical subject to first-order degradation. Subsurface ozone transport is limited by ozone reaction as it moves through the soil. The importance of ozone reaction rates on ozone transport is illustrated by considering simplified radial transport of ozone from an injection well.

Further discussion of limitations is presented later.

Model Development

The radial transport of a gaseous chemical subject to first-order degradation can be solved analytically as follows. The flow velocity (v) during radial flow is expressed as,

$$v = \frac{dR}{dt} = \frac{Q}{A} = \frac{Q}{2\pi RHnS_g} \qquad (41)$$

Where, R = radial distance from well, t = time, Q = injection rate, A = flow cross-sectional area, H = height of flow zone, n = soil porosity, S_g = gas saturation. Solving for t in equation 42 ,

$$t = \int dt = \int \frac{1}{V} dR = \int \frac{2\pi RHnS_g}{Q} dR = \frac{\pi HnS_g R^2}{Q} \qquad (42)$$

Combining equation 42 with the standard first-order decay equation yields,

$$C = C_0 e^{-kt} = C_0 e^{-[k\pi HnS_g R^2/Q]} \qquad (43)$$

Where, k = degradation constant = 0.693/half life, C_0 = initial concentration. Equation is an analytical solution for steady-state radial gas transport subject to first-order decay.

Analytical Model Results and Discussion

Application of equation 43 shows that steady-state subsurface ozone transport is highly sensitive to reaction rate (Figure 7). For 5 percent ozone injection into a 5-foot-thick zone at 5 cubic feet per minute, ozone concentrations are depleted to 1 ppm at distances of 11, 18, and 32 feet for ozone half lives of 5, 15, and 45 minutes respectively. These results represent the steady-state ozone concentration distribution at a constant degradation rate. However, this analysis yields an upper limit estimate of maximum transport distances, because it does not account for ozone mass transfer into groundwater, and the assumption of first-order kinetics is simplified. Actual ozone reaction rates are second order, i.e., a function of ozone and contaminant concentrations, which means they vary spatially and temporally. The numerical approach described below simulates unsteady state ozone transport which accounts for mass transfer and second order reactions by adapting the observed system to pseudo-first order mechanics in order to estimate time of remediation with a given dosage. A more detailed discussion of attenuation under varying ozone content can be found with Clayton (19).

Primer on Mass Balance Kinetics

For determining the expected time course of removal and daily dosage of ozone, a mass balance is normally conducted in advance of treatment. A dry cleaner spill with PCE contamination is used as an example. The following desirable site information is listed:

• Soil type		sand
• Depth to water table		12 feet (3.7m)
• Vertical depth of contamination		21 feet (6.6m)
• Aqueous mean concentration:	PCE	1,000 μg/L (ppb)
	TCE	250 μg/L (ppb)
	DCE	100 μg/L (ppb)
• Soil mean concentration:	PCE	920 μg/kg (ppb)
	TCE	230 μg/kg (ppb)
	DCE	92 μg/kg (ppb)
• Saturated treatment zone depth		26 feet (6.6m)

For this site, the ratio of PCE in aqueous phase compared to solid phase (adsorbed) was found to be 1:4. A liter of sand from the site weighed 1600 gms, containing 1300 gms solid and 300 gms of water (300 cc or 30% volume). A circular area, 21 feet (6.6m) and 21 feet deep (6.6m) was expected to be treated by a single Spargepoint® placed at 25 feet (7.6m) below water table depth (Figure 8). The radius versus depth was found from previously published data (20). For a given cylinder, the liquid volume and solid volumes are computed separately:

$$\text{Liquid volume} = V_L = \pi r^2 hn = (3.14)(6.6)^2(.30)(6.6) = 271M^3 \text{ or } 271{,}000L \quad (44)$$

$$\text{Solid volume} = V_S = \pi r^2 h(1-n) = (3.14)(6.6)^2(.70)(6.6) = 632m^3 \text{ or } 632{,}000L \text{ or } 821.6 \text{ kg} \quad (45)$$

Where: V_L = Liquid volume (liters)
V_S = Solid volume (kg)
π = 3.14, a constant
r = radius of influence (m)
h = height of cylinder (m)

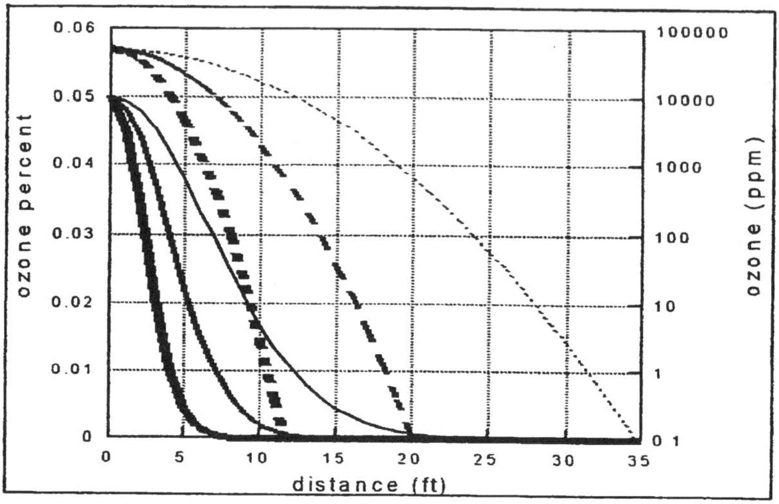

Figure 7. Steady state radial ozone transport at ozone half lives of 5, 15, and 45 minutes (coarse to fine lines). Ozone concentrations shown in percent and ppm (dashed). (Reproduced with permission from reference 18. Copyright 1972 Wiley.)

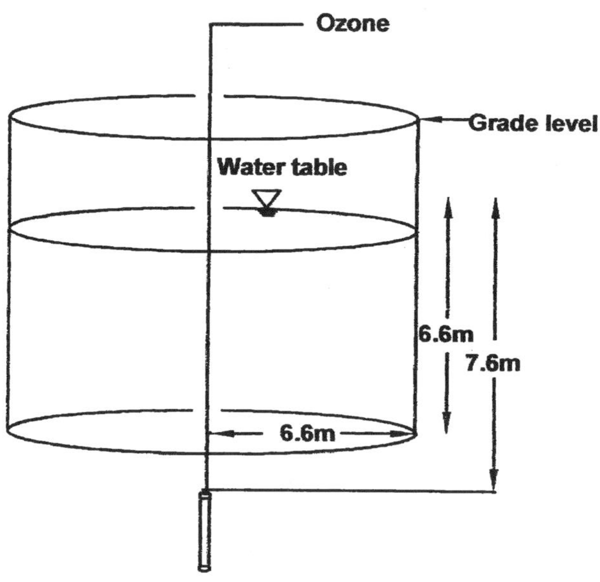

Figure 8. Treatment zone for an example Spargepoint®.

To convert to mass balance, the mean concentrations of PCE are multiplied times the respective water volume and solid volume.

Liquid content: $V_L \bullet C_{oq} = 271{,}000L \times 1{,}000\ \mu g/L = 271$ gm PCE
Solid content: $V_S \bullet C_S = 821.6\ kg \bullet 920\ \mu g/kg = 756$ gm PCE
Total Content: 1027 gm PCE

Equivalent mean concentrations for TCE and DCE were similarly computed yielding:

	PCE	TCE	DCE
Liquid content	271gm	68gm	27gm
Solid content	756gm	189gm	76gm
Total content	1027gm	257gm	103gm

Expected Oxidant Demand

The expected oxidant demand is a sum of stoichiometric chloroethene demand, soil demand, oxidizable metals demand, and background organic demand. Other volatile organics, particularly aromatic hydrocarbons, should be analyzed for, since benzene, ethylbenzene, toluene, and xylenes complete for ozone (21,22), and the oxidation demand can be estimated at over 3 mole equivalents for the benzene ring. For sandy soils, with aerobic conditions and acid pH, a 30% assumption for oxidation demand has been frequently used. The gas reaction reacts less with the soil substrate than aqueous oxidants (such as peroxide and permanganate). Bench-scale testing can be conducted on drained uncontaminated soil samples to estimate the soil demand.

The reaction of gaseous ozone with target organic chemicals in soil systems can be complicated and normally involve second-order mechanics (23,24). Masten (25) has previously detailed the reactivity of aqueous ozone in soils with TCE, pyrene, and naphthalene.

With oxidizable metals, the ferrous iron (Fe^{2+}) demand can be estimated from the dissolved iron content in acid soils. A mole of ozone can be consumed by two moles of Fe^{2+}.

$$O_3 + 2Fe^{2+} \rightarrow Fe_2O_3 \qquad (46)$$

If the mean dissolved iron content is 0.1 mg/L (ppm), then the oxidant demand would be expected to be 0.05 mg/L or .05 x 271,000 liters or 13.6 gms total for the treatment region.

Stoichiometric Demand

For PCE, .58 gms O_3 reacts with 1.0 gms PCE (C_2Cl_4) (Table II). To compute the stoichiometric demand, .58 gms is multiplied times the total PCE content for the treatment region (1027 gm) yielding 596 gms.

For TCE, .365 gms O_3 reacts with 1.0 gms TCE (C_2HCl_3) (Table II). To compute the stoichiometric demand, .365 gms is multiplied time the total TCE content for the treatment region (257 gm) yielding 93.8 gms.

For DCE, .50 gms O_3 reacts with 1.0 gms DCE ($C_2H_2Cl_2$) (Table II). To compute the stoichiometric demand, gms is multiplied times the total DCE content for the treatment region (103 gm) yielding 51.5 gms.

For vinyl chloride, .77 gms O_3 reacts with 1.0 gms VC (C_2H_3Cl) (Table II). To compute the stoichiometric demand, gms is multiplied times the total vinyl chloride content for the treatment region (0gm) yielding 0 gms.

Other Organics

Groundwater analysis at the site shows toluene is present at a concentration of 100 μg/L or ppb in the aqueous phase. Toluene will require 3 moles of ozone to decompose the benzene ring and 6 moles to convert carbon to CO_2.

$$C_7H_8 + 6O_3 \rightarrow 7CO_2 + 4H_2O \quad (47)$$

92 gm: 96 gm ratio, 1.04 gms O_3 to 1.0 gm toluene
271,000L x .1 mg/L = 27 gm C_7H_8
27 x 1.04 = 28 gms

The total ozone demand is computed in Table III.

Table III. Total Ozone Demand for Chloroethenes

	PCE	*TCE*	*DCE*	*Sum*
Stoichiometric Demand (SC) =	596 gms	94 gms	52 gms	742 gms
Soil Oxidant Demand (SOD) =				179 gms
Oxidizable Metals Demand (OMD) =				14 gms
Other Organics Demand (OOD) =				20 gms
Total Ozone Demand =				955 gms

Ozone Generation

The portable (Model 3600) C-Sparger® system normally generates from 5 to 15 gm/hr or 120 to 360 gms/day depending upon the oxygen supplied to it. Allowing for site conditions, 3 to 8 spargewells may be operational. The ozone dosage received by a single spargewell depends upon its needed mass loading and the total ozone mass generated per day. In the site example given here, 3 spargewells were operating, so each was set to receive 1/3 of the total loading or 1.7 gms/hr.

Time to Treat (Duration) Computation – Mass Basis

If ozone is supplied at 5 gm/hr to 3 spargewells, the total demand (TOD) can be divided by the spargewell oxidant supply to yield an approximation of the time-to-treat.

TOD/S = 955 gms/1.7 gms/hr = 562 hrs or 23 days

Monitoring the VOC Decay

Two procedures have been used to obtain VOC samples to perform field matchings of decay rates. Most simply, the aqueous phase can be sampled (groundwater sampling) from monitoring wells situated at varying distances from the injection spargewell. Measurements of PCE, TCE, and DCE were made at time intervals from beginning of injection of 5 gm/hr at 3 wells. The distance from the principal spargewell was 5 ft. and 35 ft. Recirculation systems were used to enhance the mixing and treatment zone. Samples obtained from peristaltic pumping opposite screen regions and field analyzed with a portable GC showed a more rapid attenuation rate than groundwater samples obtained from purging 6 volumes of the monitoring well, sampling, and then forwarding to the laboratory for EPA 8020 purge and trap followed by ms) analysis. (Table IV)

Table IV. Rates of Decay of PCE (as exponential of t) from Groundwater Samples

	5-ft. Well Laboratory Analyses (formation water)	*5-ft. Well Headspace Analyses*	*35-ft. Well Well Water Samples*
PCE	-.060t	-.138	-.092
TCE	-.050	-.092	-.087
DCE	-.035	-.077	-.069

Generally the rate of removal in the monitoring well screen region was found about double that in the formation, perhaps reflecting the capacity for screen areas to collect microbubbles at about twice the background groundwater flow mimicking a hydraulic phenomenon noted by Wheatcraft (26). The rate of removal was most rapid with PCE, followed by TCE and finally DCE, as would be expected based upon Henry's Constant. Figure 9 shows the rate of decomposition of PCE, TCE, and DCE observed during treatment at the dry cleaners. Each compound was attacked separately, without organic decomposition products, as is frequently observed with microbial decomposition under anaerobic conditions (27).

The rate of decay is presented for k values for different distances from the spargewell. For PCE removal, $C = Co\ e^{-.138t}$ was obtained for the 5 ft(1.5m) distance and $C=Co\ e^{-.092t}$ for the rate of decay at the 35 ft. (10.6m) distance. The mean rate of decay needs to be computed based upon ozone density applied.

Time to Complete

Figure 10 plots the linear decay rates on a logarithmic set of axes to depict the time to bring core region concentrations to 10 μg/l-ppb. The observed attenuation rates (k) ranged from a high of .14 to a low of .07. The highest observed value (close to 600 ppb, solid squares) is paired with the slowest attenuation rate to give the slowest removal. The most rapid rate is paired with the lowest concentration (200 ppb, open squares) to determine the most rapid removal expected. With PCE, the time to reach 10 ppb ranged from 13 to almost 60 days. Some caution is necessary. Prolonged tests or full remediation has often shown a doubling of time over short-term (20-day) pilot test (12).

In-Situ Decomposition of Vapors

Third-party review of vapor emissions from pilot-testing has verified the capacity to adjust ozone concentration to reach breakpoint to yield no VOCs in vapor control systems placed just above water table depths (12 ,28) or in spring discharges (13). During the first two days of operation, rapid depletion of ozone during injection may cause some transport of VOCs, but after adjustment of ozone content, the VOC content can be titrated to zero. Often a receiving drum (35 gal) of activated carbon is used for assurance, but VOC loading to the carbon is negligible.

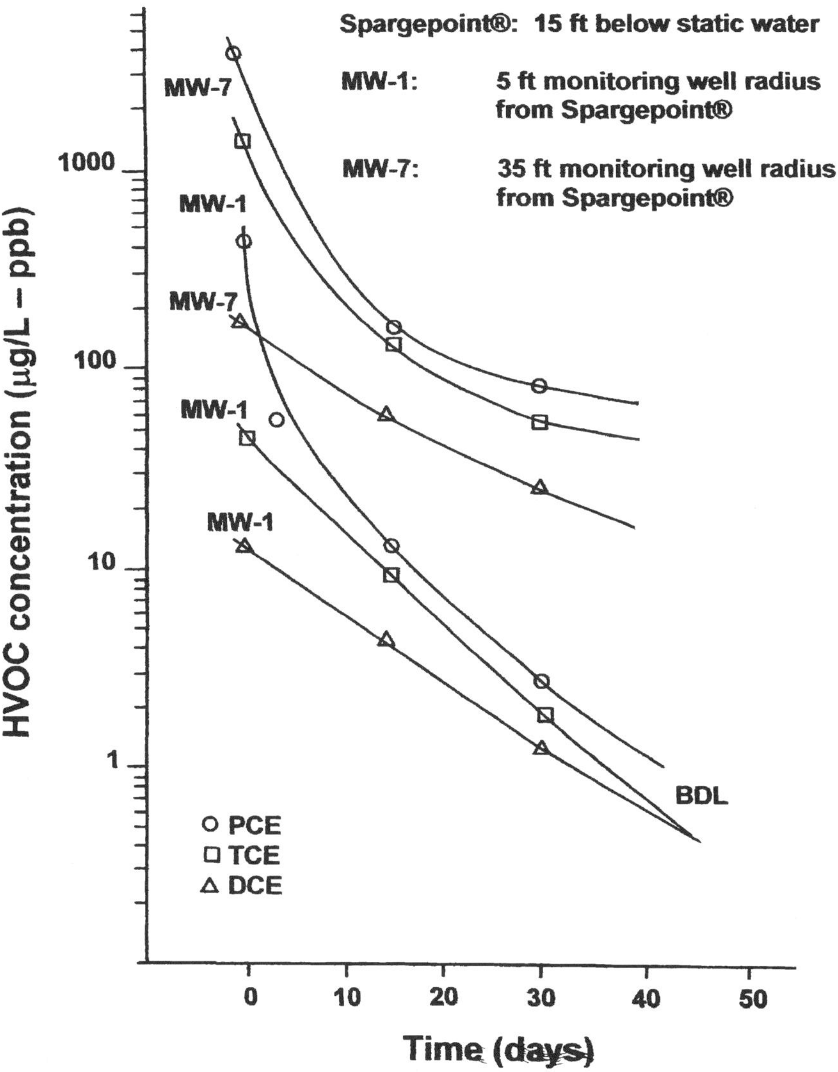

Figure 9. Rate of removal of PCE, TCE, and DCE during treatment.

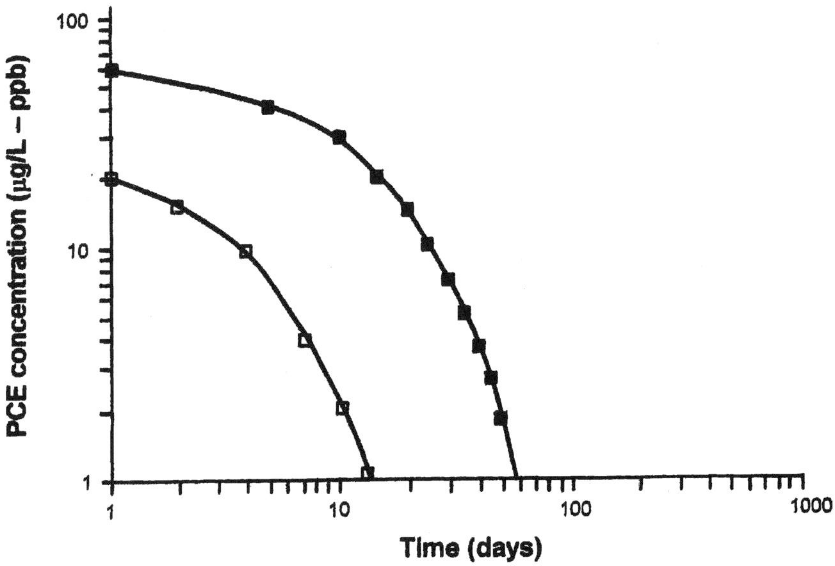

Figure 10. Projection of time-course of attenuation.

Conclusions

The use of fine bubbles encourages in-situ stripping and chemical oxidation of chloroethenes. The increase of efficiency of reaction with high surface-to-volume ratios and Criegee oxidation allows a low ozone content (50 to 500 grams/day) to be used in field applications involving removal of mg/L groundwater concentrations of chloroethenes.

The stoichiometric ozone demand for a gram equivalent of the chloroethenes PCE, TCE, DCE, and vinyl chloride was found to be .58 gms, .365 gms, .50 gms, and .77 gms. A pilot test is recommended to verify radius of influence (ROI) and estimate the time for remediation. An approximation of time-to-treat can be found from ozone mass supplied versus oxidation demand.

References

1. Kerfoot, W.B.; McCulloch, W.: Connors, J. In *The Tenth National Outdoor Action Conference and Exposition,* National Groundwater Association, Columbus, Oh, 1996; pp 77-95.
2. Kerfoot, W.B. In *Contaminated Soils;* Kostecki, P.; Calabrese, E.J.; Bonazountas, M., Eds.; Amherst Scientific Publishers: Amherst, MA, 1997; Vol. 2, pp 275-284.
3. Masten, S.J. Ph.D. thesis, Harvard University, Cambridge, MA, 1986.
4. Bailey, P.S. *Ozonation in Organic Chemistry,* Olefinic Compounds; Academic Press: New York, NY, 1978; Vol. 1.
5. Ehleringer, B.E.; Kerfoot,W.B In *The First International Conference on Oxidation and Reduction Technologies for In-Situ Treatment of Soil and Groundwater,* Niagara Falls, Ontario, Canada, 2001; pp 124-125.
6. Dreiling, D.N.; Henning, L.G.: Jurgens, R.D.; Ballard, D.L. In *Physical, Chemical and Thermal Technologies, Remediation of Chlorinated and Recalcitrant Compounds;* Wickramanayake, G.B.; Hinchee, R.E., Eds.; Battelle Press, Columbus, OH, 1999; pp 247-253.
7. Kerfoot, W.B. U.S. Patent 5,855,775, 1999.
8. Kerfoot, W.B. U.S. Patent 6,083,407, 2000.
9. Dowideit, P.; vonSonntag, C. *Environ. Sci. Technol.*, **1998**, 32, pp 1112-1119.
10. Hoigne, J.; Bader, H. *Water Research,* 1983, Vol. 17, pp 173-183.
11. Hoigne, J.; Bader, H. *Water Research,* 1983, Vol. 17, pp 185-194.
12. *Pilot-Scale Demonstration Report – C-Sparge™ In-Well Stripping System for HVOC-Contaminated Groundwater,* Ecology and Environment, Inc., U.S. EPA Superfund Technical Assessment and Response Team (START), Overland Park, KS, 2000, 300 pp.

13. *Final Rowe Spring Pilot Test Report,* U.S. Army Environmental Center, SFIM-AEC-ETT, Building E4430, Aberdeen Proving Ground, MD, 2000, 104 pp.
14. Shia, C.; Delp, D.; Kerfoot, W.; Hoke, B.; Stone, P.; Hill, S. In *Contaminated Soils, Sediments and Water*, Amherst Scientific Publishers, Amherst, MA, 2000, Vol. 5.
15. Mathias, E.; Sanhueza, E.; Hisatsune, I.C.; Heicklen, J. *Canadian Journal of Chemistry*, 1974, vol 52, 3852-3862.
16. *Technical Plan for Rowe and Rocky Springs Pilot Studies, Letterkenny Army Depot, Chambersburg, PA*, U.S. Army Environmental Center, Aberdeen Proving Ground, MD, 1999, 100 pp.
17. Henley, E. J.; Seader, J.D. *Equilibrium-State Separation Operations.* In *Chemical Engineering,* Chapter 16, John Wiley and Sons, New York, NY, 1981.
18. Levenspiel, O. *Chemical Reaction Engineering,* Chapter 3, John Wiley and Sons, New York, NY, 1972.
19. Clayton, W.S. In *Physical, Chemical, and Thermal Technologies, Remediation of Chlorinated and Recalcitrant Compounds;* Wickramanayake, G.B.; Hinchee, R.E., Eds.; Battelle Press, Columbus, OH, 1998, pp 389-395.
20. Kerfoot, W.B. In *Contaminated Soils*; Calabrese, E.J.; Kostecki, P.; Bonazountas, M., Eds; Amherst Publishers, Amherst, MA, 1996; Vol. 1, pp 1923-1929.
21. Bailey, P.S. *Ozone reactions with organic compounds.* American Chemical Society, Washington, DC, 1972, 129 pp.
22. Staehelin, I.; Hoigne, J. *Environ. Sci. Technol.,* **1982,** 16, p 676.
23. Hsu, M.I.; Masten, S.J. *Environ. Eng. Science*, **1997,** Vol. 14, pp 207-208.
24. Hsu, I.; Davies, S.H.R., Yao, J.; Masten, S.J. Proc. 4th Annual WERC Technology Development Conf., Las Cruces, NM, **1994,** *The Use of Ozone for the Removal of Contaminants from Unsaturated Soils,* 11 pp.
25. Masten, S.J. *Ozone Sci. Eng.,* **1991,** 13, pp. 287.312.
26. Wheatcraft, S.W.; Winterberg, F. *Water Resources Research, 1985,* Vol. 12, no. 12, pp 1923-1929.
27. Cirpka, O.A.; Bisch. In *Bioremediation and Phytoremediation, Chlorinated and Recalcitrant Compounds,* Wickramanayake, G.B.; Hinchee, R.E., Eds., Battelle Press, Columbus, OH 1998, pp 129-135.
28. Kerfoot, W.B.; Schouten, C.J.J.M.; vanEngen-Beukeboom, V.C.M. In *Contaminated Soils*, Vol. 4, Kostecki, P.T.; Calabrese, E.J.; Bonazountaus, M., Eds. Amherst Scientific Publishers, Amherst, MA, 1999, pp 99-112.

Chapter 7

Estimating Chlorinated Ethene Bioattenuation Rates from a Single Well

Brian S. Aiken and Laurie LaPat-Polasko

Parsons Engineering Service, Inc., 4801 East Washington Street, Phoenix, AZ 85034

The reductive dechlorination rates of PCE and TCE were estimated at a VOC site near Memphis, TN by assuming plug flow reactor type conditions and conservation of VOC mass. The VOC plume was not at steady state and anaerobic groundwater was partially aerated along the flowpath, making other estimating methods unusable. The resulting rates provided an accurate RT3D model calibration, and could be used to identify the source location.

Introduction

Natural attenuation of chlorinated ethenes continues to receive increasing attention as a remedial alternative. Often, the natural bioattenuation rates must be estimated for the sequential reductive dechlorination of tetrachlorethene (PCE) or trichloroethene (TCE) to cis-1,2-dichloroethene (DCE), vinyl chloride (VC) and ethene. Two common methods for estimating rates are Buscheck and Alcantar (1) and Moutoux et al. (2). Although popular, each of these methods have limitations in their applicability since both require steady state conditions, uninterrupted flowpaths, only calculate the overall attenuation rate and can be sensitive to boundary conditions (3). Site-specific bioattenuation rates are often instrumental in developing a successful dynamic groundwater model. Since models are generally used to predict concentrations far in the future (often an order of magnitude longer than the time period used to calibrate the model), inaccurate input parameters can have a large impact.

The subject site has a VOC plume that was being considered for monitored natural attenuation. Initial attempts to estimate the natural bioattenuation rates were compromised because a groundwater remediation system interrupted the continuity of the available flowpath wells. Figure 1 shows the flowpath wells.

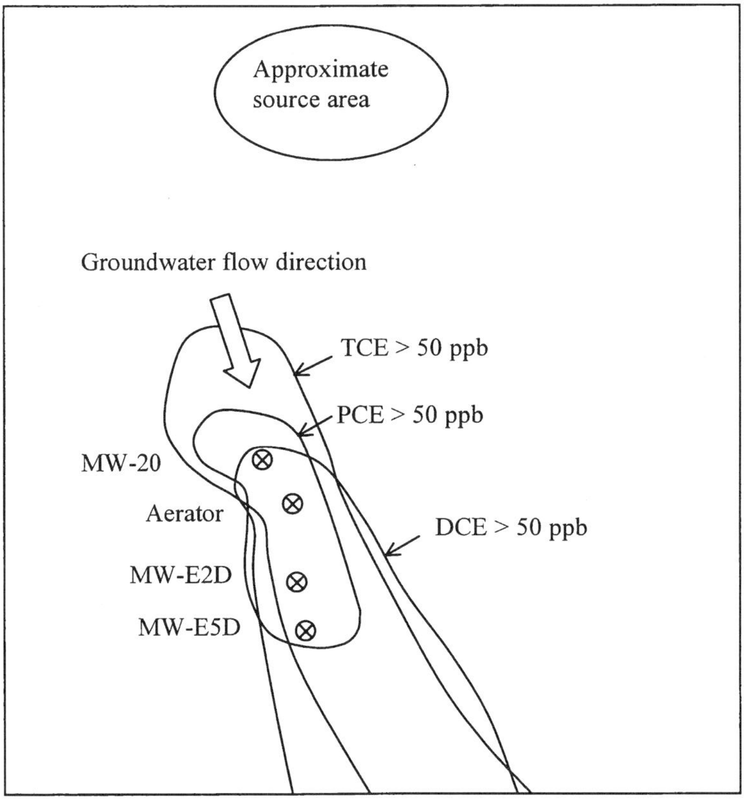

Figure 1. Well and plume locations.

An RT3D model was used at the site; however, after calibrating the model, the evaluation team concluded that the model input parameters (which included bioattenuation rates) were not reasonable, and must be incorrect. The method presented herein, using data from a single well to estimate rates, was developed to establish natural bioattenuation rates and provide better input parameters to begin model calibration. The resulting rates led to an accurate RT3D model calibration.

Site History and Overview

The subject site is an abandoned manufacturing facility north of Memphis, Tennessee. The source of PCE was thought to be a waste disposal pit, shown in Figure 1. Several hundred feet upgradient of the source is a fertilizer manufacturing facility, which is a source of ammonium chloride and possibly some VOCs. The chloride coming from the fertilizer manufacturer proved to be useful as a conservative tracer. After VOCs were discovered in downgradient wells, the soils underlying the disposal pit were excavated and removed. The mean groundwater velocity was estimated to be about 15 meters per year. The age of the source was not known, but the plume length (~500m) suggests the plume is 30 to 40 years old. The depth to groundwater is about 2m below the ground surface (bgs). Lithology in the subsurface is primarily a mixture of silt, fine sands and some gravels. There is a semi-confined upper aquifer layer extending from the groundwater surface to about 5m bgs. A lower semi-confined aquifer extends from the bottom of the upper layer to about 20m bgs. The VOCs migrate from the source area in the upper aquifer until just upgradient of well MW-20 (shown on Figure 1), where they migrate down to the lower aquifer layer. Most of the VOC contamination is then confined to a gravely portion of the lower aquifer layer, about 17 – 20m bgs.

An in-well aerator was installed just downgradient of well MW-20 (Figure 1). The in-well aerator caused a local change in the groundwater chemistry, which temporarily slowed reductive dechlorination. Redox values showed a clear indication of this change (although the aerated groundwater was re-introduced into the aquifer, residual dissolved oxygen was not observed in downgradient wells). The normal redox values for the aquifer were -85 to -55 mV; however, wells MW-E2D and MW-E5D, immediately downgradient of the in-well aerator, were +16 mVand -5 mV, respectively.

The change in groundwater chemistry slowed reductive dechlorination and the production of DCE. As shown in Figure 2, a reduction in the DCE production rate in well MW-E2D coincides with the arrival of aerated groundwater after the in-well aerator was started. Although well MW-20, MW-E2D and MW-E5D would otherwise have been ideal flowpath wells to calculate VOC biodegradation rates, the localized conditions were not representative of the overall plume.

Before the effect of the in-well aerator was realized, the natural attenuation evaluation team used the flowpath wells to estimate the bioattenuation rates by assuming first order decay along the flowpath wells. (Note: Chloride was present in the plume at a level sufficient to use it as a conservative tracer, and corrected VOC values were used to calculate bioattenuation rates). These initial rate estimates are shown in Table 1. The rate calculations were based on a plume that was known to be at a non-steady state condition; however, the calculated rates were considered a reasonable starting point to calibrate the RT3D model.

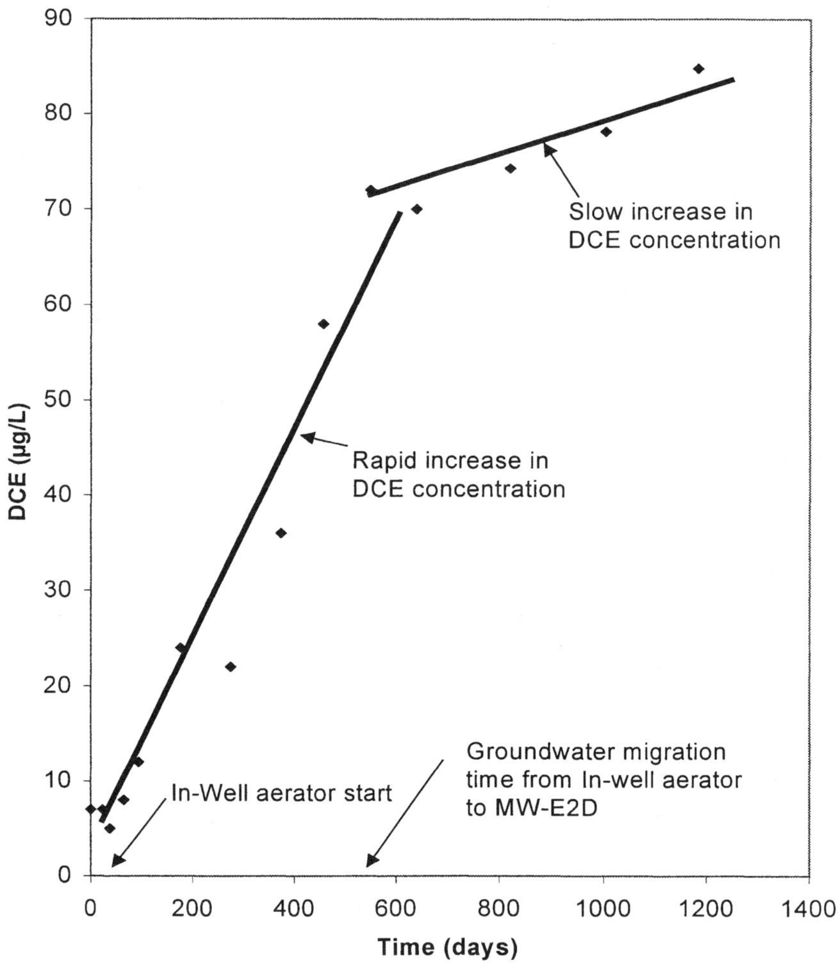

Figure 2. DCE concentration in well MW-E2D.

Table 1. Initial estimated rates and corresponding retardation factors.

Constituent	Degradation rate day^{-1}	Retardation factor
PCE	0.00015	2.7
TCE	0.00030	3.4
DCE	0.00001	4.6

In order to achieve model calibration, the retardation factors for PCE, TCE and DCE were adjusted. The initial model output was made to accurately fit known data; however, the retardation factors (also shown in Table 1) were not reasonable. The retardation factor for DCE was much higher than expected, and the value trend was reversed (TCE should have a lower retardation factor than PCE and DCE should be lower than TCE). The natural attenuation evaluation team could not reconcile the retardation factors and still obtain reasonable model results. Therefore, several additional steps were taken prior to re-evaluating the site, including: 1) soil samples were obtained and the organic carbon content was analyzed in order to be able to directly calculate retardation factors, and 2) the method described herein was developed to provide a better estimate of the natural attenuation rates.

The results of the revised estimated natural attenuation rates are shown in Table 2. The calculated site-specific retardation factors are also shown in Table 2. The RT3D model was re-initiated using the input parameters from Table 2. Although the new model results overpredicted the concentrations of TCE and DCE, the trends in PCE, TCE and DCE were well represented by the model. The overprediction of TCE and DCE were consistent with the change in groundwater chemistry caused by the in-well aerator.

Table 2. Final estimated bioattenuation rates and calculated retardation factors.

Constituent	Degradation rate day^{-1}	Retardation factor
PCE	0.00080	3.2
TCE	0.00087	2.8
DCE	0.00002*	2.7

* Estimated, not calculated

The local effect of the in-well aerator was not included in the final RT3D model used to project the future plume migration. A comprehensive evaluation of the aquifer conditions showed that the in-well aerator only affected downgradient redox conditions to approximately well MW-E5D. It also appeared that the influence of the in-well aerator decreased with time due to fouling, which caused short-circuiting within the well. The initial influent PCE concentrations in the in-well aerator were about 500 μg/L PCE, which was similar to upgradient well MW-20. After several years of operation, the in-well aerator influent concentrations dropped to about 5μg/L even though the PCE concentrations in well MW-20 continued to increase.

VOC Bioattenuation Rate Calculations

Many methods used to calculate VOC attenuation rates include both bioattenuation and abiotic dilution and dispersion. In the method presented herein, the calculated rate approximates the bioattenuation rate. We consider groundwater flow to be comparable to a non-dimensional plug-flow reactor. The parent compound (in this case PCE) is released from the source, and undergoes bioattenuation, dilution, dispersion and retardation. At any given point in the aquifer, the VOCs present have migrated from the source along approximately the same flow path. The simplifying assumption was made that dilution and dispersion affect PCE and the byproducts TCE and DCE equally. Since rate calculations are based on concentration ratios, loss in mass due to dilution and dispersion does not affect the result. There is also some error introduced by assuming that the migration rate is the same for each compound. Because this is not true in the case of PCE, TCE and DCE, the resulting PCE degradation rate constant would be overestimated during a period of increasing VOC concentrations (which was the case at this site).

The key criterion for applying this rate calculation method is the conservation of VOC mass. Figure 3 shows the total VOC concentration in wells MW-20, MW-E2D, MW-E5D, and T2-4M (well T2-4M is not shown on Figure 1, and is approximately 400m downgradient of well MW-E5D). As shown on the figure, the total VOC mass does not change significantly as the plume migrates. The conservation of VOC mass along the flowpath also confirms the assumption that dilution and dispersion is similar for VOCs and the tracer compound (chloride). Additional evidence of VOC accumulation were also present at the site, including the presence of ferrous iron and sulfate, and lack of the daughter products VC, ethene and ethane. There was also a second VOC plume on the site that showed all the same indicators of DCE accumulation. Conservation of VOC mass can be described by equation 1 - equations are from Levenspiel (4):

$$[PCE]_o = [PCE] + [TCE] + [DCE] \qquad (1)$$

Where: $[PCE]_o$ is the concentration of PCE released from the source and [PCE], [TCE], and [DCE] are the measured downgradient concentrations.

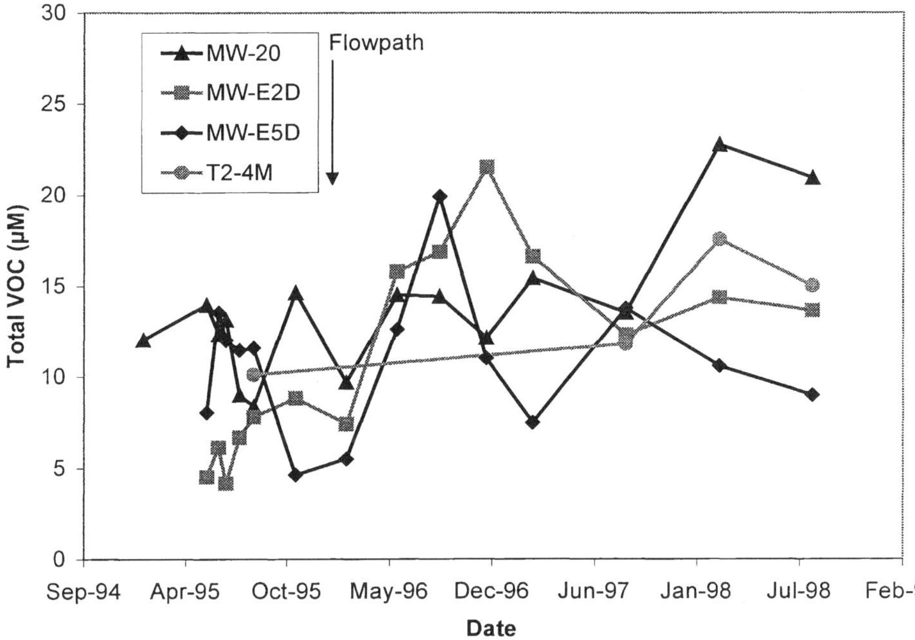

Figure 3. Total VOC concentration (PCE + TCE + DCE) in flowpath wells.

As stated previously, migration of VOC through the aquifer was considered to be similar to plug flow. The governing equation for a plug flow reactor is:

$$\frac{[PCE]}{[PCE]_o} = e^{-kt} \quad (2)$$

The initial time, t_i, can be arbitrarily chosen, and the time (t) for each subsequent sampling event is measured from t_i. For this site, t_i was selected as the date when semi-annual monitoring began (this was about 1 1/2 years after the plume was discovered). The "initial" sampling event, in January of 1995, was assigned t=0. Based on equation 2, the bioattenuation rate for PCE was estimated by plotting $[PCE]/[PCE]_o$ versus time (figure 4). The resulting curve yielded the

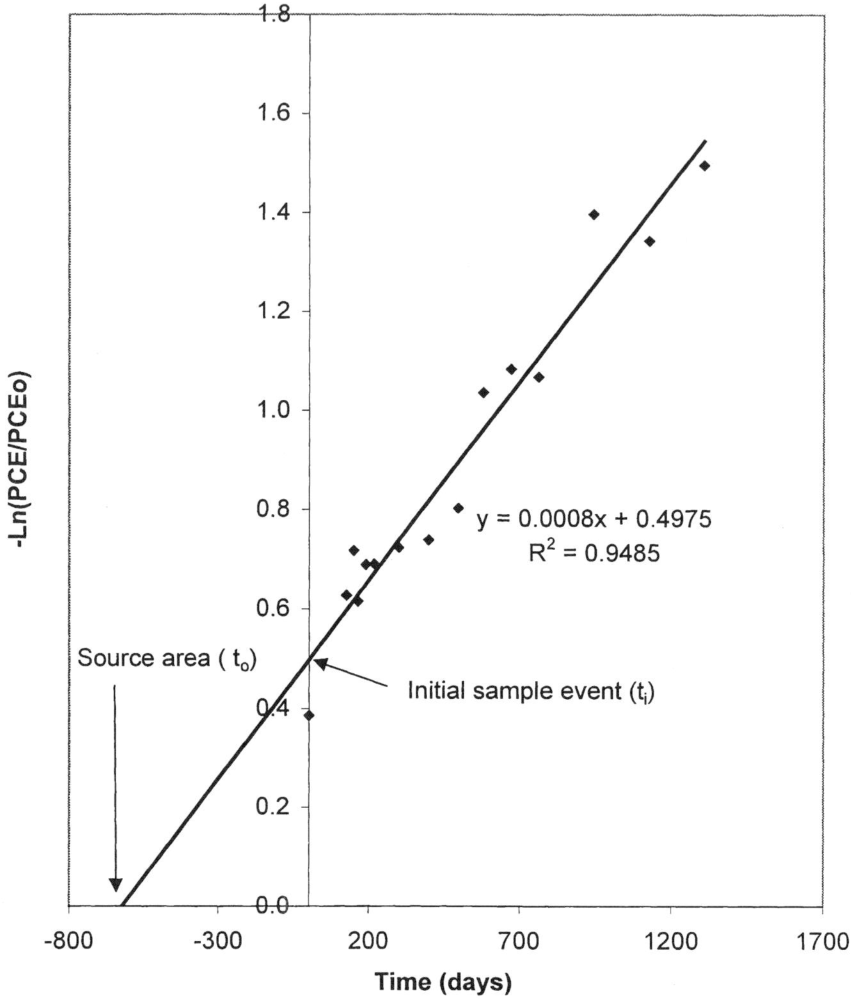

Figure 4. K(PCE) calculation.

PCE bioattenuation rate of 0.0008/day (half life = 2.4 years).

The overestimation of the decay rate caused by the differing migration rates would increase with larger time intervals, resulting in a non-linear decay curve. The PCE decay curve had a high correlation coefficient, indicating that the time interval used for the analysis is small enough not to have been significantly affected by the difference in migration rates.

By setting $[PCE]/[PCE]_0 = 1$, the decay curve should predict an approximate travel time from the source. Using the known groundwater flowrate and estimated PCE retardation factor, this method accurately predicted the source area location.

Next, the TCE natural bioattenuation rate was estimated. The governing rate equation for the second step of a two-step sequential reaction in plug flow is:

$$\frac{[TCE]}{[PCE]_o} = \frac{k_{PCE}}{k_{TCE} - k_{PCE}}(e^{-k_{PCE}t} - e^{-k_{TCE}t}) \quad (3)$$

Where: k_{PCE} is the reductive dechlorination rate of PCE; and,
k_{TCE} is the reductive dechlorination rate of TCE.

Since K_{PCE} was calculated previously, the only unknown from equation 3 is K_{TCE}. Equation 3 is not amenable to a simple log-linear plot; therefore, K_{TCE} was calculated directly for each time point by inputting the corresponding PCE_o, TCE and t value. The initial time is not arbitrary, and must be calculated with respect to time t_o, which is shown on Figure 4. Linear regression analysis showed $t_i - t_o = 622$ days. The average and standard deviation of the calculated k_{TCE} values were 0.00089 ± 0.00024/day (total of 15 sampling events). Data used to perform rate calculations are available from the author.

Conclusion

Traditional methods of estimating natural bioattenuation rates are not always applicable to site conditions. The method presented here to estimate rates, using data from a single well, was useful at this particular location. In this case, there was a change in groundwater chemistry along the flowpath, and VOC mass was conserved (DCE accumulated). DCE (or other daughter product) accumulation is a necessary prerequisite to using this method. Differing migration rates of PCE, TCE and DCE results in error in using this method; however, the error was not apparent using data from this site. The resulting estimated natural attenuation rates improved the accuracy of the RT3D model employed at the site, and accurately predicted the source location. Initial RT3D calibration results could be made to fit the data known at that time; however, the projection using the initial input parameters would have led to an inaccurate estimate of the future plume development. The mathematical approach used here was not rigorously developed, and should not be applied beyond the intended use - as a starting point for groundwater model calibration. In this case, subsequent data collected after the natural attenuation evaluation showed the updated RT3D model, which used the bioattenuation rates calculated by the single well method presented here, more accurately predicted plume development without further refinement of the estimated bioattenuation rates.

Literature Cited

1. Buscheck, T. E.; Alcantar, C. M. *Intrinsic Bioremediation*, Battelle Press, Columbus, Ohio, **1995**, 109-116.
2. Moutoux, D.E.; Benson, L.A.; Swanson, T.H.; Wiedemeier, T.H.; Lenhart, J., Wilson; J.T., Hansen, J.T. *Proceedings of the 1996 API/NGWA Conference on Petroleum Hydrocarbons and Organic Chemicals in Ground Water,* Houston, Texas, **1996,** 100-105.
3. McNab, W.W.; Dooher, B.P. *Ground Water*, **1998**, 36 (6), 983-987.
4. Levenspiel, O.; *Chemical Reaction Engineering*; John Wiley and Sons, New York, New York, **1972**, 172.

Chapter 8

Enhanced Bioremediation for Treatment of Chlorinated Solvent Residual Source Areas

K. S. Sorenson, Jr.

North Wind Environmental, Inc., P.O. Box 51174, Idaho Falls, ID 83405

Recent studies have demonstrated enhanced mass transfer of chlorinated solvents from the nonaqueous phase to the aqueous phase during bioremediation. In addition to two previously identified mechanisms for enhanced mass transfer, a third mechanism observed during studies at the Idaho National Engineering and Environmental Laboratory is described. Laboratory measurements demonstrated that sodium lactate (used as an electron donor for biostimulation) can decrease interfacial tension with nonaqueous TCE by 26% when the concentration is increased from 0.1% to 3% by weight, and by 47% when the concentration is increased to 30%. A field study showed sodium lactate injection resulted in a factor of 23 increase in trichloroethene (TCE) concentrations due to enhanced mass transfer. The TCE was then stoichiometrically converted to *cis*-1,2-dichloroethene and ultimately was completely biologically reduced to ethene.

Introduction

Chlorinated solvents are the most common class of contaminants in ground water at hazardous waste sites in the U.S. In 1993, the Agency for Toxic Substances and Disease Registry (ATSDR) compiled a list of the top 25 contaminants detected at hazardous waste sites on the National Priorities List (NPL). The ATSDR ranking identified eight of the top 20 contaminants as chlorinated solvents and their intrinsic degradation products (1). The ranking was updated by the ATSDR based on 1996 data with similar results. Of particular significance is the identification of trichloroethene (TCE) and tetrachloroethene (PCE) as the first and third most common contaminants at NPL sites in both surveys.

The prevalence of chlorinated solvents is due both to their widespread use and to their longevity in the environment. Their longevity is partly due to the hydrophobic nature that makes them such good solvents, as well as their relatively oxidized states that prevent them from serving as electron donors in microbial metabolism. At many sites, solvent sources are present in the subsurface as dense nonaqueous-phase liquids (DNAPLs). DNAPLs are hydrophobic liquids with a density greater than water. Pertinent to their longevity is the fact that the solubility of the common chlorinated solvents (PCE, TCE, TCA, and carbon tetrachloride) ranges from about 200 to 1400 mg/L at 25°C (2). These relatively low solubilities play a significant role in limiting mass transfer to the aqueous phase once the solvents contaminate ground water.

DNAPL Mass Transfer Limitations

Interphase mass transfer (dissolution) of a solvent NAPL into ground water is governed by the difference between the aqueous solubility of the compound and the actual concentration in ground water. Sale (3) provides an excellent discussion of fundamental interphase mass transfer from DNAPLs. At typical ground water velocities, the aqueous concentration of the solvent in the immediate vicinity of the ground water-NAPL interface approaches the solubility within the first few centimeters of flow along the interface (3). Because ground water flow is generally laminar, very little mixing of the water near the interface occurs with water even a few centimeters from the interface.

The lack of mixing characteristic of laminar flow has at least two important implications. First, it explains why ground water concentrations of chlorinated solvents greater than 10% of their solubility are rarely measured, even at contaminated sites with large quantities of DNAPL. Second, the attainment of concentrations approaching solubility within a few centimeters of ground water flow along the interface effectively prevents mass transfer out of the DNAPL for the remainder of flow along the interface. For example, if ground water flows across a pool of DNAPL (or through an area of residual saturation) several meters long in the direction of flow, mass transfer into the aqueous phase will be insignificant along all but the first few centimeters of the flow path. The result is that chlorinated solvents persist in ground water for many decades, or perhaps even centuries.

Influence of Mass Transfer on Remediation of NAPLs

Cleanup of chlorinated solvent sources in ground water is often considered technically (or economically) impracticable because of their density and hydrophobicity, often compounded by subsurface heterogeneity. As a result, many sites have resorted to pump and treat or other containment technologies. Operations and maintenance costs of such systems become very large over time, however, because of the longevity of the subsurface sources discussed above. While significant progress has been made in addressing solvent source areas, parties responsible for cleaning up sites with chlorinated solvent residual source areas in ground water are still faced with several technologies with significant capital costs, secondary waste streams, the involvement of hazardous materials or energy, and the potential for additional worker or environmental exposure. A more ideal technology would involve lower capital costs, would not generate secondary waste streams, would be non-hazardous to workers and the environment, would destroy contaminants in situ, would be low maintenance, and would minimize disturbance of the site.

In situ bioremediation satisfies all of the characteristics of an ideal technology listed above, but it has traditionally been viewed as very passive with respect to source area remediation. That is, conventional wisdom suggests that bioremediation is limited by the rate at which nonaqueous contaminants dissolve or diffuse to where bacteria can degrade them. If that were true, bioremediation would have all the benefits of an in situ technology regarding low capital cost, lack of secondary waste streams, low maintenance, minimal site disturbance, etc., but would not be much different than pump-and-treat in terms of cleanup times. Recent advances have shown however, that mass transfer rates of chlorinated solvents from the nonaqueous phase to the aqueous phase (where they are bioavailable) can be substantially increased during bioremediation. This paper provides an overview of mechanisms that have been shown to contribute enhanced mass transfer during bioremediation, then presents field data demonstrating the effect in a full-scale application of in situ bioremediation.

Mechanisms for Enhanced Mass Transfer

Microbial reductive dechlorination of chlorinated ethenes has been well documented (4-13); however, the stimulation of this process for source-zone restoration has not been rigorously evaluated. Investigations into biological dechlorination-based source zone restoration have begun only recently both in the laboratory (14-16), and in the field (17). Previously, the applicability of bioremediation for source-zones has been thought to be limited by two concerns. The first is the potential toxicity of high contaminant concentrations on microbial populations. Several studies have confirmed recently, however, that biological dechlorination can occur at aqueous saturation concentrations of PCE and high concentrations of TCE (13, 16, 18-20). It has been speculated that halorespiring organisms might actually have an advantage in environments of high PCE and TCE concentrations, such as NAPL source zones, where organisms normally competing with them for energy sources are not able to thrive (16, 21-24).

The second hypothetical limitation of bioremediation in source zones is the issue of mass transfer discussed above. If mass transfer rates from the NAPL phase to the aqueous phase are the same for bioremediation as for pump and treat or other containment technologies, cleanup times would be no different. Recent research has demonstrated, however, that mass transfer is significantly enhanced during biological dechlorination. It has been shown that rapid rates of biological dechlorination in NAPL-containing source areas can dramatically reduce the length of time that a NAPL will continue to be a source of chlorinated solvent contamination (14-16). In addition to this work, field and preliminary laboratory results from the Idaho National Engineering and Environmental Laboratory (INEEL) presented herein have revealed that electron donor solutions used to stimulate biological dechlorination can dramatically enhance bioavailability of the chlorinated solvents. The enhanced mass transfer observed in these different studies is attributable to at least three distinct mechanisms, which are discussed below:

1. Dechlorinating bacteria are capable of living in close proximity to the NAPL/water interface. Thus their metabolic activity increases the driving force for mass transfer (i.e., the concentration gradient). Under these conditions, mass transfer from the NAPL to the aqueous phase can occur along the entire length of a flow path over a pool or through a residual source area, as opposed to only in the first few centimeters of an abiotic system.
2. The metabolic products of dechlorination have increasingly higher aqueous solubilities with each step of the dechlorination process. The total mass of chlorinated compounds in solution is therefore much greater during dechlorination than in the abiotic dissolution case.

3. Specific electron donor solutions can be used that not only stimulate dechlorination, but also decrease the interfacial tension between the aqueous and nonaqueous phase, thereby increasing the effective solubility of the NAPL (patent pending).

The first two mechanisms have been documented in laboratory tests. In glass-bead column studies, Cope and Hughes observed a 16-fold increase in PCE removal from a NAPL in biotic systems as compared to abiotic "washout" (15). This result was similar to the observation of a 14-fold increase in PCE removal rates from a NAPL in a continuous-flow stirred-tank reactor (14). If similar results were obtained in field systems, a 100-year source of PCE would be present for only 6.25 years. In independent batch microcosm studies, Yang and McCarty (16) found PCE dissolution was enhanced by up to a factor of five during reductive dechlorination as compared to the abiotic case.

The third mechanism for enhanced mass transfer of chlorinated solvent NAPLs during bioremediation is different in that it relies not on the biological activity itself, but on the electron donor used to stimulate the activity. Interfacial tension is a property that is often used to screen co-solvents and surfactants for use in DNAPL remediation. It measures the force per unit length along the interface between two liquid phases arising from the surface free energy. The higher the interfacial tension between two liquids, the less likely one is to dissolve into the other, and the more difficult it is for one to be transported within the other. Of particular importance for electron donor solutions that decrease interfacial tension is the fact that the enhanced mass transfer of the NAPL is occurring into the solution that facilitates its degradation. This is an ideal scenario from a mixing standpoint for halorespiration because the electron donor and electron acceptor are comingled in the aqueous phase.

The interfacial tension between nonaqueous TCE and sodium lactate electron donor solutions were measured as a function of the sodium lactate concentration using the pendant drop method (25, 26) coupled with real-time video imaging (27). As shown in Figure 1, interfacial tension decreased by about 26% when sodium lactate was increased from 0.1 to 3%. When sodium lactate was increased to 30%, the interfacial tension decreased to 47% of the value measured at a concentration of 0.1%. The data shown represent the mean values of a minimum of ten measurements. The sodium lactate concentrations are very high relative to concentrations used in most laboratory studies, even when electron donor concentrations were considered excessive (10); however, as discussed below, high electron donor concentrations have been shown to be effective in the field both for enhanced mass transfer, and for achieving complete reductive dechlorination.

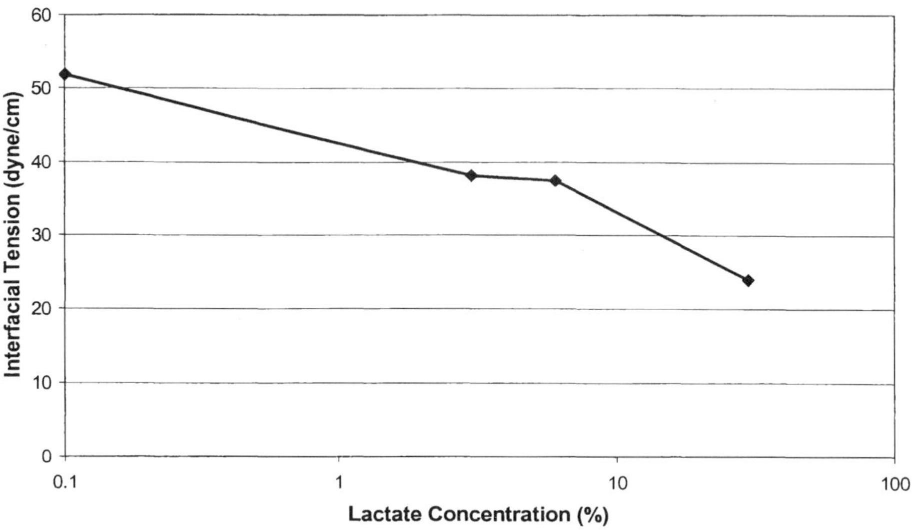

Figure 1. Impact of sodium lactate concentration on DNAPL/water interfacial tension.

Field Results and Discussion

A full-scale field study of bioremediation of a chlorinated solvent source zone at the INEEL began in January 1999. TCE and other contaminants were injected with sewage sludge into a fractured basalt aquifer 60-90 m below land surface for a 15-20 year period at the INEEL facility known as Test Area North. This combined sludge comprises a residual source of contamination extending approximately 30 m radially from the injection well (17). High concentrations of a sodium lactate solution ranging from 3 to 60% by weight were injected into the former injection well to stimulate reductive dechlorination of TCE. Transformation of TCE to ethene was observed as far as 15 m from the injection well within 5 months (17). Figure 2 shows the chloroethene concentrations at monitoring wells within 15 m of the injection location for the first 14 months of the test. By the end of this period, ethene was by far the dominant compound detected at all three wells. Dechlorination was strongly correlated to redox conditions and electron donor distribution (17, 28).

While these results were very exciting, it was of particular interest that strong evidence indicated the lactate solution significantly enhanced the mass transfer (and hence the bioavailability) of TCE in the residual source area, thereby accelerating degradation of the source material. The bioremediation performance observed in the field was consistent with the results of the laboratory studies discussed above, and showed substantial evidence that all three mechanisms were at work. Specific aspects of the field data that demonstrated enhanced mass transfer are discussed below.

Generation of Transformation Products

As shown in Figure 2, the quantity of *cis*-1,2-dichloroethene (*cis*-DCE) generated exceeded the amount of TCE initially present in the aqueous phase at TAN-25 and TAN-31 (compare the thickness of the *cis*-DCE band in March and April 1999, respectively, to the initial thickness of the TCE band). The most likely source of TCE for the "excess" *cis*-DCE is TCE associated with the organic sludge that comprises the residual source at the site. This effect was not observed at TAN-26 because this well is completed deeper than the residual source, while the other two wells are within the residual source (17). The observed effect is consistent with enhanced transfer resulting from Mechanisms 1 and 2 above. The increased dissolution due to increasing concentration gradients throughout the residual source zone and the increasing solubility of TCE's reduced products would be expected to cause both an increase in the *cis*-DCE generated relative to the initial TCE present, and a gradual increase in the total moles of chlorinated solvents in solution.

Another effect that might play a role is the degradation of sorbed-phase TCE. A potentially analogous process is the use of iron-III from mineral surfaces by iron-reducing bacteria to produce aqueous iron-II. The solubility of iron-III is extremely low (less than 150 μg/L) and it is bound in hydrous iron oxides and other minerals associated with basalt surfaces. In spite of this, dissolved iron-II concentrations in excess of 3000 μg/L were commonly observed following the onset of iron reduction at TAN (17). Jordan et al. (29) discussed some of the possible mechanisms for microorganisms to increase the bioavailability of surface-bound substrates.

An additional effect that may contribute to the "excess" of degradation products is the disruption of the equilibrium between sorbed and aqueous contaminants. When all of the PCE and TCE is removed from the aqueous phase, the reversible partitioning reaction between the solid and aqueous phases will shift toward the aqueous phase. This is analogous to Mechanism 1 and creates a flux of contaminants into the aqueous phase that is in addition to the flux from the secondary source, thereby providing an additional source of the parent compounds. In all likelihood, a combination of these effects is probably responsible for the production of more *cis*-DCE and other transformation products than would have been predicted from the initial aqueous TCE concentration. In any case, the result is significant because it indicates that accelerated degradation of nonaqueous as well as aqueous TCE was achieved.

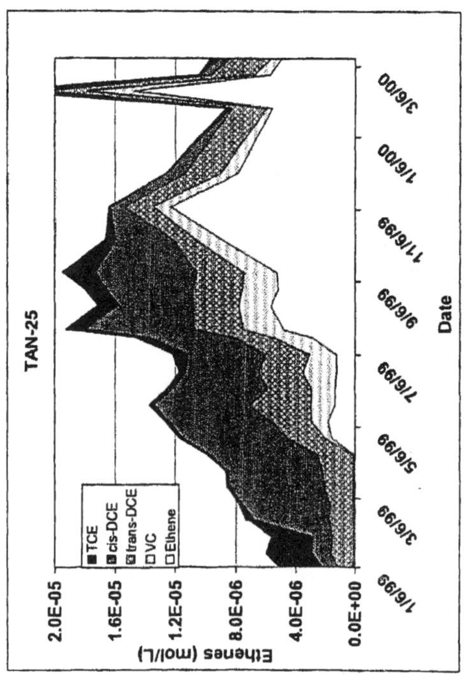

TAN-25
TCE
cis-DCE
trans-DCE
VC
Ethene
Ethenes (mol/L)
2.0E-05
1.6E-05
1.2E-05
8.0E-06
4.0E-06
0.0E+00
1/6/99
3/6/99
5/6/99
7/6/99
9/6/99
11/6/99
1/6/00
3/6/00
Date

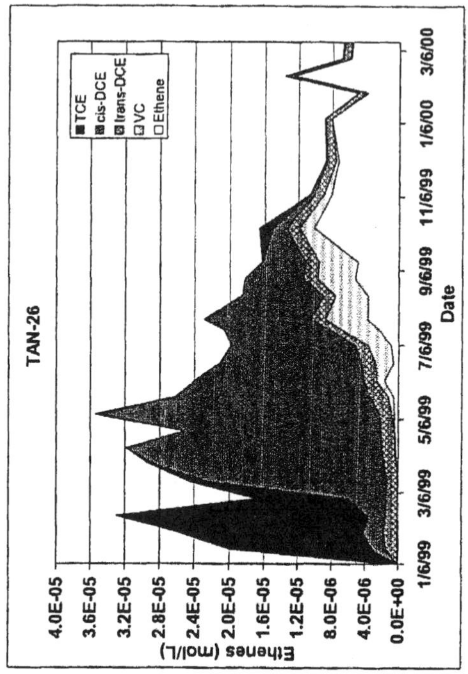

TAN-26
TCE
cis-DCE
trans-DCE
VC
Ethene
Ethenes (mol/L)
4.0E-05
3.6E-05
3.2E-05
2.8E-05
2.4E-05
2.0E-05
1.6E-05
1.2E-05
8.0E-06
4.0E-06
0.0E+00
1/6/99
3/6/99
5/6/99
7/6/99
9/6/99
11/6/99
1/6/00
3/6/00
Date

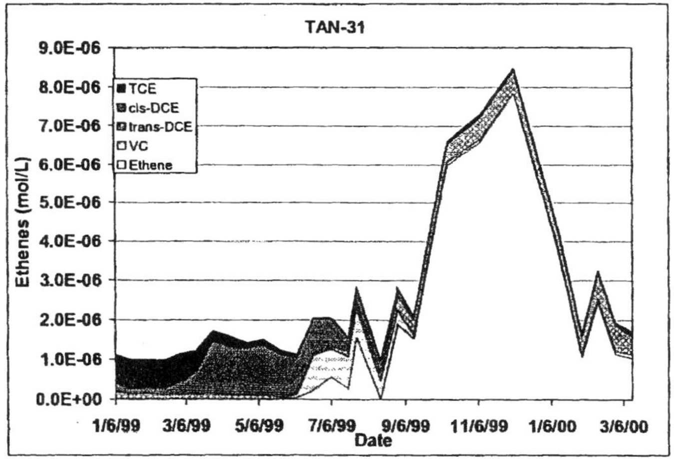

Figure 2. Reductive dechlorination of TCE at 3 wells within 15 m of the lactate injection well.

Abiotic Enhanced Mass Transfer and Bioavailability

The large increase of TCE and total ethenes in Well TAN-26 (Figure 2) appears to be due primarily to Mechanism 3 identified above. The TCE increase occurred simultaneously with the arrival of the highly concentrated lactate solution, but before fermentation of the lactate to acetate and propionate began (17). Thus, the effect had an abiotic origin. In addition, the peak TCE concentration was actually significantly higher than historical measurements for TAN-26. These facts strongly suggest that the transport of TCE to Well TAN-26 was associated with the downward migration of the electron donor (the solution was initially injected with a specific gravity of about 1.4). This could occur through two mechanisms. One possibility is that the electron donor simply pushed secondary source material along in front of it as it migrated out from Well TSF-05 through the secondary source and down toward Well TAN-26. This would explain the large and rapid increase in TCE concentrations; however, examination of Figure 3 rules out this possibility. These data show that while TCE and transformation products increased significantly, tritium concentrations were unaffected. Because tritium is also in the residual source material (30), physical displacement of the material would have caused tritium to increase along with TCE.

A second possible explanation for TCE increase in Well TAN-26 is that the electron donor preferentially facilitated transport of the organic contaminants. Two hypotheses that might explain the facilitated transport are: (1) that the electron donor enhanced mass transfer from the nonaqueous phase to the aqueous phase by decreasing interfacial tension, similar to a mild cosolvent or surfactant, or (2) that the solution, because of its high concentration, displaces sorbed chlorinated ethenes, driving them into solution. Both of these mechanisms would result in the transport of the chlorinated ethenes in intimate contact with the electron donor solution. This "mixing" with the electron donor solution would make the chlorinated ethenes eminently bioavailable.

The behavior of the TCE in TAN-26 after the peak concentration demonstrates that it was, in fact, extremely bioavailable. The drop in TCE concentration from the peak concentration to less than the detection limit of 10 μg/L occurred with a TCE half-life of less than 20 days (assuming first-order kinetics for illustration). Just as important for the facilitated transport hypothesis, cis-DCE increased to a peak concentration within 20% of the peak TCE concentration (Figure 2), indicating an excellent mass balance, and the total ethene compound concentration remained elevated near that peak concentration as long as the density of the injected solution remained high (Figure 3). This is consistent with expectations for Mechanism 3 because the electron donor was injected on a weekly basis through September 1999, and would be expected to continue dissolving and transporting the organic contaminants with it as long as it migrated downward through the secondary source.

The drop in the total ethene compound concentration after several months was expected and intentional because the electron donor solution concentration had been reduced several times by June 1999, with a final concentration that was a factor of 20 less than the initial 60% sodium lactate solution. These changes reduced the density of the solution significantly, so less electron donor, and therefore less total ethenes, was transported to Well TAN-26. In general, the enhanced mass transfer observed makes large quantities of the chlorinated ethenes available for biodegradation that otherwise would remain associated with the secondary source.

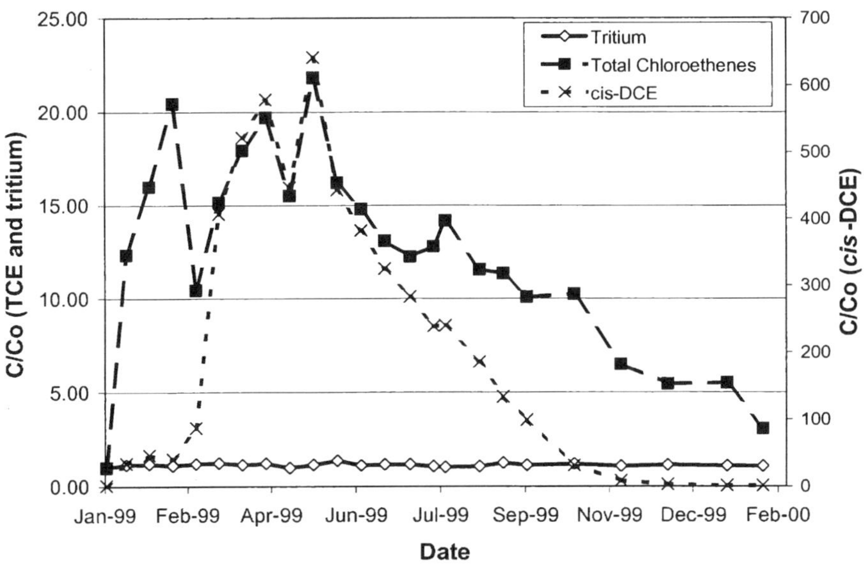

Figure 3. Concentration changes of tritium, total chloroethenes, and cis-DCE over time as measured by the ratio of the concentration at a given time (C) to the initial concentration (Co).

Conclusions

The biological mechanisms responsible for reductive dechlorination of aqueous chloroethenes have been described and are becoming relatively well understood. However, the interactions of biological communities and added electron donor solutions with nonaqueous (liquid or sorbed) residual chloroethenes in situ are only beginning to be studied. The results of the studies at TAN, as well as the other laboratory studies discussed, suggest these interactions may be very important in terms of the overall effectiveness of enhanced in situ reductive dechlorination as a remediation strategy. In fact, it appears that enhanced in situ bioremediation may be far more attractive than previously thought for remediation of chlorinated solvent residual source areas. Not only does it have the benefits of low capital cost, low maintenance, lack of worker or environmental risk associated with bringing contaminants above ground or using hazardous substances, and minimal site disruption, it also has the benefit of accelerated source removal, currently only achieved by more expensive and hazardous technologies. At sites with large volumes of DNAPL in the subsurface, however, it may prove beneficial to combine enhanced bioremediation with the more aggressive technologies. The aggressive technologies, such as thermal or surfactant technologies, could be tightly focused on areas with large volumes at high DNAPL saturations, while bioremediation could be used both as a polishing step for those areas and as a stand-alone remedy for areas with lower DNAPL saturations. The Idaho National Engineering and Environmental Laboratory has a pending patent on the use of high concentrations of electron donor solutions to enhance bioavailability of chlorinated solvents during biostimulation.

References

1. National Research Council (NRC). *Alternatives for Ground Water Cleanup;* National Academy Press: Washington D.C., 1994.
2. Pankow, J. F.; Cherry J. A. *Dense Chlorinated Solvents and Other DNAPLs in Groundwater;* Waterloo Press, Inc.: Portland, OR, 1996.
3. Sale, T. Ph.D. thesis, Colorado State University, Ft. Collins, CO, 1998.
4. Bouwer, E. J.; McCarty, P. L. *Appl. Environ. Microbiol.* **1983**, *45*, 1286-1294.
5. Parsons, F.; Wood, P. R.; DeMarco, J. *Journal of the American Water Works Association* **1984**, *76*, 56-59.
6. Vogel, T.; McCarty, P. *Appl. Environ. Microbiol.* **1985**, *49*, 1080-1083.
7. Fathepure, B.; Boyd, S. A. *FEMS Microbiol. Lett.* **1988**, *49*, 149-156.
8. Freedman, D. L.; Gossett, J. M. *Appl. Environ. Microbiol.* **1989**, *55*, 2144-2151.
9. DiStefano, T. D.; Gossett, J. M.; Zinder, S. H. *Appl. Environ. Microbiol.* **1991**, *57*, 2287-2292.
10. deBruin, W. P.; Kotterman, M. J. J.; Posthumus, M. A.; Schraa, G.; Zehnder, A. J. B. *Appl. Environ. Microbiol.* **1992**, *58*, 1996-2000.

11. DiStefano, T. D.; Gossett, J. M.; Zinder, S. H. *Appl. Environ. Microbiol.* **1992**, *58*, 3622-3629.
12. Fennell, D. E.; Gossett, J. A.; Zinder, S. H. *Environ. Sci. Technol.* **1997**, *31*, 918-926.
13. Carr, C. S.; Hughes, J. B. *Environ. Sci. Technol.* **1998**, *32*, 1817-1824.
14. Carr, C. S.; Garg, S.; Hughes, J. B. *Environ. Sci. Technol.* **2000**, *34*, 1088-1094.
15. Cope, N.; Hughes, J. B. *Environ. Sci. Technol.* **2001**, *Accepted for publication.*
16. Yang, Y.; McCarty, P. L. *Environ. Sc. Technol.* **2000**, *34*, 2979-2984.
17. Sorenson, K. S. Ph.D. thesis, University of Idaho, Moscow, ID, 2000.
18. Sharma, P. K.; McCarty, P. L. *Appl. Environ. Microbiol.* **1996,** *62*, 761-765.
19. Isalou, M.; Sleep, B. E.; Liss, S. N. *Environ. Sci. Technol.* **1998,** *32*, 3579-3585.
20. Nielsen, R. B.; Keasling, J. D. *Biotechnol. Bioeng.* **1999,** *62*, 160-165.
21. Holliger, C.; Schraa, G.; Stams, A. J. M.; Zehnder, A. J. B. *Appl. Environ. Microbiol.* **1993,** *59*, 2991-2997.
22. Holliger, C.; Hahn, D.; Harmsen, H.; Ludwig, W.; Schumacher, W.; Tindall, B.; Vazquez, F.; Weiss, N.; Zehnder, A. J. B. *Arch. Microbiol.* **1998**, *169*, 313-321.
23. Neumann, A.; Scholz-Muramatsu, H.; Diekert, G. *Arch. Microbiol.* **1994,** *162*, 295-301.
24. Scholz-Muramatsu, H.; Neumann, A.; MeBmer, M.; Moore, E.; Diekert, G. *Arch. Microbiol.* **1995,** *163*, 48-56.
25. *Structure/Performance Relationships in Surfactants;* Rosen, M. J., Ed.; American Chemical Society: Washington D.C., 1984; pp. 329.
26. Bagnall, R. D.; Arundel, P. A. *J. Phys. Chem.* **1978,** *82,* 898.
27. Herd, M. D.; Lassahn, G. D.; Thomas, C. P.; Bala, G. A.; Eastman, S. L. Proceedings Paper No. SPE/DOE 24206; Eighth Symposium on Enhanced Oil Recovery; Tulsa, OK, 1992; pp. 513-519.
28. Sorenson, K. S.; Peterson, L. N.; Ely, R. L. In *Groundwater 2000;* Bjerg, P. L.; Engesgaard, P.; Krom, Th. D., Eds.; A. A. Balkema Publishers: Rotterdam, Netherlands, 2000; Vol. 2, pp 379-380.
29. Jordan, R. N.; Nichols, E. P.; Cunningham, A. B. *Water Sci.Tech.* **1999,** *39,* 91-98.
30. Sorenson, K. S.; Peterson, L. N.; Hinchee, R. E.; Ely, R. L. *Biorem. J.* **2000,** *4*, 337-357.

Chapter 9

Nanoscale and Microscale Iron Emulsions for Treating DNAPL

Cherie L. Geiger[1], Christian A. Clausen[1], Kathleen Brooks[1], Christina Clausen[1], Christian Huntley[1], Laura Filipek[1], Debra D. Reinhart[2], Jacqueline Quinn[3], Tom Krug[4], Suzanne O'Hara[4], and David Major[4]

Departments of [1]Chemistry and [2]Civil and Environmental Engineering, University of Central Florida, 4000 Central Florida Boulevard, Orlando, FL 32816–2366 (fax: 407–823–2252; cgeiger@mail.ucf.edu; clausen@mail.ucf.edu)
[3]NASA, Kennedy Space Center, Kennedy Space Center, FL 32899
[4]GeoSyntec Consultants, Inc., 130 Research Lane, Guelph, Ontario N1G 5G3, Canada

This study demonstrated the feasibility of using emulsified nanoscale and microscale iron particles to enhance dehalogenation of DNAPL free-phase. The emulsified system consisted of a surfactant-stabilized, biodegradable oil-in-water emulsion with nanoscale or microscale iron particles contained within the emulsion droplets. It was demonstrated that DNAPLs, such as TCE, diffuse through the oil/surfactant membrane of the emulsion particle whereupon they reach an aqueous interior and the surface of an iron particle where dehalogenation takes place. The hydrocarbon reaction by-products of the dehalogenation reaction, primarily ethene (no chlorinated products detected), diffuse out of the emulsion droplet. This study also demonstrated that an iron-emulsion system could be delivered in-situ to the DNAPL pool in a soil matrix by using a simulated push well technique. Iron emulsions degraded pure TCE at a rate comparable to the degradation of dissolved phase TCE by iron particles, while pure iron had a very low degradation rate for free-phase TCE. The iron-emulsion systems can be injected into a sand matrix where they become immobilized and are not moved by flowing water. It has been documented that surfactant micelles possess the ability to pull pooled TCE into emulsion droplets where degradation of TCE takes place.

Introduction

Remediation of chlorinated solvents in ground water are most commonly addressed using pump and treat technology. It is now widely recognized that due to the slow dissolution of solvents from residual or pooled dense non-aqueous phase liquid (DNAPL) sources, pump and treat primarily provides containment, rather than remediation. Because of this slow dissolution, pump and treat systems will need to be operated for long periods of time (i.e. decades or longer) in order to maintain protection of human health and the environment, incurring high operation and maintenance costs over that period.

Currently, there are no available proven technologies that can treat 100% of DNAPL sources (free-phase, residual phase, and sorbed [matrix diffused] phases). Attempts have been made to remove the DNAPL sources through heating to enhance volatilization (steam injection and radio-frequency-heating). This approach is limited because of the energy costs associated with heating the groundwater, and the exponential volume of areas that will need to be treated to ensure that the entire DNAPL source is encountered and treated. An alternative approach is to flood the source area with surfactants or oxidizing agents (1). This approach is also limited because the agents will move primarily through the most permeable zones, but considerable DNAPL mass will penetrate into areas of low permeability and will also diffuse (matrix diffusion) into geological materials. DNAPLs that exist in the matrix are dissolved or sorbed (2) and oxidizing agents can diffuse into the matrix and treat the dissolved NAPLs. Thus the very nature of DNAPL sources prevents complete removal that is required to prevent the remaining DNAPL from re-contaminating the groundwater.

There is a critical need for technologies that can effectively treat DNAPL sources in the saturated zone and result in both their destruction and containment with reduced treatment times and lower costs. This project addresses this need.

Significant work has demonstrated that zero-valent metals will reductively dechlorinate chlorinated DNAPLs such as trichloroethene (TCE) to ethene (3,4). Permeable reactive barriers (PRBs) composed of iron have been shown to be effective in treating dissolved plumes of chlorinated solvents. Although this technology is passive and requires no energy, it still relies on the DNAPL dissolution and transport of dissolved chlorinated solvents to the PRB for treatment. Thus, the length of the treatment is not shortened and there is still a requirement for long-term monitoring. Insertion of iron into source zones by use of soil mixing technologies has been attempted, but this approach is limited to accessible sources, and by the depth of soil treatment. This research project addresses these limitations through the demonstration of an innovative technique that can be applied to DNAPL pools located under existing, occupied facilities and demonstrates the feasibility of using emulsified nanoscale iron particles to enhance dehalogenation of DNAPL free-phase.

The emulsified system consists of a surfactant stabilized oil-in-water emulsion with the nanoscale iron particles contained within the emulsion droplets. The surfactant serves two purposes: 1) it makes the emulsion stable for injection into the DNAPL and 2) the micelles within the emulsion droplet aid in the delivery of TCE to the iron. DNAPLs, such as TCE, diffuse through the oil membrane of the emulsion particle whereupon they reach the surface of an iron particle where dehalogenation takes place. The hydrocarbon reaction by-products of the dehalogenation reaction diffuse out of the emulsion particle and vent to the ground surface. The iron-emulsion system can be delivered in-situ to a DNAPL pool through a series of push wells.

In the generation of the iron-emulsion system, it is desirable to use the smallest iron particles possible because the small size produces a more stable and reactive emulsion that is capable of penetrating into the smallest of pore cavities between soil particles. Thus, it is desirable to use iron particles in the nanosize or microsize range. Microscale iron can be purchased from a variety of manufacturers. Nanoscale iron must be synthesized in the laboratory.

One of the methods used to synthesize nanoscale iron particles is to slowly add an aqueous solution of $NaBH_4$ to an aqueous solution of $FeCl_3.6H_2O$. The iron particles precipitate and can be readily separated from the solution (5). These chemicals are industrial chemicals produced in large quantities and are relatively inexpensive. A similar method for preparing nanoscale iron particles has been reported in the literature (4).

The nanoscale iron that has been prepared by this technique has a surface area between 15 and 25 m^2/g. This material has been demonstrated to be active in the dehalogenation of TCE in groundwater (6). Nanoscale iron could be mixed directly into water to form a slurry that could be injected into the ground to intercept or treat a plume of contaminated groundwater. However, if this aqueous slurry was to be used to directly treat a DNAPL pool, the hydrophobic pool would reject the hydrophilic slurry and thus the iron particles would not be delivered to the TCE molecules. Thus, it is necessary to make the iron particle slurry hydrophobic.

There is a large scientific basis for the production of what is known as liquid membranes (7, 8). These liquid membranes are emulsion droplets that can be made to have a spherical hydrophobic skin and an aqueous hydrophilic interior. The iron particles are contained in the aqueous interior phase. The exterior of the droplet is hydrophobic and would be able to penetrate into a separate DNAPL phase. The chlorinated solvent would dissolve in the hydrophobic emulsion membrane and then move by diffusion into the aqueous interior. The dehalogenation reaction would then rapidly take place at the surface of the iron. A graphic rendition of what this system looks like at the microscopic level is shown in Figure 1.

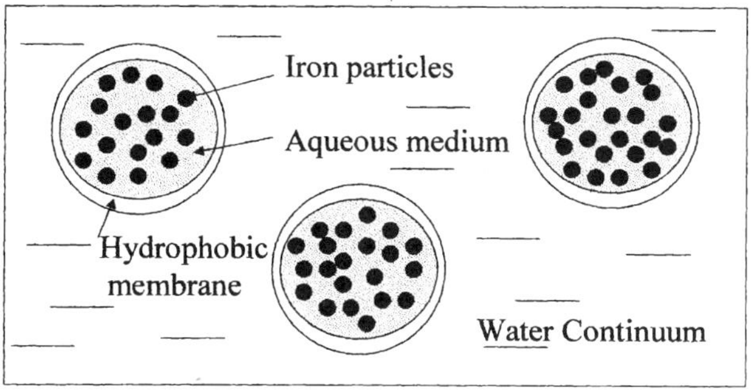

Figure 1. Nanoscale iron particles contained in emulsion droplet.

Experimental

Kinetic studies where performed to determine the degradation efficiency of the various emulsions made in the lab. In order to quantitate true destruction, ethene, the terminal product of the dehalogenation reaction, was measured by headspace analysis. Crimp-top, gas-tight vials were used for this work. All experimental work was carried out in an inert atmosphere (nitrogen) glovebox. A measured amount (based on weight of iron) of emulsion was added to the vial along with de-oxygenated, deionized water and 100-μL of TCE. Each vial had five-mL of headspace. Analysis was performed using a purge and trap concentrator and a Trimetrics gas chromatograph equipped with a flame ionization detector.

Pumping Studies

A column experiment was set-up using 2-inch o.d. x 2.0 ft. length acrylic tubing and soil obtained from a research site at launch complex 34, Cape Canaveral Air Station, Kennedy Space Center, FL. The soil used was from a lower (26-40-ft. below land surface) sandy unit of a contaminated shallow aquifer. The porosity of the column material after packing was 32%. Water and emulsion was pumped into the column (vertical up-flow) at a rate of 65-ml per hour and a water:emulsion ratio of 5:1.

Results and Discussion

The liquid membrane emulsions can be made with nonionic, cationic or anionic surfactants. During this study, more than 200 different emulsions were made and tested for TCE degradation efficiency. Micrographs of a microscale iron emulsion (a) and a nanoscale iron emulsion (b) are shown in figure 2.The micrograph shown in Figure 2(a) was taken using a slide platform of depth approximately 200 μm so the emulsion droplet wasn't flattened. Because the droplet was still spherical, some of the micro-iron appeared in focus and the iron near the top appeared out of focus. A nanoscale iron emulsion Figure 2(b) was examined using a standard microscope slide and cover slip so the spherical shaped droplet was flattened by the cover slip. It was obvious from both micrographs that the iron was contained within the interior of the droplets. Since the oil layer was the outer-most layer of the droplet, hydrophobic TCE can dissolve through that layer and into the water/iron phase where degradation takes place.

Tests narrowed the most effective emulsions to those made with three surfactants: Rhodapon LSB (anionic, sodium laurel sulfate), Span 80 (neutral, sorbitan monooleate), and Span 85 (neutral, sorbitan trioleate). These emulsions satisfied the practical need of being stable and flowable as well as being active. Emulsions were then made in the blends shown in Table I.

Table I. Composition of Emulsions Used in the Vial/Constant Headspace Kinetic Study.

Emulsion	*Nano Iron (wt%)*	*Corn Oil (wt%)*	*Water (wt%)*	*Surfactant (wt%)*
Blend #1	9.21	42.39	46.08	2.30
Blend #2	9.00	41.44	45.04	4.50
Blend #3	8.44	38.82	42.19	10.55
Blend #4	13.13	20.13	65.65	1.09
Blend #5	13.45	61.88	22.42	2.24
Blend #6	17.74	21.76	59.12	1.38

Experiments performed to test the efficiency of those blends for degradation of DNAPL TCE to ethene revealed that five emulsions gave the best results and those are shown in Table II. Rhodapon LSB (blend 4) and Span 85 (blend 5) showed the fastest reduction of TCE to ethene. Further long-term studies are needed to clarify the lifetime usefulness of the emulsions. However, these early results indicate that TCE mass in the DNAPL form can be reduced quickly using this methodology. Reduction of the source mass could significantly reduce plume remediation time and costs.

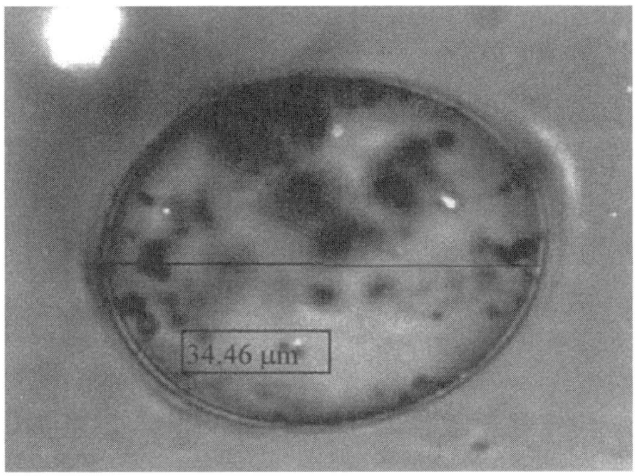

(a)

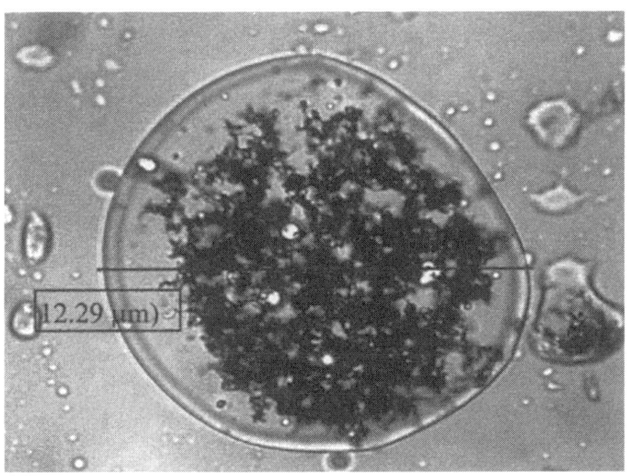

(b)

Figure 2. (a) Micrograph of a micro-iron emulsion dispersed in water. (b) Micrograph of nano-iron emulsion.

Table II. Ethene (ppm) Produced in the Constant Headspace Vials (5-mL samples) Over a 12-Day Period.

Emulsion Type	*Day 1*	*Day 2*	*Day 3*	*Day 7*	*Day 12*
Rhodapon LSB Blend #4	DL	6.15	8.91	19.32	38.06
Rhodapon LSB Blend #5	DL	2.38	2.49	5.47	13.08
Span 80 Blend #4	DL	4.03	6.21	10.29	14.69
Span 85 Blend #4	DL	3.04	4.82	6.21	9.87
Span 85 Blend #5	DL	7.04	9.92	28.94	33.88
Control	DL	DL	DL	DL	0.21

Note: DL denotes concentration below detection limit.

None of the emulsions yielded concentrations of chlorinated by-products in solution above the detection limit (DL). The only product that could be measured was ethene and some small concentrations of other hydrocarbons. This indicates that TCE entered the emulsion and degradation occured via proposed mechanisms (9). The chlorinated by-products produced during the degradation process (cis or trans-dichloroethene, vinyl chloride, etc.) remain in the emulsion and continue along the mechanistic pathways until ethene, a terminal product is formed. Ethene, which has a lower solubility in water, exits through the droplet wall where it then enters free phase water and the gaseous headspace. This is an important aspect of the use of such a potential technology since it won't contribute to the pollution problem by allowing partially de-chlorinated by-products to enter the groundwater.

Pumping Studies

In preparation for field scale tests, practical properties of the emulsions had to be determined. A primary concern was how the emulsion could be pumped into the subsurface. To investigate this concern, a column experiment was set-up using 2-inch o.d. x 2.0 ft. length acrylic tubing and soil obtained from a research site at launch complex 34, Cape Canaveral Air Station, Kennedy Space Center, FL. The soil used was from a lower (26-40-ft. bls) sandy unit of a contaminated shallow aquifer. The porosity of the column material after packing was 32%. Water and emulsion (micro iron formulation #5) was pumped into the column (vertical up-flow) at a rate of 65-ml per hour and a water:emulsion ratio of 5:1.

After a period of four hours, the emulsion had passed through the length of the column and began exiting in the effluent. Samples of the effluent emulsion were examined under a microscope and verified that the emulsion did not break down through the pumping process. The column was then dismantled by cutting it in half along its vertical length. The emulsion had saturated the bottom one-half of the column and then found preferential paths through the remainder of the soil matrix.

A similar column experiment was carried out with nano-scale iron emulsion using the same formulation as the micro-scale iron emulsion used in the previous column experiment. The nano-scale iron emulsion flowed into and through the column more effectively and saturated the entire length of the column material. The emulsion droplets exiting in the effluent were examined and found to be intact. Soil samples containing emulsion were also examined and the structure of the emulsion was retained.

Conclusions

Emulsions containing microscale and nanoscale iron have been developed and tested at the University of Central Florida Chemistry Department. The components of the emulsion, corn or vegatable oil, food-grade surfactant, water and iron particles are biodegradable and/or harmless to the environment. These emulsions have been shown to encapsulate and degrade DNAPL in both water and soil matrices. Chlorinated by-products of the degradation of TCE are not released from the emulsion in measureable quantities. Ethene, a non-chlorinated product of the reaction, can be quantified and was used to follow the rate of the degradation reaction. The emulsion can be pumped at pressures of 160-psi in laboratory experiments through fine-grained or more coarse-grained materials. A field injection test showed that a direct push injection technology is not adequate for subsurface placement of the emulsion. Other technologies are being evaluated.

A field test that is being evaluated under the US EPA SITE (Superfund Innovative Technology Evaluation) Program is scheduled for June 2002 at Cape Canaveral Air Station, Launch Complex 34. This NASA cleanup site has been extensively characterized and it is estimated that over 40,000 kg of DNAPL (TCE and cis-DCE) lie beneath a former parts cleaning facility on the site.

References

1. Reilsma, S. and M. Marshall (1990) "Experimental Study of Oxidation of Pooled NAPL" in Chemical Oxidation and Reactive Barriers. Battelle Press, Columbus, Ohio.
2. Gillham, R.W. and D.R. Burris (1992) "Recent Developments in Permeable In Situ Treatment Walls for Remediation of Contaminated Groundwater," In *Proceedings of Subsurface Restoration Conference*, June 21-24, 1992.
3. Vogan, J. L., J.K. Ealberg, B. Gnabasik, and S. O'Hannesin (1994), "Evaluation of In Situ Groundwater Remediation by Metal Enhanced Reductive-Dehalogenation - Laboratory Column Studies and Groundwater Flow Modeling," presented at the 87th Annual Meeting and Exhibition of the Air and Waste Management Association, Cincinnati, OH, June 19-24.
4. Toy, P.C., Master's Thesis, University of Central Florida, Orlando, Florida, 1998.
5. Wang, C.B. and Zhang, W.X., (1997). *Environ. Science Technol.*, Vol. 31, No. 7, p. 2154.
6. Geiger, C. L., C. A. Clausen, D. Reinhart, N, Ruiz, P. Toy, N. Lau. (1998) "The Use of Ultrasonic Energy for Regeneration of Reactive Iron Used for In Situ Remediation." Proceedings of the Air and Waste Management 91st Annual Meeting, San Diego, CA. June 14-18.
7. Cahn, R.P. and Li, N.N., (1974). *Separation Science*, Vol. 9, pp. 505-519.
8. Kim, K.S., S. J. Choi, and S. K. Ihm, (1983). *Ind. Eng. Chem. Fundam.*, Vol. 22, pp. 167-172.
9. Roberts, A. L., L. A. Totten, W. A. Arnold, D. R. Burris, T. J. Campbell. (1996) *Environ. Science Technol.*, 30,2654-2659.

Chapter 10

Reactive Dechlorination of PCE Using Zero Valent Iron Plus Surfactants

Hyun-Hee Cho and Jae-Woo Park

National Subsurface Environmental Research Laboratory (NSERL), Ewha Womans University, 11–1 Daehyon Dong Sedaemun-Gu, Seoul 120–750, South Korea

Zero valent iron (ZVI) is particularly useful as a reductant of chlorinated hydrocarbons because of its low cost and lack of toxicity. Surfactants also have been used widely for decontamination of subsurface soil and groundwater. The objective of this study is to examine the effect of various surfactants on enhancing the dechlorination of perchloroethene (PCE) by ZVI. Three surfactants—anionic, nonionic, and cationic—were chosen: sodium dodecylbenzene sulfonate (SDDBS), Triton X-100, and cetylpyridinum chloride (CPC). The rate of dechlorination of PCE using ZVI with CPC and Triton X-100 was much higher than withoutout surfactant. The combination of nonionic and cationic surfactants with ZVI increased the degradation of PCE, because PCE that bound with the surfactant in aqueous phase was readily available for dechlorination. Using SDDBS resulted in no observed dechlorination of PCE with ZVI, because no trichloroethene (TCE) was detected. However, PCE was immobilized on the sorbed SDDBS phase.

Introduction

The common occurrence in groundwater of chlorinated compounds, such as perchloroethene (PCE) and trichloroethene (TCE), is due largely to their extensive use by industry and their resistance to degradation under natural conditions (*1-3*). The release of these compounds into subsurface environments has contaminatednumerous groundwater resources.

Permeable reactive barriers, which are gaining increased use in remediating groundwater, most commonly utilize zero-valent iron (ZVI) as the primary treatment material. ZVI is particularly attractive as a reductant of chlorinated hydrocarbons because of its low cost and lack of toxicity. When ZVI is used to support in situ remediation of contaminants, the chemical reactions mediated by ZVI convert toxic compounds into non-toxic forms (*1-5*).

Previous studies have shown that the variability in first-order rates of disappearance for chlorinated compounds primarily reflects differences in reactivity of individual chemical contaminants and in available iron surface area (*6*). Factors such as pH, contaminant concentration, and flow or mixing rate also affect degradation kinetics to varying degrees (*7*).

Studies of reductive dechlorination of PCE and TCE have shown that dechlorination proceeds in a stepwise fashion, with the degradation rate slowing substantially with each dechlorination step. Hydrogenolysis and reductive β-elimination are primary pathways for degradation by ZVI (*5, 8, 9*). The reaction rate was found to be highest for saturated and perhalogenated organic oxidants, with most systems exhibiting accelerated reaction in the presence of water (*10*).

In aerobic waters, oxygen is consumed and Fe^0 oxidizes to Fe^{2+}, Fe^{3+}, and OH^- ions, as shown by the flowing reactions:

$$2H_2O + O_2 + 4e^- \rightarrow 4OH^- \quad (1)$$
$$2Fe^0 + O_2 + 2H_2O \rightarrow 2Fe^{2+} + 4OH^- \quad (2)$$

In anaerobic systems, water is the oxidizing agent, oxidizing Fe^0 to Fe^{2+}, OH^-, and H^+ ions, as shown by the reactions below:

$$2H_2O + 2e^- \rightarrow H_2 + 4OH^- \quad (3)$$
$$Fe^0 + 2H_2O \rightarrow Fe^{2+} + H_2 + 2OH^- \quad (4)$$

Although iron oxidation gives off electrons, the organic pollutants can be reduced if the electrons are received, as shown below in the equation for the reaction of iron and the chlorinated compound designated by RX:

$$Fe^0 + RX + H^+ \rightarrow Fe^{2+} + RH + X^- \quad (5)$$

Surfactants have been used widely for decontamination of the subsurface through soil-washing or contaminant immobilization (*11-15*). Surfactant molecules, which consist of hydrophilic and hydrophobic moieties, form micelles up to a certain aqueous concentration, called the critical micelle concentration (CMC) (*16*). The central part of a micelle is hydrophobic, so aqueous solubility is enhanced greatly for poorly water soluble compounds at concentrations higher than their CMCs (*17*).

Bizzigotti et al. (1997) studied the reduction of PCE under the influence of hydroxypropyl-β-cyclodextrin (HP-β-CD), which enhances the solubility of PCE in both static and flowing water systems. They found that the rate of PCE dechlorination decreased because the PCP molecules were partitioned into the hydrophobic interiors of HP-β-CD molecules (*18*). Sayles et al. (1997), who studied the influence of the nonionic surfactant Triton X-114 on the rate of 1,1,1-trichloro-2,2-bis(p-chlorophenyl)ethane (DDT), 1-dichloro-2,2-bis(*p*-chlorophenyl) ethane (DDD), and 1,1-dichloro-2,2-bis(p-chlorophenyl) ethylene (DDE) dechlorination by ZVI (*19*), found that dechlorination rates for DDT and DDE were independent of the amount of iron, with or without Triton X-114, but were higher with Triton X-114 than without.

The objective of the present study was to investigate the effects of various surfactants at low concentrations, below or about their CMCs, on the dechlorination of PCE by ZVI. Because both immobilization with surfactants and reductive dechlorination with ZVI are vaiable technologies for remediating chlorinated organics in the subsurface, we can imagine a situation in which both these technologies are applied as a combined treatment barrier or zone, in series or mixed. In such a case, the inflow of surfactant, desorbed into groundwater from the sorbed phase, is unavoidable and, even at low concentrations, could affect the reaction by ZVI. Therefore, low concentrations of surfactant were used in this study, compared to previous studies, which focused on flushing with surfactant solutions for washing rather than immobilization (*8, 9*).

Experimental Methods

Iron powder was obtained from Junsei (Japan), and 95% of the particles passed through a 100-mesh sieve. The surface area was 1.96 m^2/g based on BET adsorption using an ASAP 2000 device. PCE purchased from Fisher was 99% pure. The surfactants used were sodium dodecylbenzene sulfonate (SDDBS) from Aldrich, Triton X-100 from Sigma, and cetylpyridium chloride (CPC) from Aldrich. Table I lists the characteristics of the three surfactants. The buffer solution was 20 mM of 3- (N-morpholino) propanesulfonic acid (MOPS) with pH adjusted to 7.

To determine the rate of PCE removal, 1.0 g of ZVI (cleaned by washing in 1N HCl), 20 mL of buffer solution, and 20 mL of surfactant solution at concentrations below and above CMC were put into 40-mL vials capped with Teflon septa, then PCE was added. The concentrations of Triton X-100 were 10, 50, 100, and 150 mg/L; those of CPC were 5, 10, and 20 mg/L; and those of SDDBS were 10, 100, 500, and 1300 mg/L. The vials were mixed at 250 rpm and 25±1 °C on a shaker table. All duplicate samples were filtered through 0.1μm inorganic membrane filter (Whatman). The concentration of PCE was measured with an HPLC [high performance liquid chromatography] (Waters 515). The analyses were conducted at a UV wavelength of 220 nm and with a flow rate of 1.8 mL/min, a mobile phase of 50% water and 50% acetonitrile (Fisher, optima grade), and a 3.9×300 mm μ-bondapak C18 reverse phase column (Waters).

To evaluate the sorption isotherm for each surfactant, 1.0 g of ZVI, 15 mL of surfactant solution at initial concentrations ranging from 50 to 30,000 mg/L, and 15 mL of the buffer solution were put into centrifuge bottles having nominal volumes of 25 mL. The bottles were shaken for 48 hr at 250 rpm and 25±1 °C on a shaker table. After centrifugation, a TOC [total organic carbon] analyzer (Shimadzu, TOC-5000A) was used to analyze the supernatant for the equilibrium surfactant concentration. The sorbed mass of surfactant was calculated from the difference between the initial and equilibrium aqueous concentrations. All experiments were duplicated.

Table I. Properties of Surfactants Used

Surfactant	*Ionic type*	*M.W.*	*CMC (mg/L)*	*Structure*
Triton X-100	Nonionic	628.0	130	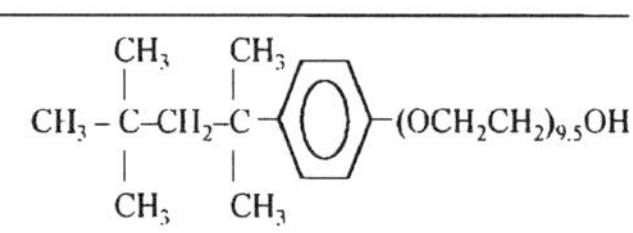
SDDBS	Anionic	348.5	1000	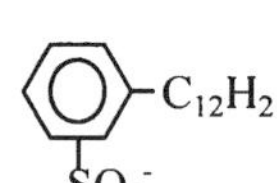
CPC	Cationic	339.5	15	$C_6H_5N^+(CH_2)_{15}CH_3Cl^-$

Results and Discussion

Reductive dechlorination of PCE using ZVI with surfactants

The removal of PCE using ZVI with Triton X-100 in time series experiments is shown in Figure 1(a). Whereas the removal of PCE rarely was observed in the control test without ZVI, the concentration of PCE decreased in the presence of ZVI both with and without surfactant. Half the initial concentration of PCE remained after 24 hours for all ZVI samples. The decrease in aqueous PCE concentration was greater when ZVI and Triton X-100 were combined than without the surfactant.

In the presence of ZVI with Triton X-100, the formation rate of TCE was higher than without Triton X-100 (Figure 1(b)). The amount of TCE formation was less than the amount of PCE removed because the transformation rate of TCE to DCEs was very fast.

Figure 2(a) shows the removal of PCE, and Figure 2(b) shows the formation of TCE, using ZVI with CPC. The removal rate of PCE using ZVI with CPC was higher than with ZVI alone. TCE formation from the degradation of PCE by ZVI was observed after 120 hours, whereas TCE formation started after 24 hours when ZVI with CPC was used. The yield of TCE was higher using ZVI with CPC than without CPC. The higher concentration of surfactant further increased PCE removal, but did not result in a corresponding increase in the formation of TCE.

Figure 3 shows the removal rate of PCE using ZVI with SDDBS. Ninety percent of the initial concentration of PCE was removed after 24 hours. The rate of PCE removal by ZVI with SDDBS was much higher than without SDDBS. With SDDBS, the anionic surfactant, the dechlorination of PCE by ZVI was not observed, because TCE was scarcely detected. Therefore, the removal of PCE might be explained by sorption of PCE on SDDBS-modified ZVI, rather than by dechlorination.

Sorption isotherms of surfactant on ZVI

Sorption isotherms of three surfactants on ZVI are shown in Figure 4. The sorption isotherms of Triton X-100 and CPC conformed to the L-type isotherm (Figure 4(a)), whereas the isotherm of SDDBS conformed to the S-type isotherm (Figure 4(b)).

The sorption data for Triton X-100 and CPC fitted to the Langmuir isotherm model (Figure 5). A Langmuir isotherm reflects a relatively high affinity between the adsorbate and adsorbent, and usually is indicative of chemisorption. An S-type isotherm suggests "cooperative adsorption," which operates if the adsorbate-adsorbate interaction is stronger than the adsorbate-adsorbent interaction. That is to say, ZVI (Fe^0) was converted to ferrous iron (Fe^{2+}) in aqueous solution, which greatly supported the interaction between iron and SDDBS. The sequence comparison of the amount of surfactant sorbed to the iron was SDDBS > Triton X-100 > CPC.

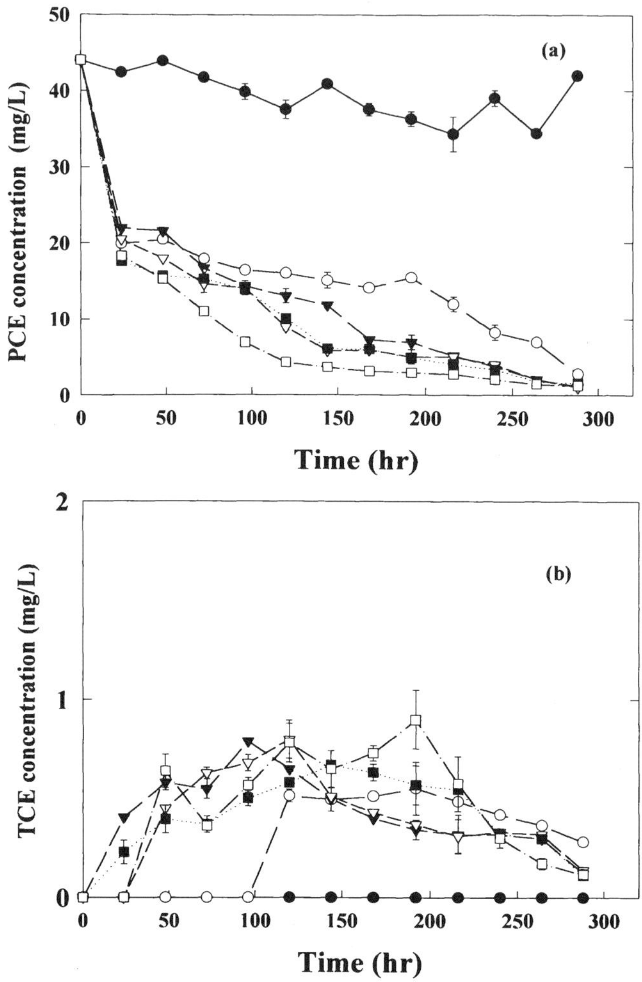

Figure 1. Dechlorination of PCE using ZVI with Triton X-100. -●- control; -○- Fe only; -▼- Fe with 10 mg/L Triton X-100; -▽- Fe with 50 mg/L Triton X-100; -■- Fe with 100 mg/L Triton X-100; -□- Fe with 150 mg/L Triton X-100.

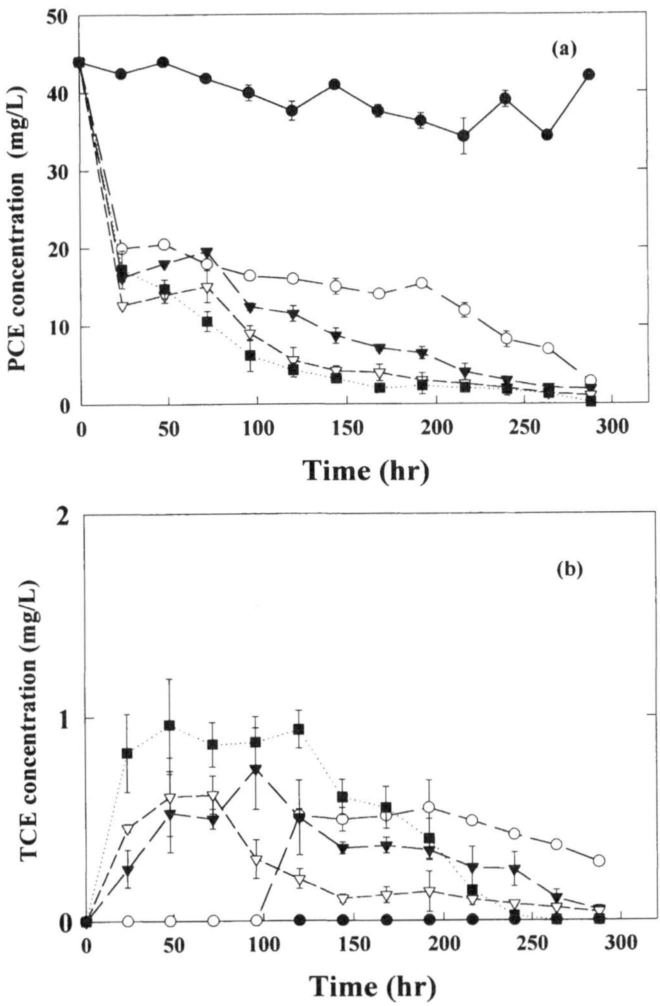

Figure 2. Dechlorination of PCE using ZVI with CPC. -●- control; -○- Fe only; -▼- Fe with 5 mg/L CPC; -▽- Fe with 10 mg/L CPC; -■- Fe with 20 mg/L CPC.

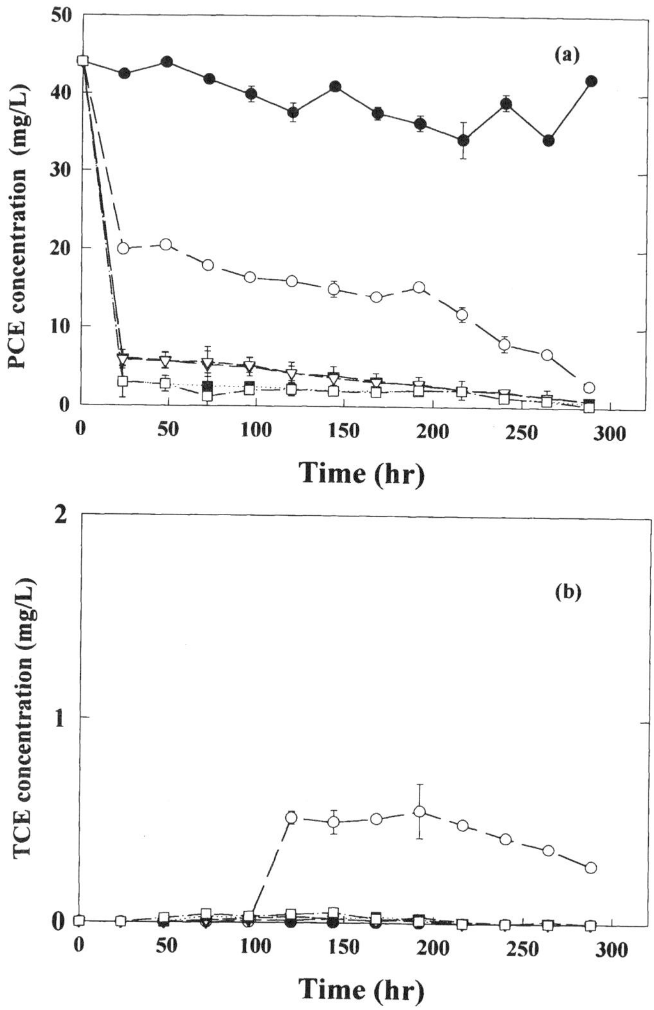

Figure 3. Removal of PCE (a) and TCE formation (b) using ZVI with SDDBS. -●- control; -○- Fe only; -▼- Fe with 10 mg/L SDDBS; -▽- Fe with 100 mg/L SDDBS; -■- Fe with 500 mg/L SDDBS; -□- Fe with 1300 mg/L SDDBS.

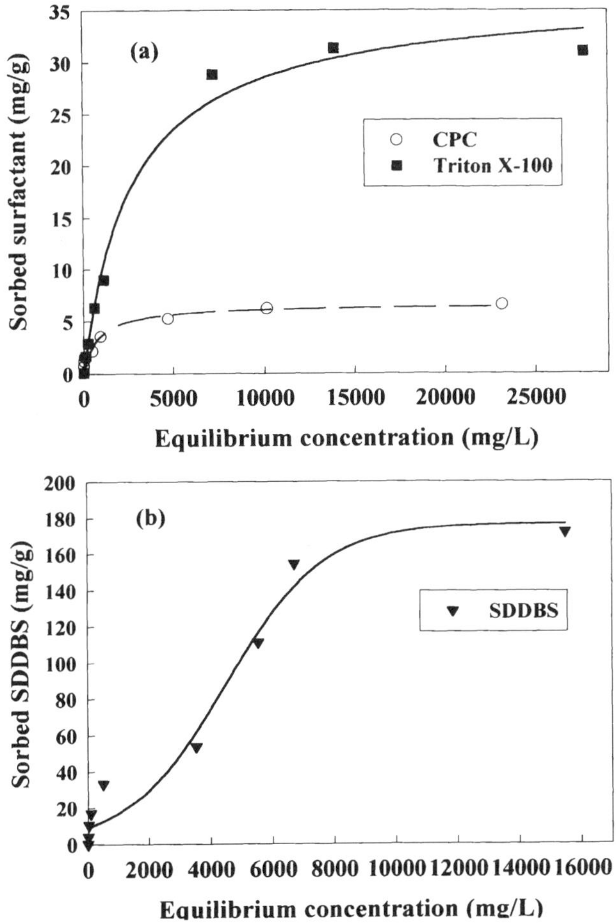

Figure 4. Sorption isotherms of Triton X-100 , CPC (a), and SDDBS (b) on ZVI.

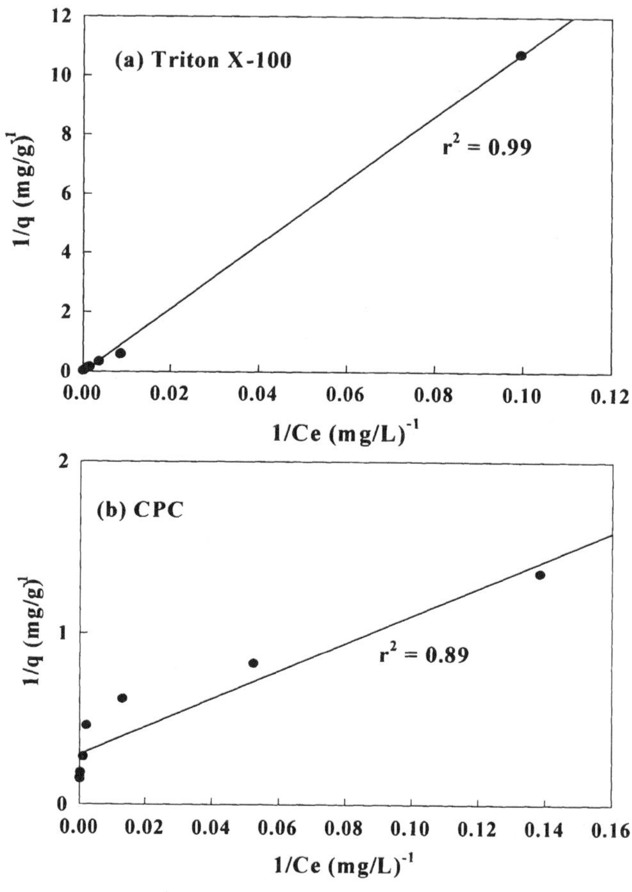

Figure 5. Plot of experimental data of 1/q versus 1/Ce. Symbols are experimental, and lines represent data predicted using the Langmuir model.

Table II shows the distribution of surfactants between sorbed and aqueous phase. Given 10 mg/L of surfactant concentration, representing less than the CMC, 78.8% of SDDBS was sorbed on the iron surface, while 17.04% of Triton X-100 and 7.64% of CPC were sorbed. The same trend can be seen with concentrations greater than the CMCs: more Triton X-100 and CPC were in aqueous phase and more SDDBS was in sorbed phase. These results show that PCE in aqueous phase was more easily available for dechlorination than was PCE in sorbed surfactant on ZVI. Therefore, PCE was simply immobilized on the sorbed SDDBS phase, although the removal rate of PCE with SDDBS was the highest of the three.

Table II. Content of Surfactants in Solid and Aqueous Phases

Surfactant concentration (mg/L)		*Surfactant type*	*Content in solid phase (%)*	*Content in aqueous phase (%)*
Below CMC	10	Triton X-100	17.04	82.96
	10	CPC	7.64	92.36
	10	SDDBS	78.80	21.20
Above CMC	150	Triton X-100	24.51	75.49
	20	CPC	8.91	91.09
	1300	SDDBS	60.19	39.91

Figure 6 shows the removal of PCE by ZVI with three surfactants at concentrations below and above the CMCs (Table II). The removal rate of PCE by ZVI with the three surfactants was higher than without them, although with SDDBS most PCE was just sorbed rather than dechlorinated, as it was with Triton X-100 and CPC.

Conclusions

The present research was performed in order to assess the potential benefit of surfactants desorbed from the immobization layers, usually at low concentrations below CMCs, on the rate of PCE dechlorination. The results of the research suggests that immobilization with surfactants and dechlorination by ZVI can be combined into one or a series of reactive zones or areas to treat a groundwater plume contaminated with chlorinated organics. The rate of PCE dechlorination was found to be faster using ZVI with Triton X-100 or CPC than without, because the TCE formation rate was higher for ZVI with Triton X-100 or CPC than for ZVI without. The combination of the nonionic or cationic surfactant with ZVI strongly increased the dechlorination of PCE because the

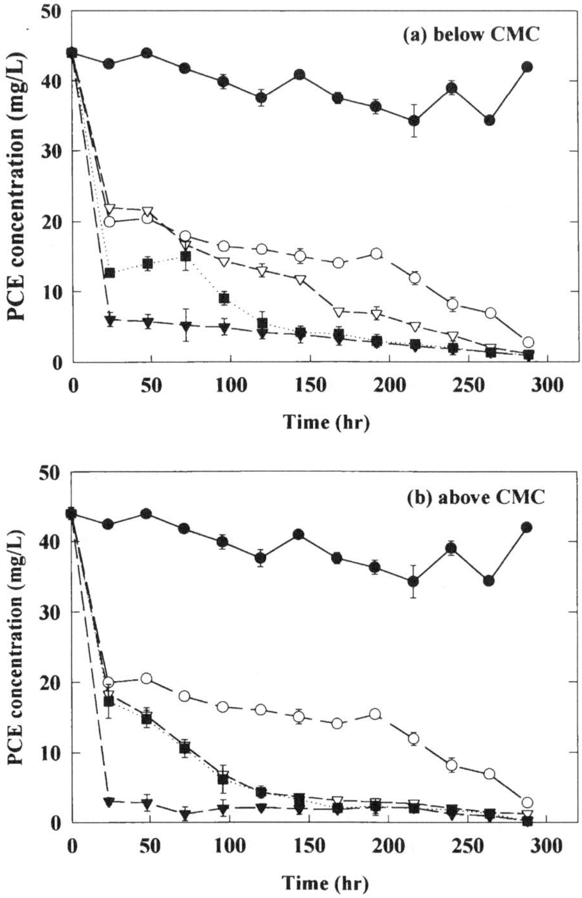

Figure 6. Removal of PCE by ZVI with three surfactants. -●- control, -○- Fe only, -▼- Fe with SDDBS, -▽- Fe with Triton X-100, -■- Fe with CPC.

PCE that bounded with surfactant in aqueous phase was available for dechlorination. With SDDBS, PCE primarily was immobilized on the sorbed SDDBS phase, rather than dechlorinated, and thus TCE was scarcely detected.

Because most soil and clay particles are negatively charged, cationic surfactants would be most useful for immobilization with surfactants, although anionic and nonionic surfactants also could be used depending on the application (*20*). According to the results of this research, cationic and nonionic surfactants, which had been sorbed in the immobilization layers and desorbed by the natural groundwater flow, would not hinder the dechlorination reaction in ZVI layers. Even with the anionic surfactant, the removal of PCE from the aqueous phase was enhanced because more PCE could be sorbed on the iron particles, even though the dechlorination under this condition could not be positively confirmed.

References

1. Burris, D.R.; Campbell, T.J.; Manoranjan, V.S. *Environ. Sci. Technol.* **1995**, 29, 2850-2855.
2. Gillham, R.W.; O'Hannesin, S.F. *Ground Water* **1994**, 32, 958-967.
3. Orth, W.S.; Gillham, R.W. *Environ. Sci. Technol.* **1996**, 30, 66-71.
4. Tratnyek, P.G. *Chemistry & Industry* **1996**, 499-503.
5. Roberts, A.L.; Totten, L.A.; Arnold, W.A.; Burris, D.R.; Campbell, T.J. *Environ. Sci. Technol.* **1996**, 30, 2654-2659.
6. Johnson, T.L.; Scherer, M.M.; Tratnyek, P.G. *Environ. Sci. Technol.* **1996**, 30, 2654-2659.
7. Johnson, T.L.; Fish,W.; Gorby, Y.A.; Tratnyek, P.G. *J. Contam. Hydrol.* **1998**, 29, 378-398.
8. Li, Z. *Advances in Environ. Res.* **1998**, 2, 244-250.
9. Loraine, G.A. *Wat. Res.* **2001**, 35, 1453-1460.
10. Matheson, L.J.; Tratnyek, P.G. *Environ. Sci. Technol.*, **1994**, 28, 2045-2053.
11. Sheng, G.; Xu, S.; Boyd, S.A. *Soil Sci. Soc. Am. J.* **1999**, 63, 73-78.
12. Sun, S.; Inskeep, W.P.; Boyd, S.A. *Environ. Sci. Technol.* **1995**, 29, 903-913.
13. Park, J.W.; Boyd, S.A. *J. Environ. Qual.* **1999**, 28, 945-952.
14. Neupane, D.; Park, J.W. *Chemosphere.* **1999**, 38, 1-12.
15. Neupane, D.; Park, J.W. *Chemosphere.* **2000**, 41, 787-792.
16. Rosen, M.J. Surfactant and Interfacial Phenomena; 2nd edn., John Wiley & Sons Inc., New York, **1989**; pp 108-162.
17. Kile, D.E.; Chiou, C.T. *Environ. Sci. Technol.* **1989**, 23, 832-838.
18. Bizzigotti, G.O.; Reynolds, D.A.; Kueper, B.H. *Environ. Sci. Technol.* **1997**, 31, 1511-1517.
19. Sayles, G.D.; You, G.; Wang, M.; Kupferle, M.J. *Environ. Sci. Technol.* **1997**, 31, 3448-3454.
20. Park, J.W.; Jaffe, P.R. *Environ. Sci. Technol.* **1993**, 27, 2559-2565.

Chapter 11

Reductive Dechlorination of Chloromethanes Using Tungsten and Molybdenum Hydrogen Bronzes or Sodium Hypophosphite

Allen W. Apblett, B. P. Kiran, and Katie Oden

Chemistry Department, Oklahoma State University, Stillwater, OK 74078

Several promising materials that can deliver hydridic hydrogen were investigated for their ability to dechlorinate chloromethanes. Molybdenum blue. $Mo_2O_5(OH)$, was found to be unsuitable for reductive dechlorination of carbon tetrachloride because coupling reactions between trichloromethyl radicals dominated to produce tetrachloroethylene. Tungsten blue, $W_3O_8(OH)$, on the other hand, cleanly converted carbon tetrachloride to chloroform. It also reacted with chloroform and methylene chloride to yield as yet unidentified products. Adsorption data suggest that CH_2Cl_2 produced from reduction of $CHCl_3$ remains bound to the substrate and reacts away at the same rate it is formed. Sodium hypophosphite was found to convert carbon tetrachloride completely to methane through the intermediates of chloroform, methylene chloride, and chloromethane.

INTRODUCTION

Reducing agents can play an important role in the destruction of environmental contaminants such as halocarbons, explosives, and heavy metals. The reductive transformation of chlorocarbons occurs via a number of processes. One interesting method of which involves the passing of chloroorganics or CFCs through a packed bed of powdered sodium oxalate, $(COONa)_2$, at 270 to 290°C (*1*). The generated reaction products are NaCl (as well as NaF in the case of CFCs), CO_2, and elemental carbon. Another approach for the reductive transformation of chlorocarbons involves the use of base metals. For example, Gillham and co-workers reported the degradation of halocarbons in groundwater in the presence of galvanized steel, stainless steel, aluminum, and iron (*2*) and Tratnyek and co-workers reported the reduction of chlorinated methanes by iron metal (*3*). Further investigation indicated that increasing the clean surface area of iron increased the rate of degradation and that increasing the pH decreased the rate of degradation. Additionally, active metals such as sodium, zinc, or aluminum are highly reactive with chlorocarbons (*4*). They operate by dechlorinating the chlorocarbon and producing a metal salt. This is highly advantageous in that the metal salt can be precipitated out of solution to allow for easy disposal of this solid waste. Unfortunately, these metals are highly reactive (in some cases, explosively) with water and air.

During an investigation of nucleophilic catalysts for hydrolytic dechlorination of chlorocarbons (*5,6,7,8*) a different type of reducing agent was discovered. In the course of the investigation of these reactions, the use of solvents other than water was investigated as a means of enhancing the solubility of chlorocarbons and, thus, their interaction with solid-phase and homogeneous catalysts. It was discovered that, the interaction of molybdenum trioxide and carbon tetrachloride in ethanol had a lengthy lag period in which chloride generation was gradual and the organic product was diethyl carbonate. However after 18 hours at 100°C, chloride generation accelerated by a factor of 50 and the organic product was found to be chloroform. Concomitant with this alteration in the dechlorination reaction was a change in color of the solid catalyst from white to blue. Subsequent analysis by X-ray powder diffraction and follow-up experiments have shown that the molybdenum trioxide had been converted to molybdenum blue, $Mo_2O_5(OH)$, via an oxidation/reduction reaction with ethanol. This reaction is catalyzed by hydrochloric acid produced in the initial dechlorination process. These results prompted an investigation of the reactions of chlorocarbons such as carbon tetrachloride, chloroform, and dichloromethane with molybdenum blue and the related tungsten blue that is reported herein. Also, initial results for reductive dechlorination of carbon tetrachloride using sodium hypophosphite will be discussed.

EXPERIMENTAL

X-ray powder diffraction (XRD) patterns were recorded on a Bruker AXS D-8 Advance X-ray powder diffractometer using copper K_α radiation. Gas chromatographic/mass spectroscopic analysis (GC/MS) was performed on a

Hewlett Packard G1800A instrument equipped with 30 m x 0.25 mm HP5 column (Crosslinked 5% PhME silicone). The temperature program used was an initial hold of 2 min at 35°C, a ramp of 5°C/min to 170°C, and a final hold of 5 min. The helium flow rate was 1 ml/min and the injection port was set at 250°C.

Preparation of Molybdenum Blue

A round bottom flask was charged with 30.00 g of MoO_3, 300 ml of n-butanol and 5 ml of concentrated HCl . The mixture was refluxed for 12 hours at which time it had turned a very dark blue color. At this point, the reaction mixture was cooled to room temperature and was filtered through a fine sintered-glass filter funnel. The dark blue solid was washed with n-butanol and was then dried in a vacuum oven at room temperature. The yield was 28.30 g (94%). The XRD pattern of the product corresponded to that of $Mo_2O_5(OH)$ in the ICDD data base.

Preparation of Tungsten Blue

Tungsten trioxide (30.0 g) and 56.8 g of granular zinc metal, 5 ml of water, and a large stirring bar were placed in a 1000 ml filtration flask. A paraffin oil-filled bubbler was attached to the flask's hose connector with a short length of Tygon tubing. Next a dropping funnel containing 200 ml of concentrated hydrochloric acid was placed in the mouth of the flask using a one-hole rubber stopper to make an air-tight connection. The flask was then placed in a water bath and the hydrochloric acid was added drop-wise to the magnetically-stirred reaction mixture (caution must be exercised with this reaction since it vigorously evolves hydrogen gas, an explosion hazard). The yellow solid rapidly turned blue. After the HCl addition was complete and hydrogen evolution had stopped, the reaction mixture was filtered through a 45 μm nylon membrane filter and was thoroughly washed with water. After drying in a vacuum oven at room temperature the yield of dark royal blue solid was 30.5 g (the excess mass is due to hydration of the product). The XRD pattern of the product corresponded to that of $W_3O_9(OH)$ in the ICDD data base.

Dechlorination Reactions

Reactions between molybdenum and tungsten blue were performed in Teflon-lined stainless steel bombs using an excess of the blue reagent (2.5 g) and 0.25 g of the chlorocarbon (CCl_4, $CHCl_3$, or CH_2Cl_2). The sealed reactors were placed in a digitally-controlled oven at various temperatures under autogenous pressure. Reactions between sodium hypophosphite (4.00 g) and carbon tetrachloride (0.25 g) were performed at 100°C in 100 ml glass media jars sealed with Teflon-lined high-temperature resistant caps. The amount of reactants and products in the reaction mixtures were determined by cooling the bombs, sampling 20 μL the headspace with a gas-tight syringe, and analyzing by gas chromatography/mass spectroscopy. Compounds were identified by comparison of their mass spectra to the NIST database. Product identity was confirmed by measuring the retention times of authentic samples of the compounds identified by mass spectroscopy.

Adsorption Equilibrium Constants

Tungsten blue (4.5 g) was placed in a 4.77 ml vial which was then sealed with a screw cap MiniInert valve. A 20 ml sample of CCl_4, $CHCl_3$, or CH_2Cl_2-saturated air was withdrawn from a container containing the liquid chlorocarbon. The needle of the syringe was inserted into a vial so that it reached as close to the bottom as possible and then the plunger was compressed rapidly so that the vial was completely flushed with chlorocarbon-saturated air. In this manner, three samples of each tungsten blue/chlorocarbon combination were prepared along with three control samples for each chlorocarbon in which empty vials were used. After 12 hours equilibration at 25°C, the headspace of the vials were sampled with a gas tight syringe and 20 μL samples were injected into a GC/MS. The concentrations of the chlorocarbons were determined from the area of the chromatogram peaks using response factors that were obtained using chlorocarbon in pentane standard solutions. In the case of methylene chloride, which co-elutes with pentane, the response factor was determined using a saturated CH_2Cl_2 sample in air and the literature value for methylene chloride's vapor pressure at 25°C. The amount of chlorocarbon adsorbed by the tungsten blue was calculated from the difference in chlorocarbon concentrations between the sample and control vials. The equilibrium distribution constants were then calculated as the ratio of the chlorocarbon concentration per gram of tungsten blue to the equilibrium concentration of chlorocarbon.

RESULTS AND DISCUSSION

The reaction of molybdenum blue, $Mo_2O_5(OH)$, with carbon tetrachloride at 150°C produced tetrachloroethene (PCE) as the major product. The formation of PCE is also observed in reductions of carbon tetrachloride with iron metal and is believed to result from coupling of dichlorocarbene or trichloromethyl radicals to give PCE or hexachloroethane (*9*). The latter compound is subsequently hydrodechlorinated to PCE. Since the interaction of $Mo_2O_5(OH)$ with CCl_4 in the presence of alcohols does not produce PCE but only $CHCl_3$, it is likely that the dimerization of trichloromethyl radicals is the reaction mechanism leading to PCE formation (Figure 1). This hypothesis is reinforced by the fact that hexachloroethane reacts with tungsten blue to give PCE as the exclusive organic product. The formation of trichloromethyl radicals on the molybdenum blue surface suggests that the first step in dechlorination is electron transfer from a Mo(V) center to CCl_4 rather than hydride attack on the CCl_4. However, a condensation reaction between the Mo^{V}-OH linkage to yield what is essentially a trichloromethoxide ion attached to a Mo(V) center is also a possibility that cannot be ruled out. Whatever the first step, it is apparent that hydrogen transfer to the trichloromethyl fragment to yield chloroform is much slower than the coupling reaction to yield tetrachloroethene.

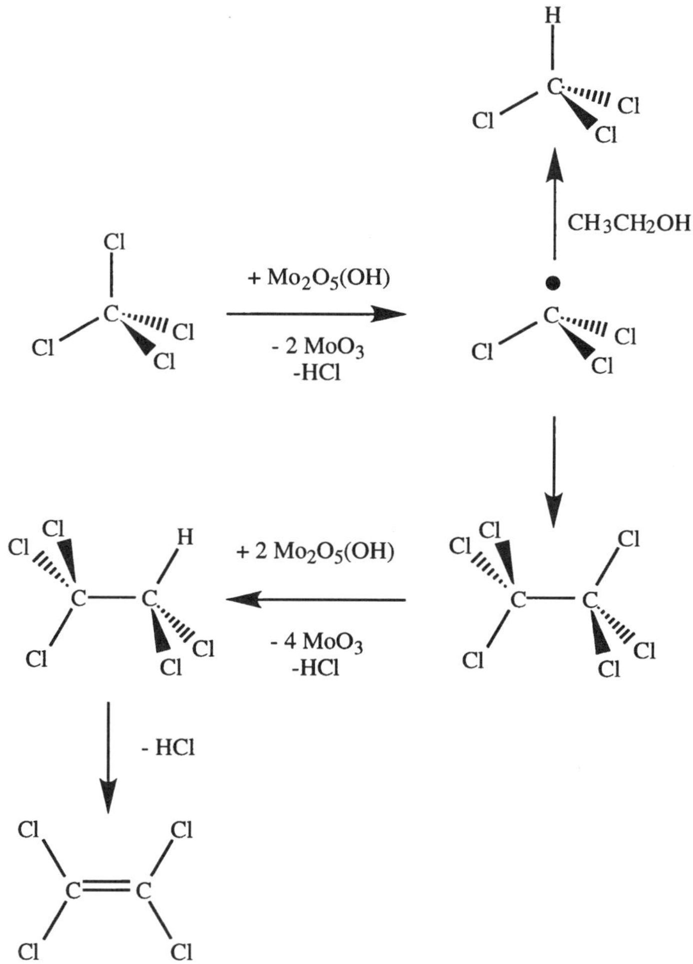

Figure 1. Reaction of Carbon Tetrachloride with Molybdenum Blue.

The formation of tetrachloroethene is unfortunate due to the fact that it is more toxic and refractory in the environment then CCl_4. In order to avoid the radical coupling reaction, a tungsten hydrogen bronze, $W_3O_9(OH)$, was used as the hydrodechlorinating reagent in place of molybdenum blue. With this substitution carbon tetrachloride was reduced to chloroform without any detectable carbon-carbon bonded side products. The effectiveness of substitution of tungsten for molybdenum for the prevention of tetrachloroethene formation remains to be determined. It may be due to electronic factors but since the materials are structurally dissimilar (molybdenum blue has a layered structure while tungsten blue has a 3-dimensional structure with the hydrogen atoms in tunnels) steric interactions may also play a role.

At 150°C, the disappearance of CCl_4 was found to be exponential and was nearly complete after 24 hours (Figure 2). A plot of the natural logarithm of the carbon tetrachloride concentration versus time, shown in Figure 3, demonstrated that the reduction was first-order in carbon tetrachloride and had a rate constant of 2.66×10^{-3} min^{-1} (from the slope of the graph). During the course of the reaction, the tungsten blue began to be converted to a yellow solid and started to display reflections in it's X-ray diffraction pattern that were typical of tungsten trioxide. Infrared spectroscopy and elemental microanalysis of the spent reagent showed no build-up of organics or chloride on the tungsten blue.

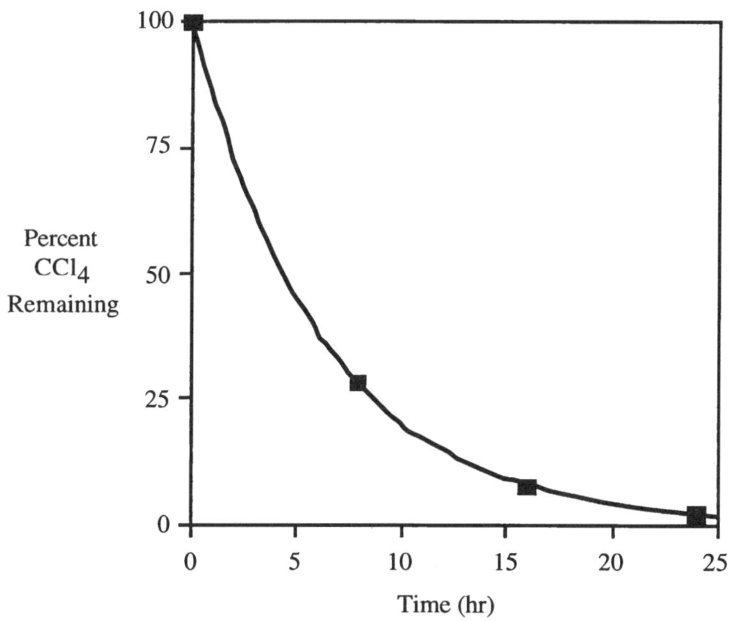

Figure 2. Concentration of CCl_4 (mole percent) versus time for reaction with tungsten blue at 150°C.

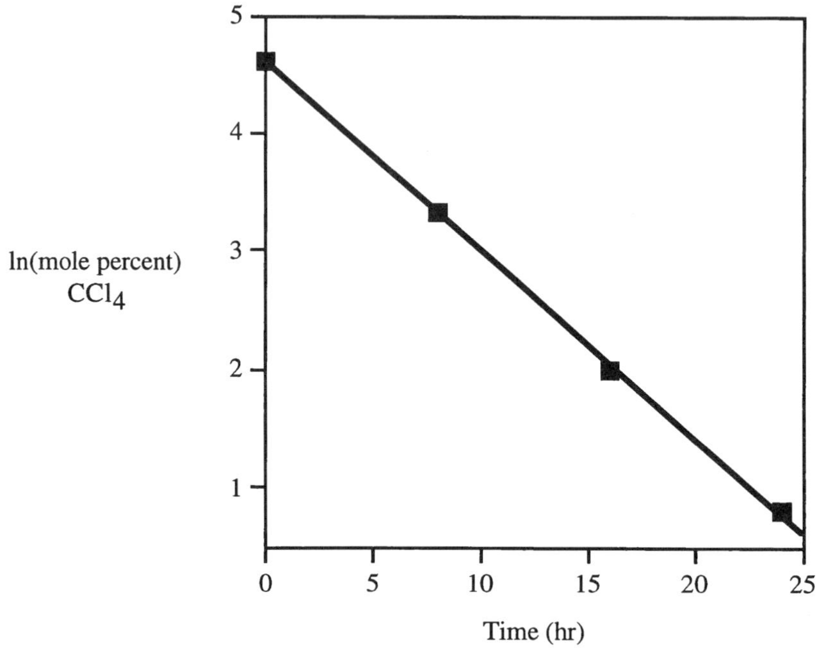

Figure 3. First order reaction kinetics (ln of concentration versus time) for reaction of CCl_4 with tungsten blue at 150°C.

$W_3O_9(OH)$ also reacts with chloroform and methylene chloride but the organic products from these reactions are either too volatile to be observed by our GC/MS instrument (for example chloromethane elutes along with the injected air) or react away at a similar rate to their formation. Significantly, CH_2Cl_2 was not observed as a product from $CHCl_3$. The pseudo first order rate constants for disappearance of $CHCl_3$ and CH_2Cl_2 at 150°C are 1.93×10^{-3} min^{-1} and 2.93×10^{-3} min^{-1}, respectively. The two rate constants are close enough to each other to expect that if methylene chloride was formed, its steady state concentration would have been detectable providing that both chlorocarbons have similar adsorption equilibria with tungsten blue (see below). Also, it is notable that the rate constant for the methylene chloride reaction is higher than that of chloroform. This is opposite what is observed in iron-mediated reduction of chloromethanes in which carbon tetrachloride is sequentially dechlorinated via chloroform to methylene chloride (3). For these iron reductions, the initial rate of each reaction step was also pseudo-first-order but became substantially slower with each step so that carbon tetrachloride typically disappeared in several hours but no significant reduction of methylene chloride was observed after one month. These dissimilarities are also suggestive of a different solid-state reaction mechanisms (including adsorption equilibria) for the $W_3O_9(OH)$

reactions with $CHCl_3$ and CH_2Cl_2. In order to try to shed some light on the mechanism of the chloroform reaction, deuterochloroform was used as a substrate to see if there was an isotope effect. It was found that $CDCl_3$ reacted at the same rate as $CHCl_3$ and this absence of an isotope effect demonstrates that the cleavage of the C-H bond is not involved in the activated complex of the rate-determining step. Therefore, a possible mechanism where the hydrogen is abstracted leading to formation of dichlorocarbene in the rate-limiting step may be ruled out.

The apparent activation energies for the reactions of methylene chloride, chloroform and carbon tetrachloride with tungsten blue were determined by measuring the pseudo first order rate constants over a range of temperatures from 140-170°C. An Arrhenius plot of ln k versus 1/T yielded straight lines with slopes that corresponded to activation energies of 104 kJ/mol and 122 kJ/mol for $CHCl_3$ and CCl_4, respectively. Methylene chloride actually reacted faster at lower temperatures and thus displayed a negative apparent activation energy. This occurs in heterogeneous reactions when the adsorption enthalpy, a negative number, is higher in magnitude than the true activation energy for the rate-controlling reaction step. In this situation, the apparent activation energy (equal to the sum of the adsorption enthalpy and the true activation energy) becomes negative. This means that adsorption equilibria play an important role in the tungsten blue/chlorocarbon reactions. For this reason, the equilibrium adsorption constants for CH_2Cl_2 ,$CHCl_3$ and CCl_4 on tungsten blue were determined by equilibrating chlorocarbon-saturated air with tungsten blue. The results of this characterization are presented in Table I and the constants are reported both in terms of per gram of catalyst and per unit surface area (the specific surface area of the tungsten blue was found to be 20.43 m^2g^{-1}).

Table I. Adsorption Data for Chlorocarbons onto Tungsten Blue at 25°C

Chlorocarbon	*Surface Concentration* *($mmol/m^2$)*	*Gas Phase Concentration* *(mmol/L)*	K_D *(L/m^2)*	K_D *(L/g)*
CH_2Cl_2	0.0297	4.21	$7.05x10^{-3}$	0.144
$CHCl_3$	$7.30x10^{-5}$	2.81	$2.60x10^{-5}$	$5.31x10^{-4}$
CCl_4	$9.67x10^{-5}$	1.42	$6.83x10^{-5}$	$1.40x10^{-3}$

It was determined that the adsorption constants for $CHCl_3$ and CCl_4 were comparable to each other with the latter chlorocarbon having a slightly higher (2.6 times) adsorption constant. The higher adsorption constants for the more chlorinated compound is in keeping with Van der Waals forces being primarily involved in binding the chlorocarbons to the tungsten blue surface. By comparison, the K_D for adsorption of CCl_4 onto UO_2, calculated from Stakebake and Goad's data (*10*), is 5.10 $x10^{-3}$ L/m^2. However, this result is for CCl_4/UO_2 equilibria in the absence of air, so the lower value found in this study is not unexpected. In this experiment, the surface coverage by CCl_4 is much less than a monolayer (2.15%) based on the theoretical area of CCl_4 being 37 Å^2. It is quite likely that the CCl_4 occupies specific sites on the tungsten blue in accordance with Machin's observation that CCl_4 is adsorbed onto metal oxides in a localized manner and is dependent on the adsorbent structure (*11*). Interestingly, methylene chloride has a much higher adsorption constant onto tungsten blue as compared to chloroform and carbon tetrachloride. The stronger binding of methylene chloride to tungsten blue is in agreement with the observation of negative activation energy for the reaction between the two materials. More importantly, the fact that K_D for CH_2Cl_2 is 271 times that of $CHCl_3$ can explain why methylene chloride is not observed in the reduction of chloroform. The higher binding of CH_2Cl_2 would allow it to be preferentially reduced providing that both chlorocarbons were competing for the same binding sites.

Another reagent for hydrodechlorination that was investigated in this study was sodium hypophosphite, NaH_2PO_2. This reagent is an inexpensive, powerful reducing reagent that is used extensively for electroless plating of metals. Since it has two hydrogen atoms attached to a phosphorus center, it undergoes a two electron oxidation to yield ortho-phosphoric acid and can deliver two hydride atoms. Thus, reaction with a C-Cl center can proceed at a single hypophosphite ion to give C-H and HCl products. In this investigation, the reaction of sodium hypophosphite with carbon tetrachloride at 100°C was found to produce the complete sequence of reductively-dechlorinated products: chloroform, methylene chloride, chloromethane, and methane (Figure 4). The methane was not observed but its presence may be surmised from the observed sequence of products and the fact that chloromethane disappears with time. At 100°C, the reaction kinetics are fairly rapid, leading to complete destruction of 0.25 g of CCl_4 by 2.5 g of sodium hypophosphite in a 24 hour period. The rate constant for the pseudo-first order disappearance of CCl_4 was found to be 1.67×10^{-3} min^{-1} . After 70 minutes reaction time at 100°C, the reaction mixture consisted of 12.5% $CHCl_3$, and 83.9 % CCl_4 and a considerable amount of CH_2Cl_2 (3.6%) indicating that chloroform is reduced at a similar rate to CCl_4. At ambient temperature, the reaction between CCl_4 and sodium hypophosphite is slow and only 0.6% conversion of CCl_4 to chloroform is observed after 48 hours (the rate constant was found to be 1.39×10^{-6} min^{-1}). However, it is quite likely that this rate can be improved by optimizing the properties of the hypophosphite reagent, especially it's surface area by placing it on a suitable support. The ability of hypophosphite to deliver two hydrides does appear to prevent formation of radical intermediates and carbon-carbon coupling reactions so that no chlorinated ethenes were observed as side products. Therefore, sodium hypophosphite appears to be an excellent reagent for dechlorination of chloroalkanes (even lightly chlorinated ones) further development of this reagent will likely be a valuable endeavor.

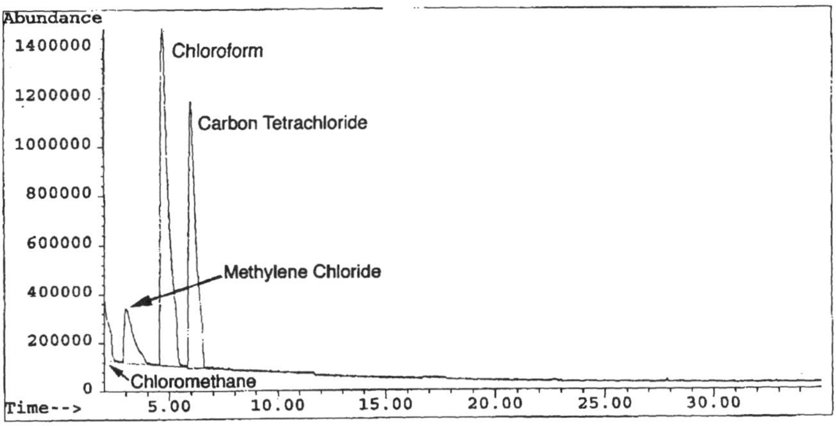

Figure 4. GC/MS trace for sodium hypophosphite/CCl₄ reaction at 100°C after 8 hours.

CONCLUSIONS

Molybdenum blue was found to be unsuitable for reductive dechlorination of carbon tetrachloride in the absence of a solvent to act as a proton donor because coupling reactions dominated to produce tetrachloroethylene. Tungsten blue, on the other hand, cleanly converted carbon tetrachloride to chloroform. It also reacted with chloroform and methylene chloride to yield as yet unidentified products. Adsorption equilibria suggest that chloroform is reduced to methylene chloride but the latter compounds is preferentially reduced further because of it's stronger binding to tungsten blue. Sodium hypophosphite was found to convert carbon tetrachloride completely to methane through the intermediates of chloroform, methylene chloride, and chloromethane. These reagents show promise for reductive dechlorination reactions but the slow kinetics at ambient temperature might be restrictive to their environmental applications. Their ability to strongly adsorb chloromethanes may be a mitigating factor that would allow sufficient contact time between the reducing agent and the substrate in aquifers to allow chlorocarbon reduction to occur. However, it is likely that manufacture of permeable reactive barriers for contaminant plumes using these reagents would likely require an auxiliary catalyst (such as a noble metal or iron) to accelerate the reaction between the reducing agent and the chlorocarbon.

ACKNOWLEDGEMENT

Oklahoma State University's Environmental Institute is gratefully acknowledged for supporting this research. The National Science Foundation, Division of Materials Research, is thanked for Award Number 9871259 that provided funds for the X-ray powder diffractometer used in this investigation. Katie Oden thanks the Freshman Scholars Program administered by Oklahoma State University for scholarship support for her research.

REFERENCES

1. Crabtree, R.; Burdeniuc, J. *Science,* **1996**, 271, 340.
2. Gillham, R.; Orth, W. S. *Environ. Sci. Technol.* **1996**, 30, 66.
3. Tratnyek, P.; Matheson, L. J. *Environ. Sci. Technol.* **1994**, 28, 2045.
4. Dickerman, J. C.; Emmel, T. E. ; Harris, G. E.; Hummel, K. E. *Technologies for CFC/Halon Destruction*; EPA Publication: Washington, DC, 1989).
5. Apblett, A. W.; Byers, L. D.; Reinhardt, L. E. *American Chemical Society, Division of Environmental Chemistry, Preprints of Papers,* **1996,** 36(2) 10.
6. Apblett, A. W.; Byers, L. D.; Reinhardt, L. E. *Extended Abstracts for Emerging Technologies in Hazardous Waste Management,* **1996,** VIII, 724.
7. Apblett, A. W.; Byers, L. D.; Reinhardt, L. E. *American Chemical Society, Division of Environmental Chemistry, Preprints of Papers,* **1997,** 37(1), 300.
8. Reinhardt, L. E. "Development of Group V & VI Oxide Chemistry: Novel Preceramic Materials and Chlorocarbon Remediation Processes", Ph.D. Thesis, Tulane University, New Orleans, LA, **1997**.
9. Hung, H.-M.; Hoffman, M. R. *Environ. Sci. Technol.* **1998**, 32, 3011.
10. Stakebake, J.L.; Goad, H.A.. *J. Catal.* **1974,** 32, 272.
11. Machin, W.D. *J. Phys. Chem.* **1969,** 73, 1170.

Chapter 12

Understanding the Mechanisms Controlling the Kinetics of Arsenate and Chromate Removal from Solution Using Zero Valent Iron

Nikos Melitas[1], Martha Conklin[2], and James Farrell[1,*]

Departments of [1]Chemical and Environmental Engineering and [2]Hydrology and Water Resources, University of Arizona, Tucson, AZ 85721

This research investigated the mechanisms controlling the kinetics of arsenate and chromate removal from water by zerovalent iron media. Batch experiments were performed to determine the kinetics of As(V) and Cr(VI) removal as a function of their aqueous concentration. The effect of arsenic and chromium on the corrosion rate of iron was investigated via analysis of Tafel polarization diagrams. Column experiments were conducted to investigate the long-term performance of zerovalent iron for arsenate and chromate removal. Arsenate removal kinetics were first-order at low aqueous concentrations, but approached zeroth-order in the limit of high concentrations. This was attributed to increasing adsorption site saturation with increasing aqueous concentration. Arsenate removal at low aqueous concentrations was limited by diffusion of As(V) through iron corrosion products to adsorption sites. At high concentrations, arsenate removal was limited by the generation rate of adsorption sites. The dividing line between the high and low concentration behavior depends on the iron corrosion rate, which is largely determined by the dissolved oxygen concentration. Rates of Cr(VI) removal could not be described by a simple kinetic model due to the effect of precipitated chromium compounds on the reactivity of the iron media. Absolute rates of chromate removal declined with increasing concentration due to passivation of the iron surfaces by adsorbed chromium compounds. This surface passivation was incomplete, and steady-state effluent concentrations from a column reactor were observed over a period spanning more than 1,200 pore volumes. The surface passivation was also reversible upon lowering the influent Cr(VI) concentration, indicating that the performance of iron media for Cr(VI) removal was hysteretic.

Introduction

Chromate

Chromate is one of the most common groundwater contaminants due to its widespread use as a metal corrosion inhibitor and chemical oxidant. In recent years there has been great interest in using permeable reactive barriers containing zerovalent iron for in situ treatment of groundwater contaminated with soluble chromium compounds. In these systems, highly water soluble Cr(VI) is reduced to Cr(III), which then precipitates as Cr(III) oxides and mixed Cr(III)/Fe(III) oxides (*1*, *2*). Although field studies have demonstrated that permeable iron barriers are effective for chromate removal (*3*, *4*), chromate is a known passivator of iron (*5*), and may contribute to decreasing reaction rates with increasing elapsed time of operation.

The kinetics of Cr(VI) removal by zerovalent iron media has been investigated in both batch and column systems (*1*, *2*, *6*). Although rapid initial rates of Cr(VI) removal have been observed, reaction rates have been found to decline or completely cease after several hours (*1*, *2*, *6*). Because iron passivation by adsorbed chromium compounds is highly dependent on the

Cr(VI) concentration, the removal kinetics determined in short-term batch tests conducted at high concentrations cannot be used to assess the long-term performance of iron for Cr(VI) removal at low aqueous concentrations. Thus, a more fundamental understanding of the mechanisms controlling Cr(VI) removal kinetics is required in order to extrapolate laboratory results to field conditions.

Arsenate

Drinking water supplies contaminated with arsenic compounds are a worldwide health concern. The problem of arsenic contaminated drinking water has received considerable media and scientific attention due to the recent discovery of high levels of arsenic contamination in developing nations (*7*). In the United States the current maximum contaminant level (MCL) of 50 μg/L for arsenic in drinking water is under review (*8*), and a level of 10 μg/L has recently been proposed (*9*). The high levels of arsenic contamination in developing nations, and the recently proposed regulatory changes have stimulated research on new methods for removing arsenic compounds from drinking water.

Several investigators have proposed that zerovalent iron filings may serve as an effective filter medium for removing As(V) (arsenate) and As(III) (arsenite) species from drinking water supplies (*10, 11, 12*). The removal mechanism involves complex formation of arsenate or arsenite with iron oxides produced as corrosion products. Although under some conditions zerovalent iron has been found to reduce As(V) to As(III) (*12*), the reduction process itself is not a removal mechanism, since As(III) compounds are highly water soluble.

The kinetics of As(V) complex formation with iron oxides are extremely rapid and occur on a time scale of milliseconds (*13*). Therefore, the slow approach to equilibrium observed in studies of As(V) complexation with iron oxides has been attributed to diffusional mass transfer limitations through the oxide phase (*14*).

The kinetics of arsenic removal by zerovalent iron have been investigated in several recent studies (*11, 12*). Su and Puls (*12*) used a first-order model to describe arsenate and arsenite removal by iron filings in batch tests. These tests were conducted at high arsenic concentrations over a short period of time, and may therefore not be useful for predicting the long-term performance of iron filings for arsenic removal at low concentrations.

Farrell et al. (*11*) proposed that arsenate removal in batch tests by corroding iron could be described by:

$$\frac{dC}{dt} = \frac{-k_0 C}{k_0 / k_1 + C} \tag{1}$$

where C is the arsenate solution concentration, t is time, k_0 is the zeroth-order rate constant, and k_1 is the first-order rate constant. This form of kinetic expression yields removal rates that are first-order in As(V) concentration in the limit of infinite dilution, and are zeroth-order in the high concentration limit. First-order removal at low arsenate concentrations arises from an abundance of unoccupied adsorption sites that eliminates competition for available sites between arsenate species. Zeroth-order removal kinetics are observed in the high concentration limit since there is competition among arsenate species for a limited number of adsorption sites. Under this condition, removal rates are limited by the rate of adsorption site generation via iron corrosion.

The objective of this research was to elucidate a mechanistic understanding of the kinetics of chromate and arsenate removal by zerovalent iron media. A mechanistic understanding will enable results from short-term laboratory tests to be used for evaluating the long-term performance of iron media for removing chromate and arsenate from drinking water supplies. Towards this end, the effect of aqueous concentration on the rates of chromate and arsenate removal by corroding iron were determined in both batch and column reactors. The effects of chromate and arsenate on iron corrosion rates were also determined. Finally, the long-term performance of zerovalent iron for arsenate and chromate removal was evaluated in column tests lasting up to one year in duration.

Materials and Methods

Batch Reactors

Batch experiments measuring soluble chromate and arsenate removal by iron wires were performed in well-stirred, sealed 0.85 L glass reactors containing potassium chromate or sodium arsenate in 3 mM $CaSO_4$ background electrolyte solutions. A schematic diagram of the batch reactors is shown in Figure 1a. Each reactor contained a single 10 cm long by 1.2 mm diameter iron wire of 99.9% purity (Aesar, Ward Hill, MA) that was used as the reactant. The reactors also contained a calomel reference electrode and a stainless steel counter electrode. The solution redox potential was measured using a platinum wire that was periodically inserted through the septum in the sampling port. Samples were taken using a 1 mL glass syringe, with and without 0.1 µm nylon syringe filters (Whatman, Clifton, NJ). The reactors were purged with ~50 mL/min of humidified nitrogen gas in order to agitate the solutions and maintain low dissolved oxygen concentrations.

Packed Column Experiments

Column experiments measuring chromate removal were performed using a 25 cm long by 0.9 cm outer diameter (o.d.) stainless steel column, while arsenate removal was measured using a 50 cm long by 2.5 cm o.d. glass column. Both columns were packed with Master Builder's Supply (Cleveland, OH) iron filings GX-27 blend. The glass column was fitted with an inlet and three intracolumn sampling ports at 0, 12.5, 25, and 37.5 cm from the influent end, as illustrated in Figure 1b. One port at each location served for taking aqueous samples and for measuring the redox potential of the solution. The other port was used to determine the free corrosion potential of the iron reactants. The corrosion potentials were measured using an iron wire permanently inserted into the column through a rubber septum at each port. The columns were operated with Cr(VI) or As(V) concentrations ranging from 100 to 23,500 μg/L in 3 mM $CaSO_4$ background electrolyte solutions. The mean hydraulic residence time in the 25 cm column was 25 min and was 20 min in the 50 cm column.

Analysis Methods

In all experiments the solution pH values were measured with test strips that were color calibrated in increments of 0.5 pH units. Aqueous chromium and arsenate concentrations were determined using a Perkin Elmer (San Jose, CA) model 4110zL graphite furnace atomic absorption (AA) spectrophotometer. The oxidation state of the aqueous arsenic species was determined via ion chromatography (IC) using an atomic fluorescence detector (PS Analytical, Kent, UK). The detection limits for the AA and IC methods (*15*) were both ~1 μg/L.

Electrochemical Experiments

To assess the effect of chromate and arsenate concentration on rates of iron corrosion, a single 10 cm long iron wire was placed in 0.75 L of a 3 mM $CaSO_4$ electrolyte solution in one of the sealed, glass reactors containing a calomel reference electrode and a stainless steel counter electrode. The solution was continuously purged with 50 mL/min of humidified nitrogen gas, and potassium chromate or sodium arsenate was added to the reactor through the vent tubing to produce aqueous concentrations ranging from 100 to 10,000 μg/L. The wire was exposed to each concentration for 1 day, at which point a Tafel scan was performed. All Tafel diagrams were produced by polarizing the wires ± 200 mV with respect to their open circuit potentials. The polarization experiments were performed using an EG&G (Oak Ridge, TN) model 273A scanning potentiostat and M270 software. Corrosion rates in the column reactors were also determined by analysis of Tafel polarization diagrams. All potentials are reported with respect to the standard hydrogen electrode (SHE).

Results and Discussion

Removal Kinetics

In a previous investigation, rates of arsenate removal in the batch reactors were found to follow equation 1 with a k_1 value of 0.10 d^{-1} and a k_0 of 120 $\mu gL^{-1}d^{-1}$, (*11*). Figure 2 shows the concentration dependence of As(V) removal rates calculated from this model, and compares them to rates of chromate removal determined in this investigation. Although both arsenate and chromate are anionic species that form complexes with iron oxides, the data in Figure 2 show that their interactions with zerovalent iron yield significantly different removal behavior. The purpose of this investigation is to provide a mechanistic basis for the removal kinetics exhibited in Figure 2.

Arsenate

The kinetics of arsenate removal show increasing deviation from first-order behavior with increasing As(V) concentration. This type of behavior is common for surface mediated reactions where there is increasing competition for reactive sites with increasing reactant concentration. The competition among arsenate species for iron oxide adsorption sites arises because arsenate does not promote iron oxidation under the conditions of these experiments. The effect of arsenate on the corrosion rate of iron is illustrated in Figure 3. Adding 100 μg/L of arsenate to the solution decreased the corrosion rate by a factor of five versus the blank electrolyte. However, further increases in the arsenate concentration from 100 to 20,000 μg/L had no effect on the iron corrosion rate.

Tafel diagrams can be used to explain the mechanism through which arsenate decreases iron corrosion rates. Figure 4 shows the Tafel plots used to

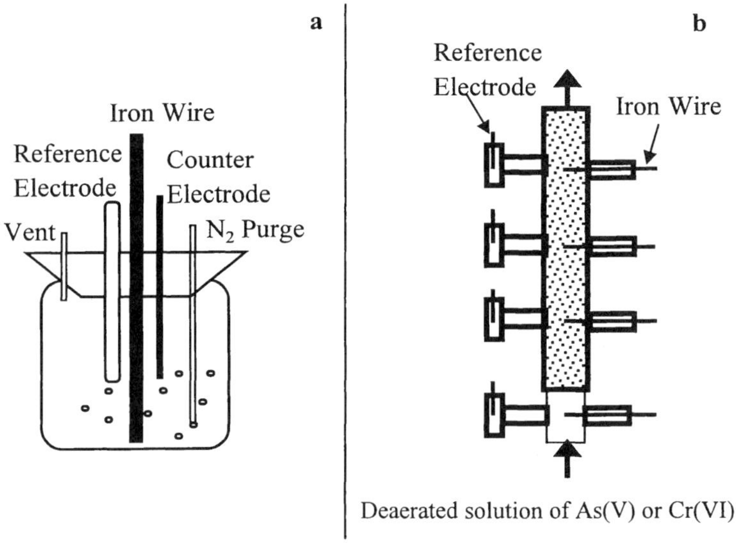

Figure 1. a) Schematic diagram of the batch reactors; b) Schematic diagram of the 50 cm long column reactor.

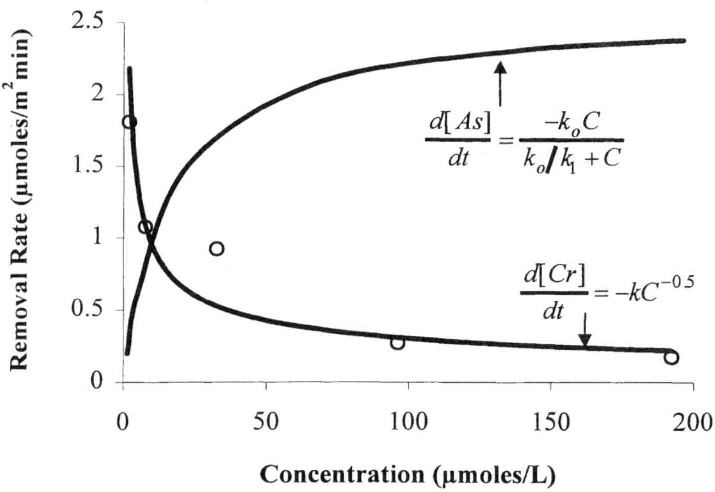

Figure 2. Arsenate and chromate removal rates by iron wires immersed in anaerobic 3 mM $CaSO_4$ electrolyte solutions as a function of the aqueous As(V) or Cr(VI) solution concentration. Arsenate removal rates were determined in a previously reported investigation (11) and were fit to the kinetic model decscribed by equation 1.

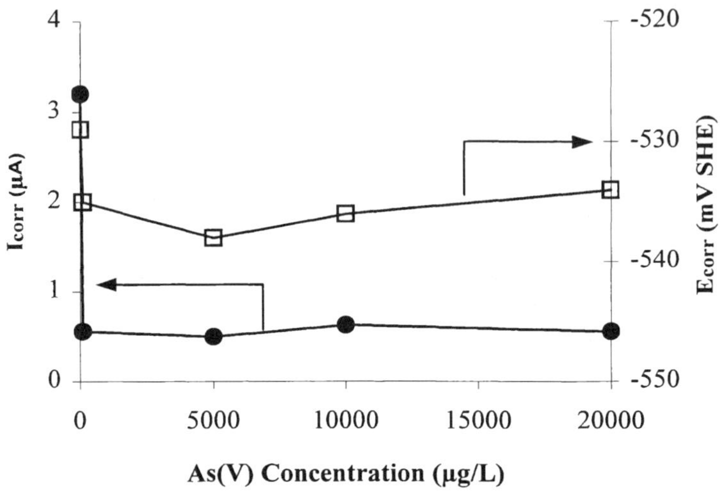

Figure 3. Corrosion currents and free corrosion potentials for a single iron wire exposed to arsenate at concentrations ranging from 100 to 20,000 µg/L after one day elapsed at each concentration.

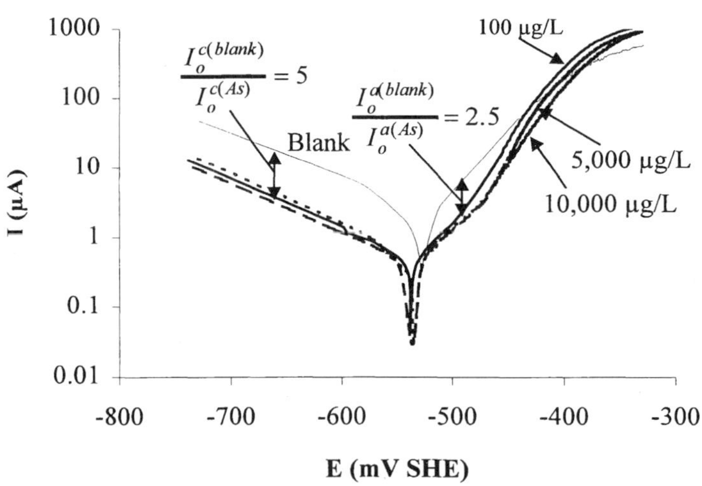

Figure 4. Tafel polarization diagrams for an iron wire in an anaerobic 3 mM $CaSO_4$ electrolyte solution and in electrolyte solutions containing arsenate at the indicated concentrations.

generate the corrosion rate data in Figure 3. The similar cathodic slopes in the blank and arsenate solutions indicate that water was the primary oxidant, and that arsenate reduction was insignificant. However, lower cathodic currents at each potential were observed in the presence of arsenate. The similar cathodic slopes and the lower cathodic currents indicate that the presence of arsenate decreased the exchange current (I_o^c) for water reduction. The I_o^c determines the magnitude of the cathodic current (I_c) at each potential, and is defined according to (*16*):

$$I_c = I_o^c\left[e^{-\beta_c(E-E_{eq}^c)}\right] \tag{2}$$

where β_c is the cathodic Tafel slope, E is the potential and E_{eq}^c is the equilibrium potential for the cathodic reaction. The exchange current is indicative of the chemical facility of the cathodic reaction, and is proportional to the oxidant concentration and the electroactive surface area. The cathodic Tafel scans in Figure 4 show that the presence of arsenate decreased the I_o^c by a factor of 5 as compared to the blank electrolyte solution. The observed decrease in I_o^c can be attributed to the blocking of cathodic sites for water reduction by adsorbed arsenate. The absence of a concentration effect for As(V) concentrations above 100 μg/L can be attributed to saturation of the adsorption sites at a concentration of only 100 μg/L. This is confirmed by the similar cathodic currents at each potential for As(V) concentrations of 100, 5,000 and 10,000 μg/L.

Arsenate also affected the iron oxidation reaction, as indicated by the anodic Tafel scans in Figure 4. The anodic scans can be divided into two regions by the potential where the slope changes at approximately -475 mV. For potentials lower than -475 mV, the anodic slope (β_a) in the arsenate solutions was similar to that in the blank solution. The β_a value of 0.017 dec/mV corresponds to the oxidation of zerovalent iron (*17*). Although the β_a values are the same, the lower anodic currents at each potential in the arsenate solutions indicate that the exchange current for the iron oxidation reaction (I_o^a) was diminished by the presence of arsenate. The I_o^a determines the magnitude of the anodic current (I_a) at each potential, and is defined according to (*16*):

$$I_a = I_o^a\left[e^{\beta_a(E-E_{eq}^a)}\right] \tag{3}$$

where E_{eq}^a is the equilibrium potential for the Fe/Fe^{2+} redox couple. The decrease in I_o^a can be attributed to the blocking of anodic sites by adsorbed arsenate. Addition of only 100 μg/L arsenate resulted in a 60% decrease in I_o^a as compared to the blank solution. Since the I_o^a is proportional to the anodic area, this suggests that there was a 60% decrease in electroactive surface area for iron oxidation.

The β_a in all arsenate solutions began to change at a potential of approximately -475 mV. This change in slope may be attributed to oxidation of iron complexed with As(V) (*5*). The greater β_a at potentials above -475 mV suggests that electron transfer for oxidation of complexed iron is more facile than electron transfer for oxidation of zerovalent iron. This more facile electron transfer for iron-arsenate complexes resulted in anodic currents that were higher than those in the blank solution at sufficiently positive potentials. This indicates that arsenate is not an effective anodic inhibitor, since it may actually increase the oxidation rate at high anodic overpotentials.

Chromate

In contrast to arsenate where the removal rate increased with increasing concentration, the rate of Cr(VI) removal actually declined with increasing concentration, as shown in Figure 2. The kinetic model for describing this effect has a rate constant (*k*) of 36×10^{-3} $\mu g^{1.5} m^{-3\ 5} min^{-1}$ and an apparent reaction order of -0.5 with respect to the aqueous Cr(VI) concentration. The decreasing removal rates with increasing concentration can be explained by iron surface passivation by precipitated chromium compounds.

The effect of Cr(VI) on the corrosion rate of iron is illustrated in Figure 5. Although Cr(VI) is a strong oxidant and will readily oxidize zerovalent iron, the iron corrosion rate actually declined with increasing chromate concentration. The drop in potential and the decrease in current between the blank and the 100 μg/L solution is consistent with adsorption of a cathodic inhibitor (*16*). The higher corrosion potentials and lower corrosion currents observed with increasing concentration between 100 and 5,000 μg/L are indicative of the action of an oxidant/passivator like chromate (*5*). The mechanisms responsible for this behavior can be elucidated by analysis of Tafel diagrams.

Figure 6 compares Tafel diagrams generated in the blank, 100, 5,000, and 10,000 μg/L Cr(VI) solutions. The similar cathodic slopes for the blank and the 100 μg/L solution indicate that water was the primary oxidant in both solutions. Although the cathodic Tafel slopes for the blank and the 100 μg/L solution were similar, lower cathodic currents at each potential were observed in the chromate solution. This can be attributed to a 45% decrease in the exchange current for water reduction. The lower cathodic Tafel slopes in the 5,000 and 10,000 μg/L solutions can be attributed to Cr(VI) reduction. In these solutions, both water and Cr(VI) reduction contributed to the rate of iron oxidation.

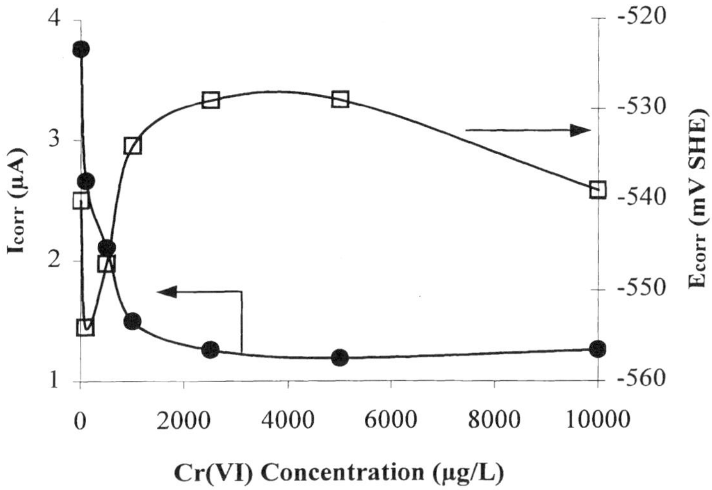

Figure 5. Corrosion currents and free corrosion potentials for a single iron wire exposed to chromate at concentrations ranging from 100 to 10,000 µg/L after one day elapsed at each concentration.

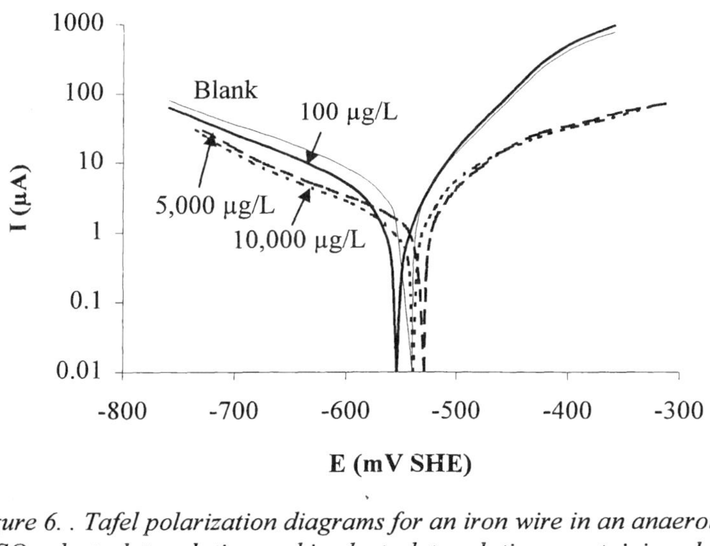

Figure 6. . Tafel polarization diagrams for an iron wire in an anaerobic 3 mM $CaSO_4$ electrolyte solution and in electrolyte solutions containing chromate at the indicated concentrations.

The presence of chromate also affected the iron oxidation reaction as shown by the anodic Tafel scans in Figure 6. The similar anodic currents in the blank and 100 µg/L solutions at low overpotentials show that anodic inhibition of iron corrosion was absent at this concentration. However, at Cr(VI) concentrations of 5,000 and 10,000 µg/L, significant anodic inhibition of iron oxidation is evident. For example, at a potential of -450 mV, the anodic current in the 5,000 and 10,000 µg/L solutions is a factor of 6 smaller than in the blank electrolyte solution.

Column Reactors

The long-term performance of zerovalent iron for chromate removal was investigated using a stainless steel column reactor. For the first 500 pore volumes the reactor was operated with a mean hydraulic residence time of 20 seconds and an influent Cr(VI) concentration of 100 µg/L. Under these conditions Cr(VI) was removed to below the detection limit of 1 µg/L. For the remainder of the test, the hydraulic residence time was increased to 25 minutes. The operational history of the column is given in Table 1.

Table 1. Influent and Effluent Cr(VI) Concentrations for the 25 cm Column Operated with a Mean Hydraulic Detention Time of 25 Minutes.

Days Elapsed	*Cumulative Pore Volumes Passed*	*Influent Concentration (µg/L)*	*Effluent Concentration (µg/L)*
0-117	6,727	100	0
117-132	7,590	300	0
132-146	8,395	600	0
146-183	10,522	1,000	0
183-193	11,097	1,500	0
193-207	11,902	5,000	0
207-252	14,490	10,000	7,000
252-267	15,352	1,000	200
267-274	15,755	shut-in	shut-in
274-296	17,020	1,000	20

Complete removal of Cr(VI) was observed for 207 days until the influent concentration was increased to 10,000 µg/L, as indicated in Table 1. An influent concentration of 10,000 µg/L resulted in a steady-state removal of 3,000 µg/L for a period of 1,200 pore volumes, as illustrated in Figure 7. Upon lowering the influent Cr(VI) concentration from 10,000 to 1,000 µg/L at 252 days elapsed, a steady-state effluent concentration of ~200 μg/L was obtained. Previously, between 146 and 183 days elapsed, a feed concentration of 1,000 μg/L resulted in complete Cr(VI) removal, as shown in Table 1. The different effluent concentrations for the same feed concentration show that the performance of the iron for removing Cr(VI) was hysteretic. Between 267 and 274 days elapsed, the column was shut-in to determine if the surface passivation was reversible. Upon resuming an influent concentration of 1,000 μg/L, a steady-state removal of 980 μg/L was obtained. This confirms that performance of the iron was hysteretic, and shows that passivation is at least partially reversible.

Arsenate removal was investigated in a 50 cm long glass column containing three intra-column sampling ports. The operational history of the column is given in Table 2. Complete arsenate removal was observed before the first sampling port for influent concentrations less than 1,000 μg/L. For influent concentrations of 2,500 μg/L and above, all sampling ports showed measurable As(V) concentrations.

Table 2. Influent and Effluent As(V) Concentrations for the 50 cm Column Operated with a Mean Hydraulic Detention Time of 20 Minutes.

Days Elapsed	*Cumulative Pore Volumes Passed*	*Influent Concentration (μg/L)*	*Effluent Concentration (μg/L)*
0-4	288	100	0
4-7	504	500	0
7-10	720	1,000	0[1]
10-13	936	2,500	470
13-16	1,152	5,000	2,100
16-19	1,368	10,000	5,500
19-23	1,656	23,500	13,500

[1]A concentration of 280 μg/L was observed at port 1.

Aqueous As(V) profiles for influent concentrations of 10,000 and 5,000 μg/L are shown in Figure 8. The faster removal between the column inlet and the first sampling port can be attributed to greater rates of iron corrosion near the influent end of the column (*18*). Throughout the investigation, corrosion rates at the inlet sampling port were at least a factor of seven times higher than those at the three intra-column ports, which all showed similar corrosion rates.

Since the k_I and k_O parameters in the equation 1 kinetic model for As(V) removal depend on the iron corrosion rate, the section of the column before the first sampling port should have different rate parameters than the remainder of the column. Therefore, the parameters in the kinetic model were independently

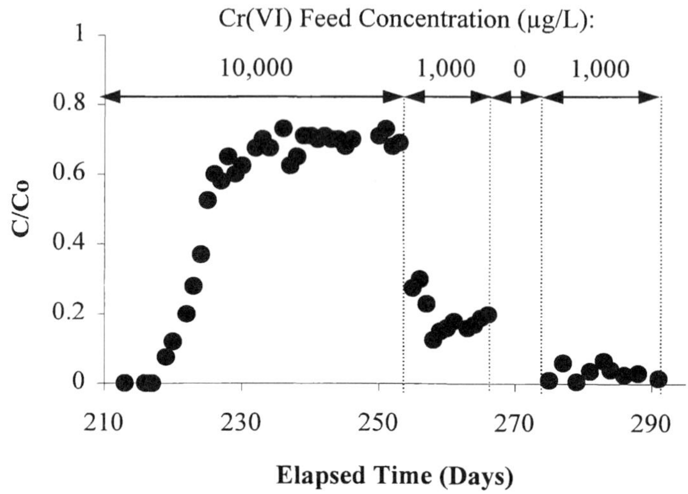

Figure 7. Effluent chromium concentrations as a function of elapsed time for a 25 cm long column packed with iron filings and operated at Cr(VI) concentrations ranging from 100 to 10,000 µg/L in an anaerobic 3 mM $CaSO_4$ background electrolyte solution. Influent and effluent chromium concentrations for the entire experiment are listed in Table 2.

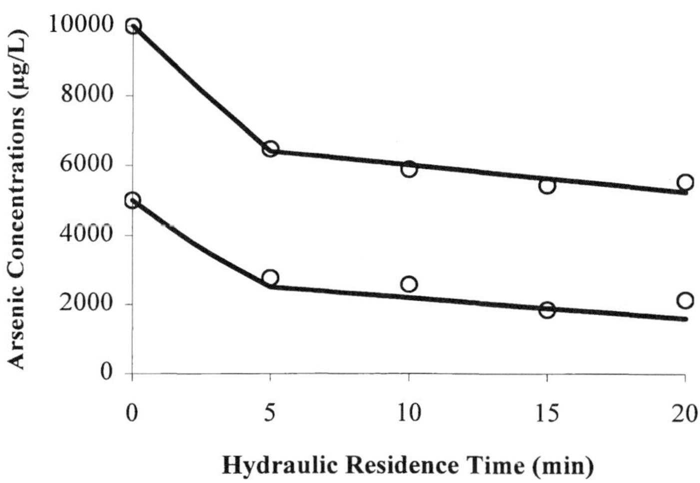

Figure 8. Aqueous As(V) profiles along the length of the 50 cm long column. The solid lines represent the equation 1 model fit with k_1=0.25 min^{-1} and k_o=1111 $\mu gL^{-1}min^{-1}$ for the section of the column between the inlet and port 1, and k_1=0.09 min^{-1} and k_o=92 $\mu gL^{-1}min^{-1}$ for the section between port 1 and the outlet of the column.

determined for the first section of the column, and for the section of the column between the first sampling port and the column outlet. Data for influent concentrations between 2,500 and 23,500 μg/L were used to determine the model parameters. Fits of the equation 1 model to the As(V) profiles are shown in Figure 8.

Conclusions

Results from this study show that the kinetics of both arsenate and chromate removal by zerovalent iron are highly dependent on their aqueous concentration. The presence of both As(V) and Cr(VI) decrease iron corrosion rates as compared to those in blank electrolyte solutions. This can be attributed to the blocking of anodic and cathodic sites for iron corrosion. The passivation of the iron surface is incomplete, and the iron continues to corrode even for high concentrations of arsenate and chromate. The steady-state effluent concentrations from the column reactors for thousands of pore volumes shows that characterizing arsenate and chromate removal in terms of a capacity is not appropriate unless all the iron has been oxidized to Fe^{3+}.

The hysteretic performance of the packed column for chromate removal shows that the kinetics of Cr(VI) removal by iron media cannot be described by a simple kinetic model. This can be attributed to chromate's behavior as both a cathodic and anodic inhibitor of iron corrosion. This behavior results in slower rates of removal with increasing Cr(VI) concentration. Increasing chromate concentration has more of an effect on the anodic reaction of iron oxidation than on the cathodic reactions of water and chromate reduction. This means that with increasing chromate concentration, the rate of chromate removal will be increasingly controlled by the ability of the iron to release electrons via corrosion. Diffusion of Cr(VI) to cathodic sites through iron corrosion products may also impact chromate removal kinetics.

The rate limiting mechanism for arsenate removal by iron media depends on both the arsenate concentration and the iron corrosion rate. At low arsenate concentrations, the corrosion rate of the iron is sufficiently fast to provide an abundance of adsorption sites. As a result, the rate of site generation is faster than the rate at which arsenate may diffuse to an unoccupied adsorption site. Therefore, the removal kinetics are approximately first-order in arsenate concentration, as would be expected for a process limited by diffusional mass transfer. At high arsenate concentrations, the iron corrosion rate is not fast enough to maintain an excess of adsorption sites. As a result, the moment a site is formed it is occupied by readily available arsenate ions. This leads to competition for unoccupied sites and deviation from first-order kinetics.

References

1. Powell, R. M.; Puls, W. P.; Hightower, S. K.; Sabatini, D. A. *Environ. Sci. Technol.* **1995**, 29, 1913.
2. Blowes, D. W.; Ptacek, C. J.; Jambor, J. J. *Environ. Sci. Technol.* **1997**, 31, 3348.
3. Blowes, D. W.; Gillham R. W.; Ptacek, C. J.; Puls, R. W.; Bennett, T. A.; O'Hannesin, S. F.; Hanton-Fong, C. J.; Bain, J. G. "An in situ permeable reactive barrier for the treatment of hexavalent chromium and trichloroethylene in ground water: Volume 1, Design and installation" Environmental Protection Agency. EPA/600/R-99/09.
4. Puls, R. W.; Paul C. J.; Powell R. M. *Appl. Geochem.* **1999**, 14, 989.
5. Uhlig, H. H.; Revie, R. W. *Corrosion and Corrosion Control, third ed.*; John Wiley: New York, 1985.
6. Gould, J. P.; *Water Res.* **1982**, 16, 871.
7. Pande. S. P.; Deshpande, L. S.; Patni, P. M; Lutade. S. L. J. *Environ. Sci. Health*, **1997**, A32(7), 1981.
8. *Research plan for arsenic in drinking water*; U. S. Environmental Protection Agency, Office of Research and Development: Washington, DC, 1998; EPA/600/R-98/042.
9. *Federal Register*, **2001**, 66, 14, 6976.
10. Lackovic, J. A.; Nikolaidis, N. P.; Dobbs, G. M. *Environ. Eng. Sci.* **2000**, 17(1), 29.
11. Farrell, J.; Wang, J. P.; O'Day. P.; Conklin, M. *Environ. Sci. Technol.* **2001**, 35, 2026.
12. Su, C.; Puls, R. W. *Environ. Sci. Technol.* **2001**, 35, 1487.
13. Grossl, P. R.; Eick, M.; Sparks, D. L.; Goldberg, S.; Ainsworth, C. C. *Environ. Sci. Technol.* **1997**, 31, 32.
14. Fuller, C. C.; Davis, J. A.; Waychunas, G. A. *Geochim. Cosmochim. Acta.* **1993**, 57, 2271.
15. Gregus, Z.; Gyurasics, A.; Csanaky, I. *Toxicological Sciences* **2000**, 56, 1, 18.
16. Bockris, J. O'M.; Reddy, A. K. N.; *Modern Electrochemistry*; Plenum Press: New York, 1970.
17. Talbot, D.; Talbot, J.; *Corrosion Science and Technology*; CRC Press: Boca Raton, 1998.
18. Melitas, N.; Wang, J.; Conklin, M.; O'Day, P.; Farrell, J.; *Environ. Sci. Technol.* in review August 2001.

Chapter 13

Trichloroethylene Reduction on Zero Valent Iron: Probing Reactive versus Nonreactive Sites

Baolin Deng[1], Shaodong Hu[1], T. M. Whitworth[2], and Robert Lee[3]

[1]Department of Environmental Engineering and [3]Petroleum Recovery Research Center, New Mexico Institute of Mining and Technology, Socorro, NM 87801
[2]Department of Geological and Petroleum Engineering, University of Missouri at Rolla, Rolla, MO 65409

Sorption and reduction of trichloroethylene (TCE) on zero-valent iron (ZVI) were examined using batch experiments in the presence and absence of cysteine and potassium sulfate. Cysteine, an amino acid containing a -SH functional group, did not significantly affect TCE adsorption at concentrations lower than 1.0 mM. However, the rates of TCE reduction and iron corrosion dramatically decreased at comparable or even much lower concentrations. The results suggest that there are two types of surface sites on the iron: the reactive sites responsible for TCE reduction and the non-reactive sites responsible for the bulk of TCE adsorption. The reactive site density was estimated to be only 2% of the total surface sites. A two-site analytical model was proposed to explain the experimental results, in which adsorption took place on both reactive and non-reactive sites, but dechlorination reactions occurred only on reactive sites. Additionally, it was observed that the rate of TCE reduction was proportional to the rate of iron corrosion. This suggested that the same types of surfaces might be responsible for both dechlorination of TCE and iron corrosion, and/or the hydrogen producing corrosion process took place prior to the dechlorination reaction.

Introduction

Chlorinated solvents, such as trichloroethylene (TCE), are commonly detected at uncontrolled hazardous waste sites in the United States, leading to groundwater contamination that can pose serious risks to water supplies. Current regulations limit the maximum contaminant level for TCE in drinking water at 0.005 ppm (1).

The prevailing method to restore groundwater contaminated with chlorinated solvents has been through pump-and-treat, i.e., the contaminated water is pumped out and treated on the surface. Although this approach is useful for some contaminated sites with simple and homogeneous geology, it is expensive and has failed to restore waste sites with complex hydrogeochemical settings and contamination history (2, 3).

Sweeny (4, 5), followed by Senzaki and Kumagai (6, 7), were among the first to report that zero-valent iron (ZVI) could be used to degrade chlorinated solvents dissolved in water. Further research was fostered by Gillham and O'Hannesin (8), when they proposed to use ZVI-based permeable reactive barriers for the in situ degradation of chlorinated organic compounds. Field studies demonstrated that 95% of the TCE and 91% of PCE in contaminated groundwater was reduced when a contaminated groundwater plume passed through such a barrier (8). The technology has quickly gone from laboratory study to pilot testing and to full field installations (9, 10).

There is compelling evidence suggesting that contact between the organic solute and the solid metal particles is required for the dechlorination reaction (11) and that sorption of the chlorinated solvents (i.e. TCE and PCE) onto the iron surface occurs. Burris et al. (12) found that a significant amount of TCE or PCE was sorbed onto the iron surface and that the bulk of the sorption was onto non-reactive sites. When PCE and TCE are both pesent, competition for sorption sites occurred. This competition, however, did not appear to influence reaction rates. Allen-King et al. (13) also observed reversible sorption for dichloroethylene (DCE) isomers on iron surfaces. Matheson and Tratnyek (14) conducted a detailed study of carbon tetrachloride (CT) degradation in systems containing granular iron and proposed that the reductive dehalogenation took place either through direct electron transfer from the surface of ZVI to CT or by reaction with dissolved Fe^{2+} or H_2, which are products of iron corrosion. Although their experimental results suggested that the direct electron transfer on the surface of the metallic iron was the most likely process, they did not exclude the possibility of TCE reduction with Fe^{2+} or H_2. Sivavec et al., (15) determined that reductive dechlorination of chlorinated solvent could occur by Fe(II)-containing minerals.

The degradation of chlorinated solvents with ZVI is accompanied by iron oxidation as represented by the stoichiometry: $Fe^0 + RCl + H_3O^+ = Fe^{2+} + RH + Cl^- + H_2O$. Elucidating the relationship between iron corrosion and the dechlorination reaction is thus desirable in order for us to obtain a mechanistic understanding of the dechlorination process. Literature on metal corrosion is abundant. Metallic iron corrosion may occur as a uniform attack or by pitting and stress cracking (16). The rate of corrosion can be inhibited by various surface treatments including: (1) adsorption of a chemical as a thin film, (2) inducing the formation of thick corrosion products, and (3) changing characteristics of the environment by either producing protective precipitates or removing and inactivating corrosion-aggressive constituents. A corrosion inhibitor is a chemical substance which, upon addition to an environment in low concentrations, effectively checks, decreases, or prevents the reaction of the metal with other ambient constituents. Deng et al. (17) examined the effects of 14 organic and inorganic chemicals (including various amino acids, phosphate, sulfate, borate, molybdate and silicate) on iron corrosion and TCE degradation under anaerobic condition. Among these chemicals, 1.0 mM cysteine appeared to totally prevent iron corrosion and TCE degradation, while sulfate at a 1.0 mM concentration level slightly enhanced the rate of iron corrosion. Detailed information on the relationship between the inhibition effect and concentrations of cysteine and sulfate, however, was not available. Solution pH, which may affect iron corrosion and hydrogen production, is another important parameter. Whitman et al. (18) found that the hydrogen ion concentration did not significantly increase the rate of iron corrosion until the pH was below 4. Reardon (19) similarly observed that pH did not appear to be a factor in governing the anaerobic corrosion rate of iron at neutral or alkaline conditions. Another set of experiments designed by Reardon (19) to examine the effect of anions on the iron corrosion demonstrated that after a reaction time of 150 hrs, the corrosion rate of iron in a 0.01 M Na_2SO_4 solution was higher than that of a comparable experiment performed in de-ionized water alone.

Optimal design of a ZVI remediation system hinges on a thorough understanding of interactions among the contaminant, iron, water, and other chemical and biological constituents. How the reductive dechlorination depends on the reactant adsorption and iron corrosion is one of the key issues that need to be addressed. In this study, the interaction between iron corrosion, as indicated by H_2 production, and reductive dechlorination of TCE, was investigated under both anaerobic and aerobic conditions in batch systems. The rate of iron corrosion was controlled by chemicals that could inhibit or enhance the corrosion reaction. The influence of the corrosion reaction on TCE degradation was examined. A two-site analytical model was developed to facilitate data interpretation.

Experimental Section

Chemicals

De-ionized Milli-Q water with 18.2 MΩ·cm resistivity (Millipore Ultra-pure water system) was used for all the experiments in this research. Trichloroethylene (TCE) (99.5+%, A.C.S reagent), palladium (II) chloride (99%), sodium hypophosphite, potassium persulfate (99+%, A.C.S. reagent) and EDTA (99%) were obtained from Aldrich. L-cysteine ($C_3H_7NO_2S$, 98%), potassium sulfate (99%) and cobalt chloride (99%) were obtained from Sigma Chemical Co. Standards of gaseous hydrogen (1.05%, N_2 balanced) and TCE (10 ppm, N_2 balanced) were obtained from Scott Specialty Gases. The iron filings (40 mesh) used in this study were supplied by Fisher Scientific Company. Prior to use, the iron filings were washed with Ar-purged 1.0 N HCl for 30 minutes, rinsed 16 times with Ar-purged water, dried at 100 °C under N_2 atmosphere, and maintained in a N_2 atmosphere. This treatment yielded a gray, granular metallic solid witt no visible oxidization products on the surface.

Adsorption of TCE and Cysteine on Metallic Iron

All experiments were carried out in borosilicate glass serum bottles with measured internal volumes of 72 ± 1 mL. In order to estimate the time for TCE to reach adsorption equilibrium, adsorption kinetics were first measured in a series of serum bottles. Each bottle contained 5.000 g of iron filings and was filled to capacity with 1.35 mM TCE solutions, so no headspace was present. After beng crimp-sealed with a Teflon-lined silicon septum, each bottled sample was first shaken by hand for 10 minutes, then mixed with an incubator shaker (Innova 4335, New Brunswick Scientific) operated at 25.0 °C and a speed of 100 rpm. TCE concentrations in the aqueous phase were measured as a function of time by sacrificing one bottle at each time point.

Experiments designed to determine TCE adsorption were performed based on the obtained equilibrium time for TCE adsorption. Each point on the TCE adsorption isotherms was developed with duplicate samples and a control. All samples contained 5.000 g of iron filings and 72 mL aqueous phase with various concentrations of TCE. For experiments examining the competitive adsorption between TCE and cysteine, the aqueous phase contained different concentrations of cysteine. The samples were sealed in bottles with no headspace and mixed in a shaker for 60 minutes at 100 rpm before sample analysis.

TCE degradation

Solutions of potential corrosion inhibitors, cysteine and potassium sulfate, were prepared by dissolving predetermined amounts of cysteine and potassium sulfate, respectively, into deionized water. Two microliters of pure phase TCE was then injected into each bottle which contained a 35 ml of cysteine or potassium sulfate solution, completing the preparation of the aqueous phases. After TCE in each bottle was completely dissolved, which took about one hour, 5.000 g of ZVI filings was added. In order to determine the role that oxygen had towards the reaction, parallel experiments were performed, with one prepared in an anaerobic chamber containing an atmosphere of 100% N_2, and the other in the ambient air. Each of the prepared samples finally contained 5.000 g of treated iron filings, 35 mL aqueous phase, and 37 mL gas phase. All sample bottles were inverted and placed in a shaker to minimize gas evaporation and mixed at a speed of 50 rpm.

In TCE degradation by ZVI, pH may influence the reaction rate directly by participating in the stoichiometric reaction and indirectly, by enhancing iron corrosion at low pH and iron hydroxide precipitation at high pH. Nevertheless, when pH values are higher than 4, the change in pH does not appear to be a factor governing anaerobic corrosion rates (18). Our preliminary tests demonstrated that un-buffered systems exhibited pH values of 6~7 and that pH changed little during the tests. Thus, most experiments in this study were carried out without the addition of a buffer.

Kinetics of TCE degradation were examined with duplicate experiments and the reaction progress was monitored by headspace analysis of TCE concentration. The solution concentration was then related to the gas phase concentration by Henry's law. Hydrogen gas (H_2) in the headspace was also measured in order to evaluate iron corrosion. It was expected that H_2 generated would dominantly partition into the vapor phase, thus the H_2 amount in the headspace would account for most of the H_2 produced (19). The hydrogen concentration was assumed to indicate the magnitude of iron corrosion in the various systems.

Analytical Methods

TCE was analyzed by gas chromatography (SRI 8610C Dual Oven GC, SRI Instruments) through either liquid or headspace injection. A silicon capillary column (MXT-Q Plot, 30 m length, Restek Co.) and flame ionization detector (FID) were employed to measure TCE. Peak areas were converted to concentrations by the standard external method using calibration curves prepared from appropriate aqueous or gas standards. Either 100.0 μl headspace or 1.0 μl aqueous phase was manually injected on-column in a splitless mode into the GC. The column temperature for TCE measurement was programmed for 60 °C for 1 minute, then raised to 200 °C at a rate of 20 °C/min with a carrier gas (N_2) flow pressure of 9 psi. Temperature for the FID was held constant at 150 °C. Hydrogen measurements were made using a molecular sieve packed column (3 feet × 1/8 inch, SRI Instruments) and thermal conductivity detector (TCD). The column temperature for H_2 analysis was held constant at 70 °C with a carrier gas (N_2) flow pressure of 7 psi. Temperature for the TCD was 100 °C.

The solution pH values were determined at the end of the experiments with an Orion 420A pH meter equipped with a gel-filled combination pH electrode. The pH electrode was calibrated at pH 4.00 and pH 7.00 with Orion application buffer solutions.

A kinetic method as reported by Garcia et al. (20) was used to determinate the aqueous phase concentration of cysteine. A 1.50 mL volume of 0.020 M Co^{III} – EDTA, 1.80 mL of Britton-Robinson buffer (pH 3.2, prepared by adding 20 mL of 0.2 M NaOH into 100 mL of acid solution with 0.04 M H_3PO_4, H_3BO_3 and CH_3CO_2H), and 48.0 μL of 5×10^{-3} M $PdCl_2$ were added in sequence into a 10 mL flasks. An appropriate volume of sample solution was also placed in the flasks to give a final cysteine concentration between 1 and 9 μM. The solution was then diluted to 8.30 mL by water and placed in room temperature (25 °C) for 10 min. A 1.70 mL of 3.0 M sodium hypophosphite was then added to the flasks and solution was mixed. The absorbance vs. time curve was then recorded at 540 nm on a spectrophotometer (Spectronic 20 Genesys™).

Model development

Surfaces of solids, such as iron, can exhibit a high degree of heterogeneity, hosting a wide variety of surface sites with different affinities to form surface complexes (21). Changes of coordination environments, as well as the presence of defects and impurities all contribute to this heterogeneity. Solute sorption, however, has often been modeled by assuming a single type of surface sites with contribution from electronic interaction (22, 23). Different reactivities on various surface sites are accommodated by introducing different adsorption constants for various surface complexes. When the reactive sites only represent a small portion of the total number of surface sites, a one-site model may not be appropriate. That is the case for the sorption and reductive dechlorination of TCE and PCE on ZVI (12, 24). Recently, Deng and Hu (25) used a numerical model to describe the reactivity of different types of sites residing on the iron surfaces and numerical solution was used to simulate TCE sorption and its reduction kinetics.

In this chapter a two-site analytical model is developed for the interpretation of experimental results on TCE reduction by ZVI in the presence of co-adsorbates. It is based on the mass balance and observed contaminant reduction kinetics applicable for the experimental system. To begin with, we designate the two types of surface sites as S_1 and S_2, on ZVI. Both types of sites can adsorb competing adsorbates, but may have different affinities. The primary distinction between the two types of sites is that site S_1 can reduce TCE and site S_2 can not, as such, the site S_1 is called the reactive site and site S_2 the non-reactive site.

First consider a system with ZVI filings and TCE solution only. If the adsorption of TCE occurs on both types of sites, the mass balance of TCE can be described by equation (1):

$$[T]_t = [T] + [S_1 - T] + [S_2 - T] + [P] \qquad (1)$$

where $[T]_t$ is the amount of TCE that has been initially added into the system, [T] is TCE concentration remaining in the aqueous phase, $[S_1\text{-}T]$ represents TCE sorbed onto the reactive site S_1, $[S_2\text{-}T]$ represents the TCE sorbed onto the non-reactive site S_2, and [P] represents the concentration of the TCE reduction products. Additionally, while it is recognized that some reaction intermediates and products during TCE reduction may be present, their sorption is considered to be negligible because of the relatively low sorption affinity for less- or non-chlorinated compounds (24). Taking the derivative of equation (1) yields:

$$\frac{d[T]}{dt} + \frac{d\{[S_1 - T] + [S_2 - T]\}}{dt} + \frac{d[P]}{dt} = 0 \qquad (2)$$

Second, adsorption reactions are considered to be in equilibrium because of the faster sorption kinetics of TCE in comparison to its reduction. Thus, the mass action relationship for the adsorption equilibrium is applicable:

$$[S_1 - T] = K_1^T [S_1][T] \qquad (3)$$

where $[S_1]$ represents the unoccupied S_1 surface site concentration.

By combining the mass balance relationships for surface sites and the mass action relationship of TCE sorption equilibrium (see ref. 31 for details), the adsorption of TCE can be described by equation (4):

$$[T]_{ads} = [S_1 - T] + [S_2 - T] = \{\frac{K_1^T[S_1]_t}{1 + K_1^T[T]} + \frac{K_2^T[S_2]_t}{1 + K_2^T[T]}\}[T] \qquad (4)$$

where $[T]_{ads}$ is the total concentration of TCE adsorbed, K_1^T and K_2^T are constants for TCE sorption onto sites S_1 and S_2, and $[S_1]_t$ and $[S_2]_t$ are the total surface site concentrations for sites S_1 and S_2, respectively. The derivative of equation (4) yields:

$$\frac{d[T]_{ads}}{dt} = \{\frac{K_1^T[S_1]_t}{(1 + K_1^T[T])^2} + \frac{K_2^T[S_2]_t}{(1 + K_2^T[T])^2}\}\frac{d[T]}{dt} \qquad (5)$$

in which $[S_1]_t$ and $[S_2]_t$ are independent of time.

Once the mass balances and equilibrium adsorption are considered, the next step is to take the degradation kinetics into consideration for the reactive system we have. It has been well established that changes of [T] as a function of time in the ZVI system can be described by a first-order kinetics (8, 26):

$$-\frac{d[T]}{dt} = k_{obs}[T] \qquad (6)$$

here k_{obs} is the observed first-order rate constant. Under the mechanistic level, we propose that the rate of product formation is directly proportional to the amount of TCE adsorbed on the reactive sites, i.e.,

$$\frac{d[P]}{dt} = k_{surf}[S_1 - T] \qquad (7)$$

where k_{surf} is the rate constant for the adsorbed TCE reduction.

Combining Eqns. (2), (3), (5), (6) and (7) yields the following equation:

$$\frac{d[P]}{dt} = (1 + \{\frac{K_1^T[S_1]_t}{(1 + K_1^T[T])^2} + \frac{K_2^T[S_2]_t}{(1 + K_2^T[T])^2}\})k_{obs}[T] = k_{surf}[S_1 - T] = k_{surf}K_1^T[S_1][T]$$

Rearranging:

$$k_{obs} = \frac{k_{surf}K_1^T[S_1]}{1 + \{\frac{K_1^T[S_1]_t}{(1 + K_1^T[T])^2} + \frac{K_2^T[S_2]_t}{(1 + K_2^T[T])^2}\}} \qquad (8)$$

This result suggests that the observed rate constant k_{obs} is determined by the adsorption of both TCE on both types of surface sites as well as the surface degradation constant on the reactive sites.

Now let's consider the situation in the presence of a co-adsorbate. The mass balance equation for TCE can be formulated by considering the sorption onto both reactive and non-reactive sites in this bi-component system:

$$[T]_t = [T] + [S_1 - T]^x + [S_2 - T]^x + [P] \tag{9}$$

here the superscript x is used to represent the bi-component system. The derivative of equation (9) yields equation (10):

$$\frac{d[T]}{dt} + \frac{d\{[S_1 - T]^x + [S_2 - T]^x\}}{dt} + \frac{d[P]}{dt} = 0 \tag{10}$$

Similar to the logic behind Eqn. (2), (3), and (7), the following equations are also applicable:

$$\frac{d[P]}{dt} = k_{surf}[S_1 - T]^x \tag{11}$$

$$-\frac{d[T]}{dt} = k_{obs}^x [T] \tag{12}$$

$$[S_1 - T]^x = K_1^T [S_1]^x [T] \tag{13}$$

Application of a mass balance in the bi-component system (see ref. 31 for details) yields the following equation for TCE adsorption:

$$[T]_{ads}^x = [S_1 - T]^x + [S_2 - T]^x = \{\frac{K_1^T [S_1]_t}{1 + K_1^A [A] + K_1^T [T]} + \frac{K_2^T [S_2]_t}{1 + K_2^A [A] + K_2^T [T]}\}[T] \tag{14}$$

where [A] is the co-adsorbate concentration in the aqueous phase. Taking derivative of Eqn. 14 yields:

$$\frac{d[T]_{ads}^x}{dt} = \{\frac{K_1^T [S_1]_t}{(1 + K_1^A [A] + K_1^T [T])^2} + \frac{K_2^T [S_2]_t}{(1 + K_2^A [A] + K_2^T [T])^2}\}\frac{d[T]}{dt} \tag{15}$$

in which [A] is considered to be a constant. Combining equation (10), (11), (12), (13) and (15), and rearranging yields:

$$k^{x}_{obs} = \frac{k_{surf} K_1^T [S_1]^x}{1 + \{\frac{K_1^T [S_1]_t}{(1 + K_1^A [A] + K_1^T [T])^2} + \frac{K_2^T [S_2]_t}{(1 + K_2^A [A] + K_2^T [T])^2}\}} \quad (16)$$

The results here illustrate that the observed degradation constant, k^x_{obs}, is not only dependent on the TCE adsorption and reactivity on the reactive sites, but also the adsorption characteristics of the coadsorbate. Taking the ratio of Eqn. (8) and Eqn. (16), we have:

$$\frac{k^x_{obs}}{k_{obs}} = \frac{[S_1]^x}{[S_1]} \times \frac{1 + \{\frac{K_1^T [S_1]_t}{(1 + K_1^A [A] + K_1^T [T])^2} + \frac{K_2^T [S_2]_t}{(1 + K_2^A [A] + K_2^T [T])^2}\}}{1 + \{\frac{K_1^T [S_1]_t}{(1 + K_1^T [T])^2} + \frac{K_2^T [S_2]_t}{(1 + K_2^T [T])^2}\}} \quad (17)$$

The concentrations of free surface sites can be expressed by:

$$[S_1] = \frac{[S_1]_t}{1 + K_1^T [T]} \quad (18)$$

$$[S_1]^x = \frac{[S_1]_t}{1 + K_1^T [T] + K_1^A [A]} \quad (19)$$

Substituting Eqn. (18) and (19) into Eqn. (17) yields:

$$\frac{k^x_{obs}}{k_{obs}} = \frac{1 + K_1^T [T]}{1 + K_1^T [T] + K_1^A [A]} \times \frac{1 + \{\frac{K_1^T [S_1]_t}{(1 + K_1^A [A] + K_1^T [T])^2} + \frac{K_2^T [S_2]_t}{(1 + K_2^A [A] + K_2^T [T])^2}\}}{1 + \{\frac{K_1^T [S_1]_t}{(1 + K_1^T [T])^2} + \frac{K_2^T [S_2]_t}{(1 + K_2^T [T])^2}\}} \quad (20)$$

Because most adsorption of TCE occurs on non-reactive sites S_2 (Ref. 12), equations (4) and (14) can be simplified as:

$$[T]_{ads} \approx [S_2 - T] = \frac{K_2^T [S_2]_t}{1 + K_2^T [T]} [T]$$

$$[T]^{x}_{ads} \approx [S_2 - T]^{x} = \frac{K_2^T[S_2]_t}{1 + K_2^A[A] + K_2^T[T]}[T]$$

As the result, Eqn. (20) can be simplified to:

$$\frac{k^{x}_{obs}}{k_{obs}} \approx \frac{1 + K_1^T[T]}{1 + K_1^T[T] + K_1^A[A]} \times \frac{1 + \dfrac{K_2^T[S_2]_t}{(1 + K_2^A[A] + K_2^T[T])^2}}{1 + \dfrac{K_2^T[S_2]_t}{(1 + K_2^T[T])^2}} \qquad (21)$$

This relationship characterizes how the coadsorbate affects the rate of TCE degradation when two types of surfaces sites are present. In order to illustrate the application of this model, let us discuss some special cases.

Case 1: The co-adsorbate prefers the reactive site S_1 to the non-reactive site S_2, so $K_1^A[A] >> K_1^T[T]$ and $K_2^A[A] << K_2^T[T]$. Under this extreme, eqn. (21) breaks down to:

$$\frac{k^{x}_{obs}}{k_{obs}} \approx \frac{1 + K_1^T[T]}{1 + K_1^A[A]} \approx 0$$

It illustrates under this circumstance, there is strong competition between TCE and co-adsorbate, therefore, the co-adsorbate could significantly reduce the reduction rate of TCE, and under extreme conditions, totally block the degradation reaction.

Case 2: The co-adsorbate prefers the non-reactive site S_2 to the reactive site S_1, i.e., $K_1^T[T] >> K_1^A[A]$ and $K_2^T[T] << K_2^A[A]$. From Eqn. (21):

$$\frac{k^{x}_{obs}}{k_{obs}} \approx \frac{1 + \dfrac{K_2^T[S_2]_t}{(1 + K_2^A[A])^2}}{1 + \dfrac{K_2^T[S_2]_t}{(1 + K_2^T[T])^2}} = \frac{(1 + K_2^A[A])^2 + K_2^T[S_2]_t}{(1 + K_2^T[T])^2 + K_2^T[S_2]_t} \times \frac{(1 + K_2^T[T])^2}{(1 + K_2^A[A])^2} \approx 1$$

Under this circumstance, the presence of a small amount of co-adsorbate does not affect the reduction rate of TCE.

The ratio, k^{x}_{obs}/k_{obs}, is therefore the key parameter for characterizing the relative activities with and without coadsorbate, and will be used for analyzing our experimental data.

Results and Discussion

Adsorption onto Metallic Iron

To understand the complex kinetic behavior and interaction among TCE, coadsorbate, and metallic iron in the system, we start by characterizing the adsorption of TCE and other coadsorbates onto the metallic iron. Accurate measurements of TCE adsorption on metallic iron is complicated by the potential reduction that may occur simultaneously. A good estimate, however, is possible by noticing that the rate of TCE adsorption is typically much faster than the rate of its reduction by the metallic iron, so the initial loss of TCE from the aqueous phase should be mainly due to sorption. As shown in Figure 1a, where the aqueous phase concentration is plotted as a function of time, $[TCE]_{aq}$ was essentially constant over time for the control without ZVI. In the sample bottles containing 5.00 g of Fe^0, $[TCE]_{aq}$ initially dropped rapidly but stabilized after approximately 30 minutes. Within a 2 hrs of experimental time period, $[TCE]_{aq}$ was decreased by about 30%. Based on literature (12) and the experimental results as presented later in this chapter, TCE degradation can be modeled by a pseudo-first order kinetics, from which we expect that less than 2% of TCE would be reduced within a 2 hrs period. The observed 30% decrease in $[TCE]_{aq}$ was, therefore, mostly due to the adsorption onto the iron surfaces. The kinetics results indicated that the sorption quasi-equilibrium was established within 1 hr for TCE. The results also indicated that approximately 0.38 mM (28%) of the TCE was adsorbed onto the iron surfaces under the experimental conditions.

Data needed to establish a TCE adsorption isotherm were collected 1 hr after the experimental setup, which is long enough to establish a TCE sorption quasi-equilibrium but short enough that no significant TCE reduction takes place. The results, as shown in Figure 1b, can be empirically described by the Langmuir adsorption isotherm:

$$[T]_{ads} = \frac{K^T[S]_t[T]_{aq}}{1 + K^T[T]_{aq}} \tag{22}$$

where K^T is the adsorption constant of TCE on surface sites, $[S]_t$ represents the total surface sites of iron, and $[T]_{aq}$ is the concentration of TCE in the aqueous phase. Using a modeling software package, Scientist for Windows (Mircomath Scientific Software, Salt Lake City, UT, USA), we obtained $[S]_T = 4.3\pm0.16\times10^{-3}$ mole/l and $K_T = 0.19\pm0.01\times10^3$ l/mole.

The adsorption isotherm of cysteine was obtained under comparable conditions and the result is plotted in Figure 2. The inset represents the data at the low concentration range. It is clear that cysteine sorbs strongly onto the

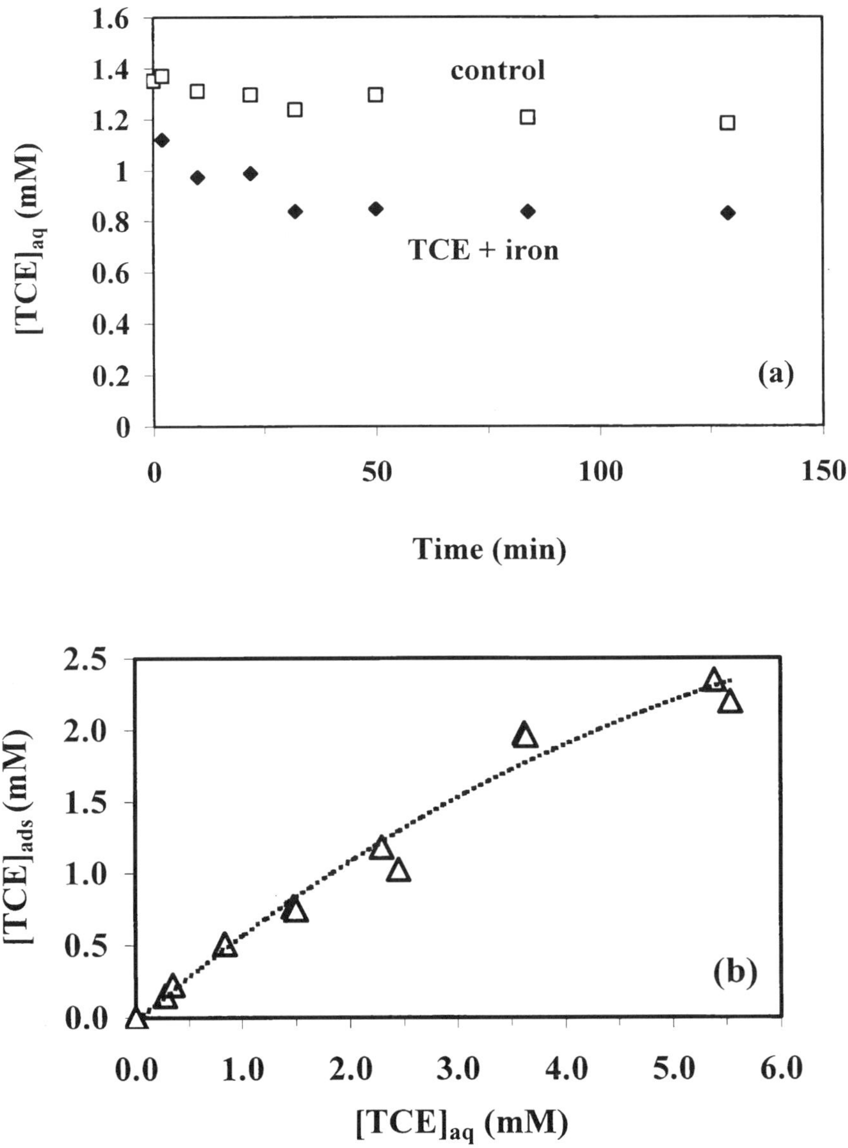

Figure 1. TCE adsorption kinetics (a) and isotherm (b) for metallic iron (69 g Fe^0 /l).

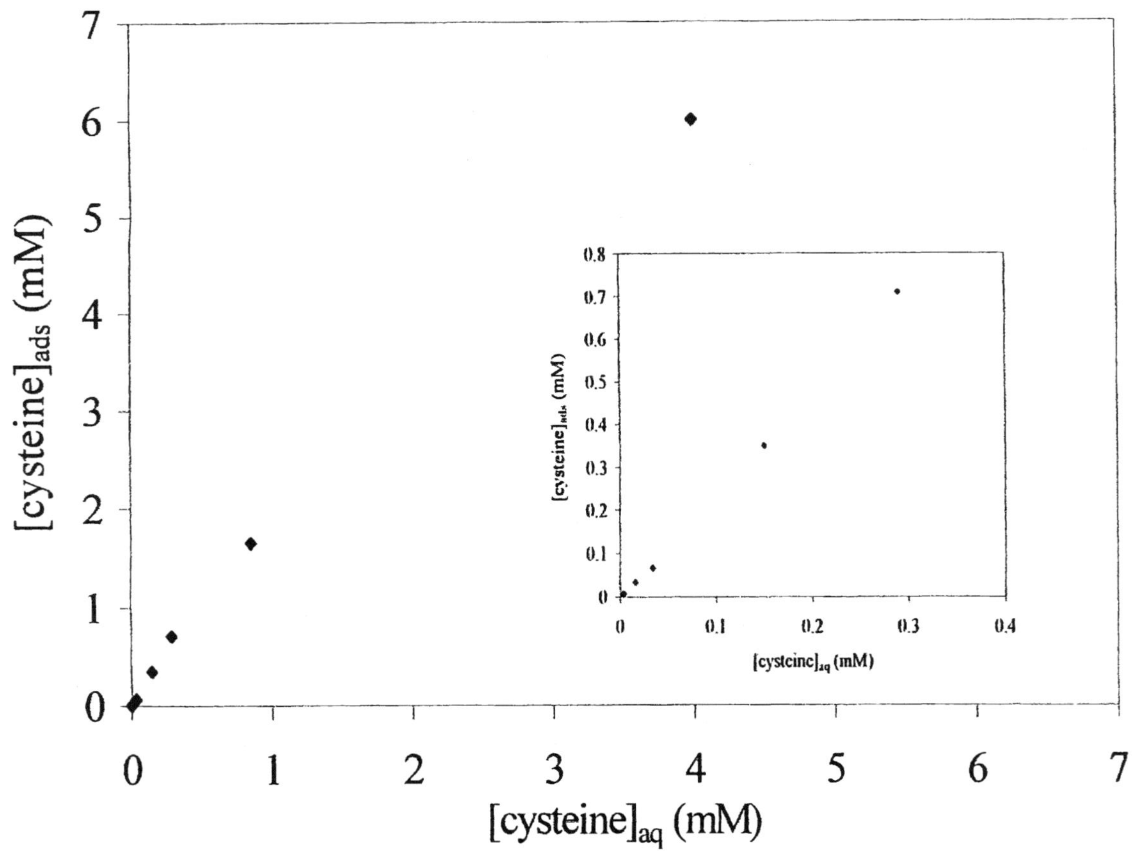

Figure 2. Sorption isotherm of cysteine onto zero valent iron.

metallic iron surface. The amount adsorbed appears to be directly proportional to the aqueous phase concentration. It is unclear whether the Langmuir adsorption model is still valid for cysteine; if it is applicable, the result suggests that the adsorption has not reached saturation with respect to the surface under the millimolar concentrations of cysteine used in this study.

The effect of cysteine and potassium sulfate on TCE adsorption was investigated by experiments under fixed TCE concentration, but each experiment used various concentrations of cysteine and potassium sulfate up to 1.00 mM. The effect of cysteine was examined in both the anaerobic chamber and the system open to air, and for sulfate, only under the anaerobic condition. The results are presented in Figure 3, with each data point representing the average value of duplicate tests and error bar representing the concentration range of the duplicates. The systems with metallic iron have lower aqueous TCE concentrations than in the control, indicating that the adsorption has occurred. For both cysteine and sulfate, however, as concentrations are increased from 0 to 1.0 mM, the measured $[TCE]_{aq}$ remains essentially the same, i.e., the effect of cysteine and potassium sulfate on the adsorption of TEC is minimal, when concentrations are lower than 1.0 mM. The results also demonstrate no significant difference in TCE adsorption between aerobic and anaerobic conditions. This suggests that although an oxide film may form on the iron surface in the presence of oxygen, the effect is not enough to change TCE adsorption behavior significantly.

Effects of Cysteine and Potassium Sulfate on TCE Degradation and Iron Corrosion

TCE concentrations in the aqueous phase ($[T]_{aq}$) were used for assessing the relative TCE degradation rates. Since analysis was performed for the headspace samples, the Henry's law and the ideal gas state equation below were used to convert the headspace TCE concentration to the aqueous TCE concentration:

$$PV = nRT$$

$$[T]_{aq} = \frac{[T]_g}{K_H}$$

where P = 1 atm; T = 298 K; K_H = 0.392 (27).

The first set of experiments with cysteine and potassium sulfate were conducted at two different concentration levels: 0.50 mM and 0.010 mM for cysteine, and 1.0 mM and 0.010 mM for potassium sulfate. The results for TCE degradation are presented in Figure 4 by the plots of ln $[TCE]_{aq}$ versus time, and concurrent H_2 production with time was shown in Figure 5. It is apparent that the rate for TCE degradation can be characterized by first order kinetics

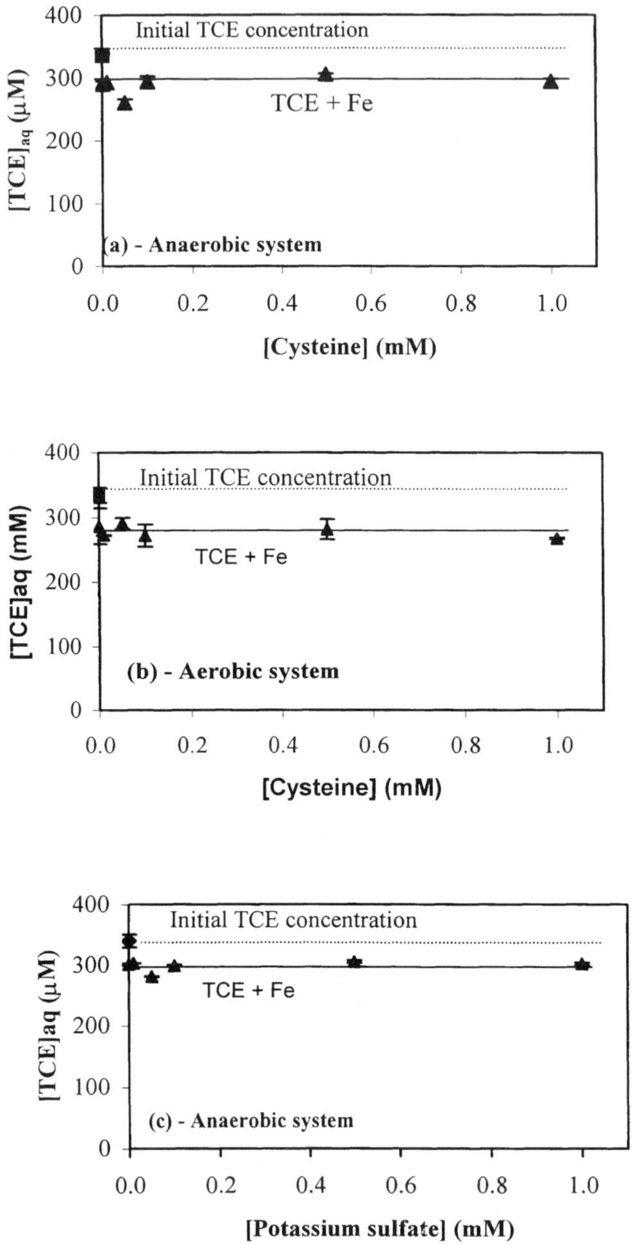

Figure 3. Effects of cysteine and potassium sulfate on the adsorption of TCE.

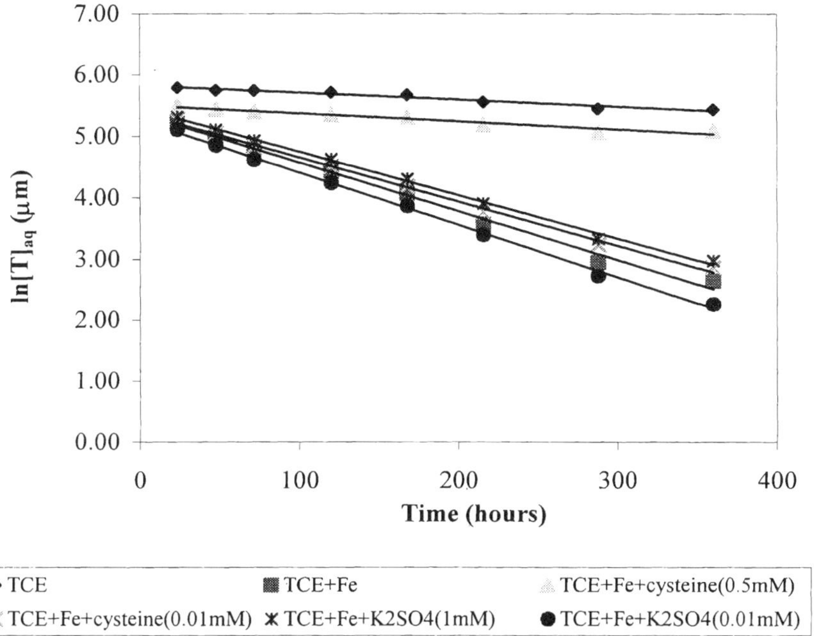

Figure 4. Effects of cysteine and potassium sulfate on TCE degradation.

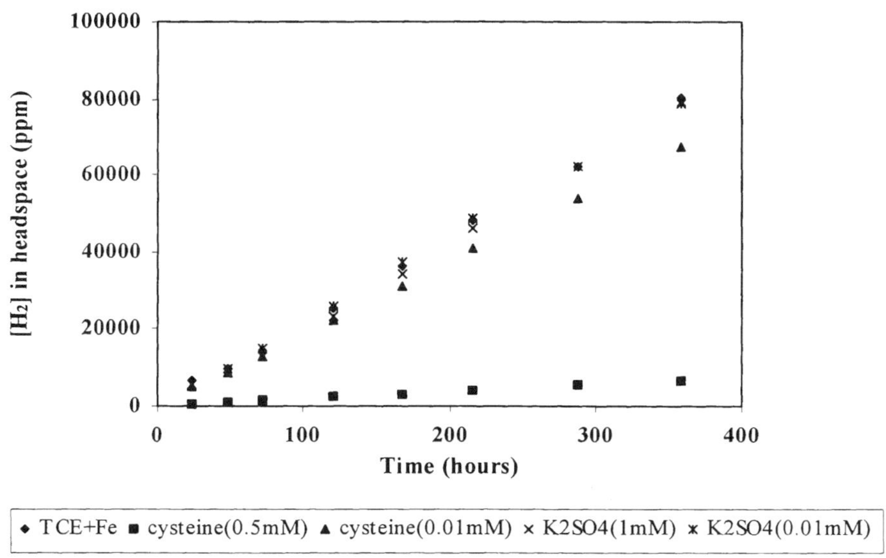

Figure 5. Effects of cysteine and potassium sulfate on H_2 production during TCE degradation.

reasonably well, based on which, the observed first-order rate constants ($^{*}k_{obs}$) were calculated as listed in Table 1. For the control system without any ZVI

Table 1. Rate constants for TCE degradation in systems containing cysteine and potassium sulfate (anaerobic condition)

System	$^{*}k_{obs}$(1/hour) ($\times10^3$)	R^2	k_{obs}(1/hour) ($\times10^3$)	k^{x}_{obs}/k_{obs}
[a]TCE	1.1 ± 0.1	0.9606	-	-
TCE	1.1 ± 0.1	0.9612	-	-
[a]TCE+Fe^0	8.2 ± 0.2	0.9926	7.1 ± 0.2	-
TCE+Fe^0	8.1 ± 0.5	0.9871	7.0 ± 0.5	-
[a]TCE+Fe^0+K_2SO_4(1 mM)	7.2 ± 0.2	0.9970	6.1 ± 0.2	0.86
[a]TCE+Fe^0+K_2SO_4(0.01 mM)	8.9 ± 0.2	0.9978	7.8 ± 0.2	1.10
[a]TCE+Fe^0+cysteine(0.5 mM)	1.3 ± 0.1	0.9447	0.2 ± 0.1	0.03
TCE+Fe^0+cysteine(0.5 mM)	1.2 ± 0.1	0.9305	0.1 ± 0.1	0.01
TCE+Fe^0+cysteine(0.25 mM)	1.5 ± 0.1	0.9659	0.4 ± 0.1	0.06
TCE+Fe^0+cysteine(0.1 mM)	2.9 ± 0.1	0.9959	1.8 ± 0.1	0.26
TCE+Fe^0+cysteine(0.05 mM)	4.7 ± 0.2	0.9973	3.6 ± 0.2	0.51
[a]TCE+Fe^0+cysteine(0.01 mM)	7.6 ± 0.2	0.9965	6.5 ± 0.2	0.92
TCE+Fe^0+cysteine(0.01 mM)	7.5 ± 0.4	0.9963	6.4 ± 0.4	0.91

[a] the first set of experiments

being present, there was still slight decrease in TCE concentration with time due to system leaking, which was reflected by a non-zero $^{*}k_{obs}$ that equaled (1.1± 0.1) $\times10^{-3}$ hr^{-1}. The real rate constants for TCE degradation, k_{obs}, therefore, were calculated after correction by subtracting the control (Table 1). The effect of cysteine and potassium sulfate on TCE degradation could then be assessed by k^{x}_{obs}/k_{obs}, the ratio of rate constants in the presence and absence of the coadsorbate (see eq. 21). In order to confirm the experimental reproducibility and obtain more detailed information on the effect of cysteine towards TCE degradation, a second set of experiments were performed using five different concentrations of cysteine (0.50, 0.250, 0.10, 0.050, and 0.010 mM). The kinetics was analyzed in a similar way and the rate constants were listed in Table 1.

The results show that at the lowest tested concentration of 0.010 mM, cysteine only had a minor effect on TCE reduction with $k^{x}_{obs}/k_{obs} = 0.92$. As cysteine concentration was increased, however, the rate of TCE degradation was significantly decreased, and when the concentration reached 0.50 mM, cysteine dramatically decreased TCE degradation with $k^x_{obs}/k_{obs} = 0.03$, i.e., the rate of TCE degradation was decreased by approximately 97%.

This trend is clearly illustrated by a plot of k_{obs} versus cysteine concentration (Figure 6). The effect was most pronounced when the concentration of cysteine was increased from 0.010 mM to 0.10 mM, and increase in cysteine concentration beyond 0.10 mM could further inhibit TCE degradation until a total shut-off of the reaction occurs. The two trend lines in the figure intercepted at a cysteine concentration around 0.13 mM. The result suggested that when the concentration was lower than 0.13 mM, free reactive sites were still available, so additional sorption of cysteine at a higher concentration could significantly decrease the rate of TCE degradation. When cysteine concentration in the system was higher than 0.13 mM, majority of the reactive sites had already been occupied, therefore, the effect of adding more cysteine was less dramatic. Assuming that the adsorption isotherm of cysteine measured in the absence of TCE (Figure 2) was also applicable for system with both cysteine and TCE, we could estimate that the amount of cysteine adsorbed was 0.093 mM, should there be 0.13 mM total cysteine in the system. In comparison to the total surface sites reflected by the maximum TCE adsorption of 4.3 mM (Figure 1), the reactive sites estimated this way account for approximately 2% of the total surface sites, that is, the reactive sites for TCE reduction are only a very small fraction of total surface sites for TCE sorption.

It is believed that the presence of reactive versus nonreactive sites arises from the heterogeneous nature of the surfaces. Burris et.al (24) found that sorption of PCE onto electrolytic iron was quite low, and the adsorption of TCE and PCE onto cast iron was much higher due to the presence of non-reactive sites. The source of this non-reactive sorption appears to be the exposed graphitic nodules or flakes embedded in the cast iron (28).

Cysteine similarly decreased the production of H_2 as shown in Figure 5. In the system without cysteine, H_2 concentration in the headspace reached approximately 80,000 ppm or 8% at 360 hrs. The concentration was slightly decreased to 6.9% with 0.010 mM cysteine and dramatically to 0.69 % with 0.50 mM cysteine. The rate of H_2 production was constant during the reaction as indicated by the linear $[H_2]$ versus t plots. Under other cysteine concentrations, similar straight lines were observed (data not shown). The slopes of these straight lines, or the reaction rate constants, were plotted as a function of cysteine concentration (Figure 7). Compared to Figure 6, it was clear that the effect of cysteine on H_2 production or iron corrosion was very similar to the effect on TCE degradation.

Experimental results on the effect of potassium sulfate were also presented in the Figures 4-5 and Table 1. Increasing ionic strength may affect TCE reduction in two ways. The hard base anions, such as Cl^-, Br^-, and SO_4^{2-} are known to be aggressive towards the passivated oxide film because they diffuse

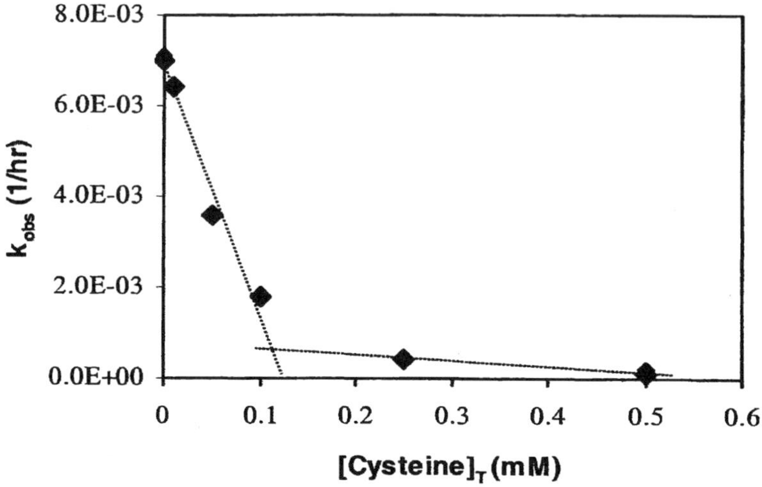

Figure 6. Rate constants for TCE degradation under various concentrations of cysteine.

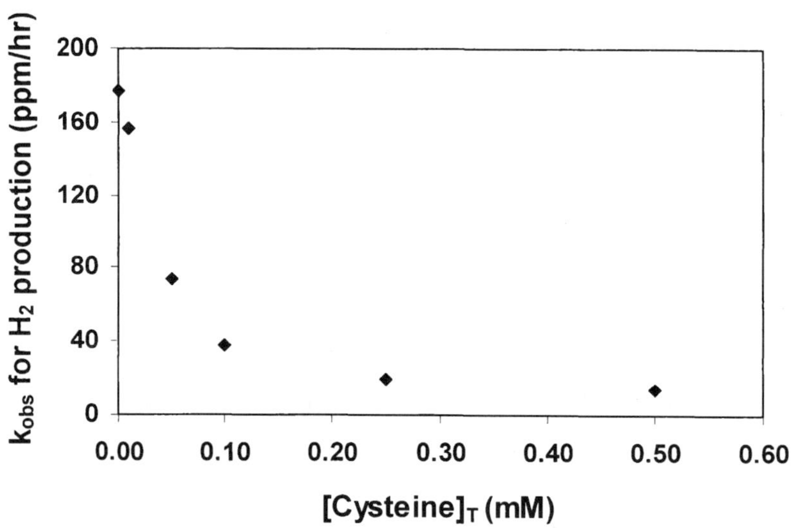

Figure 7. Rates constants for H_2 production under various concentrations of cysteine.

readily into the film and form strong complexes with iron centers (29, 30). These complexes enhance dissolution of the protective oxide film and often induce pitting, which ultimately increases the iron corrosion rate. On the other hand, at high concentrations of potassium sulfate (>0.5 mM), adsorption of electrolyte ions (K^+ and SO_4^{2-}) on the iron surfaces could occur. As a result, TCE adsorption on reactive sites may be decreased due to competition. Since the TCE degradation rate is directly proportional to TCE sorption on the reactive sites of the iron, the rate of TCE reduction would decrease if this competition for the surface sites is significant. When examining the effect of anions on iron corrosion rate, Reardon (19) found that 0.01 M SO_4^{2-} increased the iron corrosion rate after 150 hours of reaction time.

The results demonstrated that low concentration of sulfate (0.010 mM) did not affect TCE degradation and iron corrosion, but at a concentration of 1.0 mM, it slightly inhibited the TCE degradation and iron corrosion (Figures 4-5 and Table 1). This is different from the previously reported results (17, 19). The difference may be due to the different types of iron used. In this study, it appeared that iron passivation was not an important process, so no passivation layer was to be destroyed by sulfate for rate enhancement. On the contrary, at high sulfate concentration, weak competition for the surface sites did take place so as to slightly retard the reactions.

Oxygen is expected to be an important parameter for ZVI system, due to its redox sensitive nature. Thus, parallel experiments prepared in the ambient air were performed at two concentration levels to assess the effect of cysteine and potassium sulfate. The data for TCE degradation and H_2 production are shown in Figures 8 and 9, respectively. The rates for TCE degradation were analyzed by the model discussed previously, with the resulting rate constants listed in Table 2.

Table 2. Rate constants for TCE degradation in systems containing cysteine and potassium sulfate (aerobic condition)

System	k^*_{obs}(1/hour) (x10^3)	R^2	k_{obs}(1/hour) (x10^3)	k^x_{obs}/k_{obs}
TCE	0.8 ± 0.1	0.9471	-	-
TCE+Fe^0	10.9 ± 0.5	0.9876	10.1 ± 0.5	-
TCE+Fe^0+K_2SO_4(1 mM)	9.3 ± 0.4	0.9992	8.5 ± 0.4	0.84
TCE+Fe^0+K_2SO_4(0.01 mM)	10.6 ± 0.6	0.9971	9.8 ± 0.6	0.97
TCE+Fe^0+cysteine(0.5 mM)	1.7 ± 0.1	0.9848	0.9 ± 0.1	0.09
TCE+Fe^0+cysteine(0.01 mM)	9.8 ± 0.4	0.9943	9.0 ± 0.4	0.89

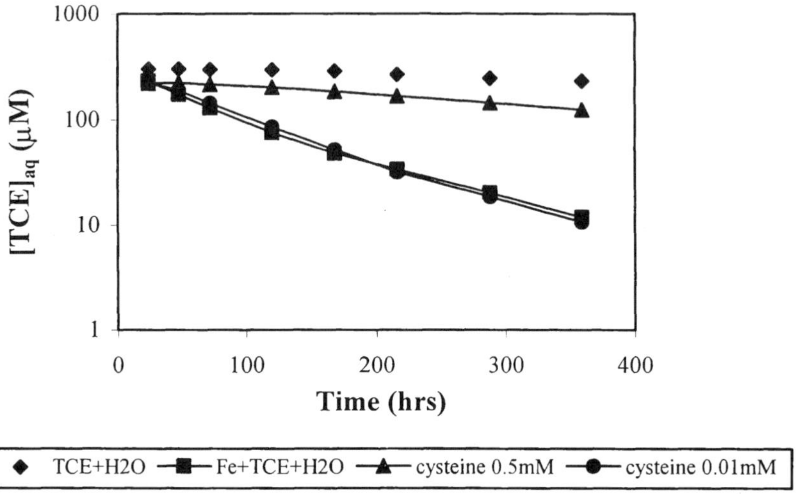

Figure 8. Effects of cysteine and potassium sulfate on TCE degradation when experiments were conducted in the ambient air.

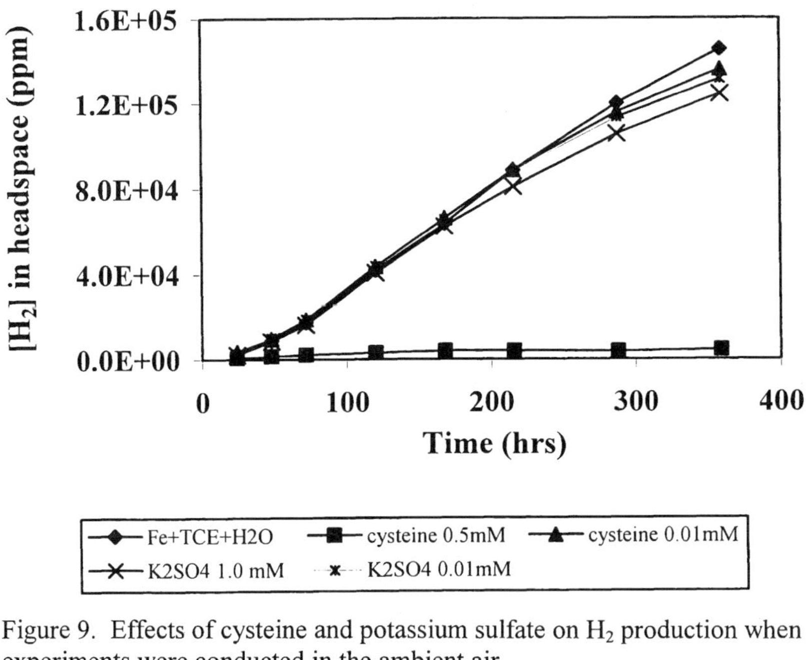

Figure 9. Effects of cysteine and potassium sulfate on H_2 production when experiments were conducted in the ambient air.

As indicated by Figures 8 and 9, as well as Table 2, the general characteristics for TCE degradation and iron corrosion in the presence of oxygen were essentially the same as those for the anaerobic systems. Cysteine decreased TCE degradation by 11% at the 0.010 mM concentration level, and in the system with 0.50 mM cysteine, the TCE degradation rate was reduced by 91%. Low concentration of sulfate (0.010 mM) didn't affect the reaction but high concentration of sulfate (1.0 mM) could decrease the TCE degradation rate by 16%. Comparing the k_{obs} in Table 1 and Table 2, it was clear that TCE degradation in the presence of oxygen was slightly faster than those under anaerobic conditions.

By comparing the results of H_2 production in Figures 5 and 9, we can see that the amounts of hydrogen produced from the aerobic systems could be 80% more than those produced from the anaerobic systems. This agrees with the observation reported in the literature (32), and the rationale is that the solubility of oxygen in water is limited and not sufficient to result in iron passivation. It is not clear whether the hydrogen produced from iron corrosion participated in the reaction of TCE dechlorination. However, the higher rate of H_2 production does proportion to the TCE degradation rate. Blocking iron corrosion by cysteine in both anaerobic and aerobic systems also blocked the dechlorination process.

Conclusions

Metallic iron surfaces exhibit two types of sites, a small fraction of reactive sites (presumably iron) for the adsorption and reduction of TCE, and a larger fraction of non-reactive sites (likely graphite inclusions in the cast iron) for the bulk of TCE sorption. We estimated that the reactive site density is approximately 2% of the total surface sites for the iron sample examined. The rate of TCE degradation appears to be correlated to the rate of iron corrosion, as indicated by H_2 production. When the corrosion is blocked by cysteine, TCE degradation is also shut off. With the higher rate of H_2 production in the aerobic system and in the presence of sulfate, the TCE degradation rate is also higher. It is likely that both H_2-production corrosion reactions and TCE degradation reactions are taking place on the same reactive sites. Another possibility is that H_2 and/or atomic hydrogen is the required precursor for TCE reduction, but this can't be proved or disproved from our batch experimental systems.

Our original goal for this research was to identify chemicals that could inhibit iron corrosion but not the reduction of chlorinated solvents. Our results demonstrate that none of the chemicals we examined posseses such properties.

Acknowledgments

Financial supports by New Mexico Waste-management Education & Research Consortium and the New Mexico Institute of Mining and Technology are greatly appreciated.

References

(1) U.S. EPA *Federal Register* **1987,** *52*, 12866-12874.
(2) Mackay, D. M.; Cherry, J. A. *Environ. Sci. & Technol.* **1989,** *23*, 630-636.
(3) National Research Council. *Alternatives for Ground Water Cleanup*; National Academy Press: Washington, D. C., 1994.
(4) Sweeny, K. H. Amer. Water Works Assoc. Res. Found. **1979**, 1487-1497.
(5) Sweeny, K. H. *Am. Inst. Chem. Eng. Ser.* **1981**; 67-71.
(6) Senzaki, T.; Kumagai, Y. *Kogyo Yosui* **1988,** *357*, 2-7.
(7) Senzaki, T.; Kumagai, Y. *Kogyo Yosui* **1989,** *369*, 19-25.
(8) Gillham, R. W.; O'Hannesin, S. F. *Ground Water* **1994,** *32*, 958-967.
(9) Focht, R.; Vogan, J.; O'Hannesin, S. *Remediation* **1996**, 81-94.
(10) US EPA, *Permeable Reactive Barrier Technologies for Contaminant Remediation;* Environmental Protection Agency: EPA/600/R-98/125, 1998.
(11) Weber, E., J *Environ. Sci. & Technol.* **1996,** *30*, 716-719.
(12) Burris, D. R.; Campbell, T. J.; Manoranjan, V. S. *Environ. Sci. & Technol.* **1995,** *29*, 2850-2855.
(13) Allen-King, R. M.; Halket, R. M.; Burris, D. R. *Environ. Toxicol. and Chem.* **1997,** *16*, 424-429.
(14) Matheson, L. J.; Tratnyek, P. G. *Environ. Sci. & Technol.* **1994,** *28*, 2045-2053.
(15) Sivavec, T. M.; Horney, D. P. *Preprints of ACS annual meeting-Division of Environmental Chemistry*, San Francisco, CA, April 13-17, 1997; pp. 115-117.
(16) Schaschl, E. In *Corrosion Inhibitors*; Nathan, C. C., Ed.; National Assoc. Corrosion Engineers: Houston, 1973, pp 28-41.
(17) Deng, B.; Hu, S.; Burris, D. R. In *Physical, Chemical, and Thermal Technologies*; Wickramanayake, G. B., Hinchee, R. R., Eds.; Battlle Press: Columbus, OH, 1998, pp 341-346.
(18) Whitman, W.; Russel, R.; Altieri, V. *Ind. Eng. Chem.* **1924,** *16*, 665-679.
(19) Reardon, E. J. *Environ. Sci. & Technol.* **1995,** *29*, 2936-2945.

(20) Garcia, M. S.; Sanchez-pedrenot, C.; Albero, M. I. *Analyst* **1990,** *115*, 989-992.
(21) Van Riemsdijk, W. H.; DeWit, J. C. M.; Koopal, L. K.; Bolt, G. H. *J. Colloid Interface Sci.* **1987,** *116*, 511-522.
(22) Schindler, P. W.; Stumm, W. In *Aquatic Surface Chemistry*; Stumm, W., Ed.; Wiley-Interscience: New York, 1987, pp 83-110.
(23) Stone, A. T.; Torrents, A.; Smolen, J.; Vasudevan, D.; Hadley, J. *Environ. Sci. & Technol.* **1993,** *27*, 895-909.
(24) Burris, D. R.; Allen-King, R. M.; Manoranjan, V. S.; Campbell, T. J.; Loraine, G. A.; Deng, B. *J. of Environ. Eng.* **1998,** *124*, 1012-1019.
(25) Deng, B.; Hu, S. In: *Physical Groundwater Remediation*; Burns, S., Smith, Eds., Kluwer Academic/Plenum Publisheres, NY, **2001**; p139-159.
(26) Johnson, T. L.; Scherer, M. M.; Tratnyek, P. G. *Environ. Sci. & Technol.* **1996,** *30*, 2634-2640.
(27) Gossett, J. M. *Environ. Sci. & Technol.* **1987,** *21*, 202-208.
(28) Deng, B.; Campbell, T. J.; Burris, D. R. *Environ. Sci. & Technol.* **1997,** *31*, 1185-1190.
(29) Bockris, J. O. M.; Khan, S. U. M. *Surface electrochemsitry, A molecular level approach*; Plenum: New York, 1993.
(30) Aramaki, K.; Tomihari, M.; Furuya, S.; Yamaguchi, M.; Nishihara, H. *Corr. Sci.* **1994,** *36*, 1133-1141.
(31) Hu, S. *Reductive Dechlorination of Trichloroethylene by Metallic Iron: Effects of Iron Corrosion and Corrosion Inhibition,* M.S. thesis, New Mexico Institute of Mining and Technology, Socorro, NM, 1999.
(32) Fontana, M. G. *Corrosion Engineering,* McGraw-Hill, New York, 1986.

Chapter 14

Preparation and Regeneration of Nickel–Iron for Reduction of Organic Contaminants

L. Gui and R. W. Gillham

Department of Earth Science, University of Waterloo, Waterloo, Ontario N2L 3G1, Canada

Abstract: Factors affecting nickel-iron (Ni-Fe) performance for TCE reduction were examined in column experiments. The Ni-Fe materials were prepared by plating 0.25 wt% nickel onto acid-washed (clean) and unwashed (oxide-covered) granular iron using both displacement and electroless methods. For a short, initial period (300 PV), acid-washed Ni-Fe degraded TCE faster than unwashed Ni-Fe. As deactivation proceeded, there was no significant difference in TCE reduction rates in acid-washed and unwashed Ni-Fe columns. Both methods plated nickel onto acid-washed iron effectively. Displacement plating, however, could not deposit nickel onto unwashed iron. Particle size was another important factor affecting Ni-Fe performance. The source of iron (e.g., Fisher vs. Connelly), however, had only limited effect. Methods for regenerating Ni-Fe were tested using *N*-nitrosodimethylamine. A partial recovery of the Ni-Fe reactivity was achieved by (1) flushing the column with water, (2) flushing the column with 0.01 N H_2SO_4, and (3) stopping the flow to the column periodically. Of the methods tested, flushing with acid proved to be the most effective in restoring the Ni-Fe activity.

Introduction

In the past decade the use of granular iron as a reactive material to treat contaminated groundwater has been demonstrated as a cost-effective technology. Laboratory and field tests have shown that granular iron can successfully degrade a wide range of contaminants, including both reducible organic and inorganic components. In particular, chlorinated solvents were reductively dechlorinated to mainly hydrocarbons, with half-lives ranging from minutes to a few hours (1, 2).

Plating a small amount of a more noble metal, such as nickel or palladium, onto the iron results in a significant increase in degradation rates (e.g., 3, 4, 5, 6). For instance, degradation rates of TCE and *N*-nitrosodimethylamine (NDMA) were increased one and two orders of magnitude, respectively (3, 5). The increase in activity could have important benefits in reducing the amount of material required for in situ applications.

The observed enhancement, however, is not consistent among different researchers since different sources of granular iron with various particle sizes were used and different plating methods were employed. In most cases, bimetals were prepared using a displacement plating method, in which the granular iron was first acid-washed to remove thick surface oxide layers / passive films. The acid-cleaned iron was then placed in a metal salt solution of the desired concentration under certain reaction conditions (e.g., time and temperature). Though successful, it is not practical and rather costly to acid-wash tons of iron for field applications. Plating onto unwashed iron surfaces by the displacement method is usually inefficient and sometimes infeasible since the reductant, Fe^0 in this case, is covered by a thick layer of passive film. A modified nickel electroless plating method, used at the University of Waterloo, avoids the acid-wash step by using hypophosphite as a reducing agent to facilitate nickel deposition onto the unwashed iron surface (4).

Furthermore, removal of passive films on the iron surface by acid-washing results in fast iron dissolution (corrosion) which produces a large quantity of hydrogen gas and mineral precipitates. Although the iron corrosion process is related to contaminant reduction, the relationship between corrosion rate and reduction rate, especially catalytic reduction rate, is not clear. Since most of the research has been conducted in the laboratory for short periods of time, the effects of gas and precipitates on long-term reactivity of the bimetal materials are not known. Limited information obtained from short-term laboratory experiments suggests some losses of reactivity of the bimetals with time. The mechanisms of deactivation of bimetals are not clear and regeneration methods for bimetals are not available.

The present study was conducted to systematically examine the factors affecting performance of nickel-iron (Ni-Fe) and to test ways of reactivating the Ni-Fe material. Specifically, the following factors that may affect Ni-Fe reactivity were examined: (1) plating method (displacement vs. electroless), (2) condition of the base metal (clean vs. oxide film covered surfaces) and (3) particle size. In addition, three possible ways of regaining reactivity of Ni-Fe were evaluated in column tests: (1) flushing the column with distilled water, (2) flushing the column with 0.01 N sulfuric acid and (3) stopping the flow to the column periodically.

Experimental

Materials

Two sources of granular iron were used in this study. One was Fisher iron (Fisher Scientific) and the other was Connelly iron (Connelly-GPM, Inc., Chicago, IL). Fisher iron contained >85% metallic iron and 3.09% carbon (7). The particle sizes were relatively uniform with an average of 40 mesh. The surface area, determined by the BET test before plating, was 2.65 m^2/g. Connelly iron, in comparison, contained 89.8% metallic iron and 2.85% carbon (Data provided by supplier, 1998). The particle sizes varied in a wide range, from −8 to +50 mesh with about 35% in the fraction of 40 mesh. The average surface area was 1.80 m^2/g. Both irons when received were coated by thick layers of iron oxides including goethite, hematite, maghemite and magnetite, as determined by Raman microspectroscopy (8). These irons were plated with 0.25 wt % nickel either as received or after acid washing.

To acid wash the iron, 6 N HCl was added to the iron in a beaker at a ratio of 1 g iron to 2.5 mL acid (6) and reactions were allowed to take place for 10 min while stirring. After repeating this step six times the iron was rinsed with distilled water six times. This cleaning procedure removed the majority of the surface oxides as confirmed by normal Ramam spectroscopy. The acid-cleaned material was plated immediately without drying.

Two plating methods were used in this study, namely displacement and electroless plating. For the displacement method, 50 mL of $NiCl_2$ solution (5.5 g/L) was added to 50 g of iron in a beaker which was placed in a water bath of 60 °C. With frequent stirring, the plating procedure lasted for 2 hours. After the supernatant was removed, the plated iron particles were rinsed with distilled water six times, using 250 mL each time. The electroless plating followed the same procedure except the plating solution consisted of 5.5 g/L $NiCl_2$, 17 g/L $NaH_2PO_2 \cdot H_2O$, 15 g/L H_3BO_3, and 5 g/L NaF. The hypophosphite is the reducing agent and NaF is the stabilizer. Details of this method are given in Odziemkowski et al. (4). Scanning Electron Microscope (SEM) and Energy Despersive X-ray (EDX) spot analysis were performed to confirm deposition of nickel onto the iron surface. Dissolved nickel concentrations in the removed supernatant and in the rinse waters of known volumes were analysed to determine plating efficiency. The resulting Ni-Fe materials prepared by both methods were packed into glass columns immediately without drying. The remaining materials were dried with pentane and methanol and were used for surface analyses (SEM, EDX and BET test).

TCE column experiments

A series of column tests was conducted to examine the factors that affect Ni-Fe performance, using TCE as a chemical indicator (Table 1). Control columns consisted of iron immersed in the electroless plating solution without $NiCl_2$ for two hours at 60 °C. The length and the diameter of the columns were 7.5 cm and 1.6 cm, respectively. A layer of Ni-Fe (3.5 cm) was sandwiched between layers (2 cm) of organic free glass beads (1 mm diameter) to prevent preferential flow in the columns. The average porosity of the columns was 0.65. TCE solution was prepared by sparging distilled water with nitrogen gas (oxygen free) for several hours and then spiking with a TCE stock solution of 500 mg/L. The solution was siphoned into a double-layered Teflon bag with N_2 gas filling the space between the bags to prevent oxygen diffusion. The initial concentration of TCE ranged between 12 and 15 mg/L and oxygen content was maintained below 0.5 mg/L. The solution was pumped through the columns from the bottom using a multichannel peristaltic pump at an average flow rate of 0.6 mL/min. To minimize oxygen diffusion, all the column fittings and tubing were stainless steel except the pump tubing.

Periodically, effluent samples were taken to determine concentrations of TCE, chlorinated breakdown products, hydrocarbons, chloride and nickel. Detailed analytical procedures for the organic compounds are given in ref. 9. Chloride concentration was determined using an ion-selective electrode. Nickel concentration was determined by Atomic Absorption Spectroscopy (AAS). The volume of gas formed in the columns was collected and measured by connecting a inverted, water-filled burette at the effluent end.

NDMA column experiment

One larger column (50 cm x 3.81 cm ID) with 7 sampling ports along the central axis of the column was used to examine means of reactivating the Ni-Fe, with NDMA as a test compound. Details of the column construction are given in Gui et al. (5). The Ni-Fe material used in this test was Connelly iron plated with 0.25% Ni by the electroless method and dried before use. The porosity of the column was 0.59. NDMA solution (5 to 28 mg/L) without deoxygenation was pumped through the column from the bottom. The flow rate varied between 0.2 and 10 mL/min. Concentrations of NDMA and its reduction product dimethylamine (DMA) were measured along the column periodically to monitor reactivity of the Ni-Fe material. During the distilled water and acid washes, effluent samples were collected for NDMA, DMA, dissolved nickel and iron and pH analyses. The analytical procedures are given in ref. 5. The organic compounds were analysed by HPLC and the dissolved metals were measured by AAS.

Table 1. Column tests to determine factors affecting Ni-Fe performance[1].

Source	*Base Iron Condition*	*Particle Size (mesh)*	*Plating Method*	*Plating Efficiency (%)*
Connelly	Acid-washed	-8 to +50	Electroless	99.4
	Unwashed	-8 to +50	Electroless	95.6
	Acid-washed	40	Electroless	99.7
	Unwashed	40	Electroless	96.1
	Acid-washed	-8 to +50	Displacement	97.1
	Acid-washed	-8 to +50	Control	NA
	Unwashed	-8 to +50	Control	NA
Fisher	Acid-washed	40	Electroless	99.7
	Unwashed	40	Electroless	96.5
	Acid-washed	40	Displacement	98.9
	Acid-washed	40	Control	NA
	Unwashed	40	Control	NA

Note: [1] Plating nickel onto unwashed iron particles using displacement method was unsuccessful and therefore was not included in column tests.

Results and Discussion

Surface analysis of the Ni-Fe materials

The surface of the dry Ni-Fe materials was examined by SEM equipped with an EDX analyzer. Images obtained by SEM showed that the surfaces of acid-washed Ni-Fe were smoother than the unwashed Ni-Fe. This was consistent with the results of the BET test which showed that the surface area of the acid-washed materials was lower than the unwashed. Quantitative EDX spot analysis of more than 10 grains of each Ni-Fe material showed that the Ni and Fe contents were higher in the acid-washed Ni-Fe mateirals than in the unwashed materials. However, the oxygen and carbon contents showed an opposite trend. The higher Ni content in the acid-washed Ni-Fe samples suggests better plating efficiency under this condition. The higher atomic ratio of Fe/O on the acid-washed surfaces indicates that the iron oxide films were partially removed by the acid washing procedure.

The acid-washed Ni-Fe plated by electroless and displacement methods contained similar amounts of Ni on the surfaces. This indicates that both plating methods were effective in depositing Ni on clean iron surface. The electroless-plated Ni-Fe also contained a small amount of phosphorous on the grain surfaces because when hypophosphite is used as the reducing agent in the electroless plating bath the base metal (iron or steel in this case) is plated as a Ni-P alloy (4, 10). EDX spot analysis also indicated that displacement plating could not deposit Ni onto unwashed iron.

Analysis of residual concentrations of Ni showed that less dissolved Ni was detected in the plating solutions and rinse waters of the acid-washed samples than the unwashed. Based on the amount of Ni remaining in the dissolved phase, the plating efficiency was estimated and the results are summarized in Table 1. In accordance with the EDX spot analyses, solution analysis also showed that deposition of Ni onto clean iron surfaces was more efficient (nearly 100%) than onto the oxides covered surfaces (96%) although the difference is not pronounced.

EDX mapping at 5000 magnification and 20 kV gun voltage revealed that Ni deposits were randomly distributed over the surface of the grain as small patches in sub-micron sizes.

Factors affecting Ni-Fe performance

TCE degradation in columns containing various Ni-Fe materials was used to examine factors affecting Ni-Fe performance (Table 1). First the effect of the base metal condition was studied in columns packed with acid-washed and unwashed Ni-Fe plated by the electroless plating method. Figure 1 summarizes the results obtained from the Connelly Ni-Fe materials. The Fisher Ni-Fe materials showed a similar trend (not shown).

For a short initial period (about 300 PV), acid-washed Ni-Fe degraded TCE at a faster rate than unwashed Ni-Fe although they both were substantially more active than the controls (Figure 1a). During this time the acid-washed Ni-Fe produced greater amounts of hydrogen gas than the unwashed Ni-Fe. The greater activity of acid-washed Ni-Fe is expected since the galvanic coupling between Ni and Fe has greater influence than the coupling between Ni and oxide-covered Fe on rates of iron corrosion and on H_2 evolution on Ni deposits. Previous Raman spectroscopic study of unwashed Ni-Fe material showed that Ni was deposited on the Fe_3O_4–oxide film of the unwashed iron (4). In addition, a small amount of Ni was detected in the initial effluent samples of the unwashed Ni-Fe columns but not in the acid-washed columns (Figure 2). The Ni concentration decreased rapidly from as high as 30 mg/L to below the detection limit (0.05 mg/L) during the initial 600 PV. This suggests that a small amount of Ni on the unwashed iron surface was incorporated into the surface oxides in its cationic form, which has no catalytic activity. This plus the differences in plating efficiency between acid-washed and unwashed iron as discussed above (Table 1) indicates that a few more deposits of catalytically active Ni° on the acid-washed iron surface also contribute to the initially faster reduction rates of TCE occurred in the acid-washed Ni-Fe columns.

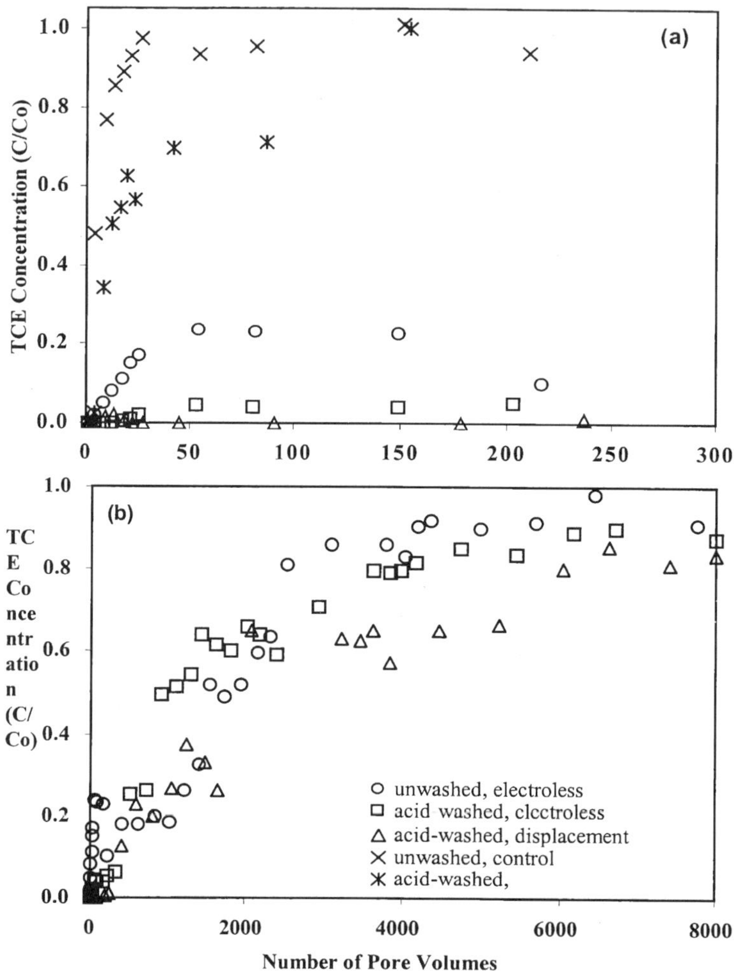

Figure 1. Normalized TCE concentrations in the effluent of columns packed with electroless-plated acid-washed and unwashed Ni-Fe, displacement-plated acid-washed Ni-Fe and acid-washed and unwashed controls during (a) initial 300 pore volumes and (b) 8000 pore volumes. Results presented here are from Connelly Ni-Fe only.

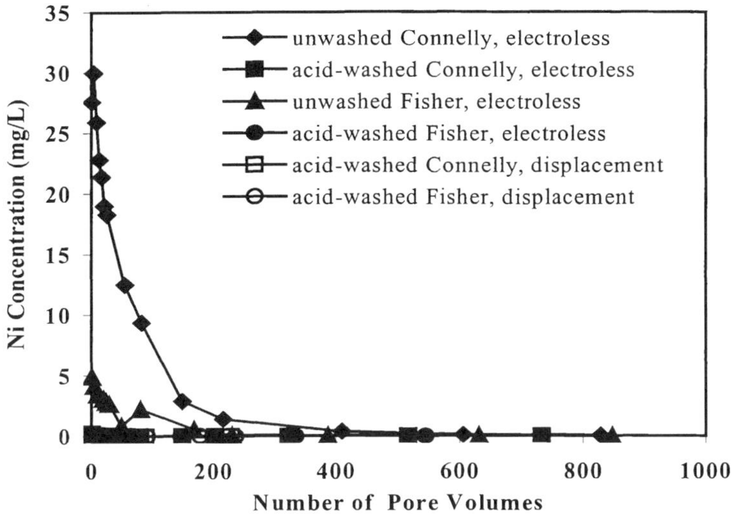

Figure 2. Dissolved Ni concentrations in the effluent samples when TCE solution passed through Ni-Fe columns.

The activity of all the Ni-Fe materials, however, decreased over time (Fig. 1b). About 90% and 50% of TCE broke through all the Connelly Ni-Fe columns (Fig. 1b) and Fisher Ni-Fe columns, respectively, after passing 6000 PV of TCE solution through. No significant difference in TCE reduction in acid-washed and unwashed Connelly and Fisher Ni-Fe columns was observed after 500 PV (Fig. 1b) and 1000 PV, respectively. The results suggest that the greater initial activity of acid-washed Ni-Fe causes faster corrosion which forms more precipitates on the Ni-Fe surfaces. Consequently, faster deactivation of the material was observed.

The effect of particle size on Ni-Fe performance was examined in columns packed with acid-washed Connelly and Fisher Ni-Fe of different particle sizes. Comparison of TCE reduction in these columns (Figure 3) indicates that particle size is also an important factor affecting Ni-Fe performance. Connelly iron in a finer particle size (40 mesh) degraded TCE faster than the coarser material. At 8000 pore volumes, about 50% of the initial TCE broke through the 40 mesh columns but 90% appeared in the effluent of the –8 to +50 mesh column. This is most likely due to the differences in surface area. Figure 3 also shows that the source of iron (Fisher vs. Connelly) had no significant effect on TCE degradation rates.

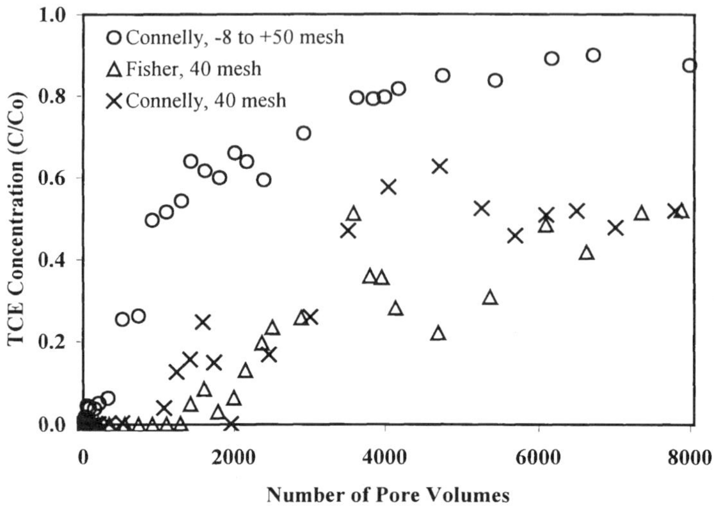

Figure 3. Normalized TCE concentrations in the effluent of columns packed with acid-washed, electroless-plated Ni-Fe using –8 to 50 mesh Connelly, 40 mesh Fisher and 40 mesh Connelly.

Analyses of breakdown products confirmed that Ni-Fe transformed TCE into innocuous products. Greater than 98% of chlorine mass balance was achieved, in which chloride alone accounted for more than 95% of total TCE degraded. Initially ethane was the predominant product that accounted for more than 90% of the TCE degraded. As the column experiment proceeded the amount of ethene increased to approximately 50% while ethane decreased to less than 50% of the TCE degraded (6000 PV). In the acid-washed and unwashed control columns, however, ethene was the predominant product that accounted for 80 to 90% of the TCE degraded. This further confirms that Ni-Fe deactivated over time. Furthermore, the initial differences in product distribution suggest that TCE reduction processes are different when using iron and Ni-Fe materials.

Regeneration of Ni-Fe materials

The above discussion demonstrated that Ni-Fe deactivation proceeds over time. Therefore three possible regeneration methods were tested. Results of the sequential experiments for regaining the reactivity of Ni-Fe are summarized in Fig. 4, which shows the time required for 5 mg/L of NDMA removal versus total number of pore volumes. Figure 4 indicates that first, recovery of Ni-Fe reactivity was achieved after each treatment. The extent of the recovery, however, decreased after each treatment with the exception of the acid wash.

Second, the rate of reactivity loss increased over time as indicated by an increasing trend in steepness between treatments. Third, of the three treatments, acid-wash was the most effective in regaining activity of the Ni-Fe.

It was observed that after shutting the column off for a few days, the brownish patches associated with Fe(III) at the bottom part of the column had disappeared. This indicates that the three-valent iron oxide on the metal surface was reduced, most likely to magnetite, through an autoreduction process (8). Magnetite is a semiconductor which can transfer electrons from the Fe core to the surface. Magnetite may also participate in TCE reduction processes.

It was found that washing the column with water and dilute acid removed adsorbed reduction product, DMA. The acid solution also dissolved some of the surface iron oxide films as indicated by high concentration of Fe (250 mg/L) in the effluent. Consequently, an increase in accessible active sites on the metal surface and thus a nearly full recovery of the lost reactivity was achieved. Acid washing, however, also resulted in dissolution of about 1% of the nickel. Therefore, repeated acid-wash should proceed with caution since it might remove a significant amount of Ni from the iron surface and cause permanent losses of reactivity.

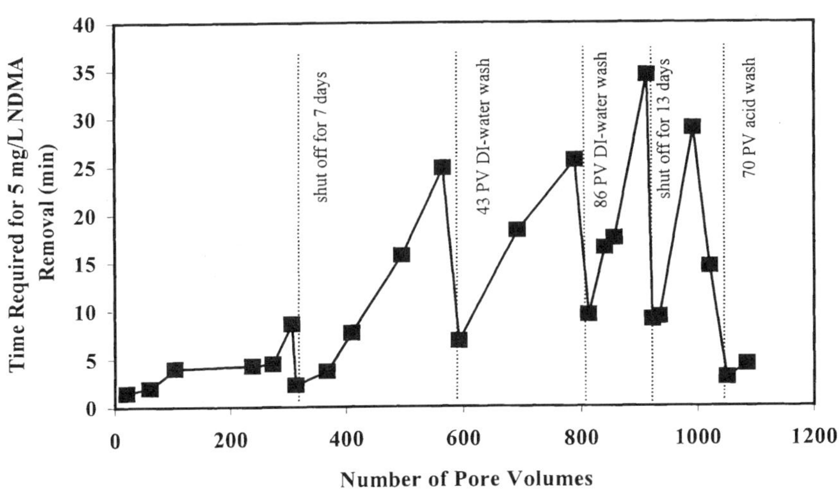

Figure 4. Recovery of lost reactivity of Ni-Fe after following treatments: (1) shutting the column off for 7 days after 308 PV and 13 days after 912 PV, (2) flushing with distilled water for 86 PV after 789 PV and (3) flushing with H_2SO_4 solution for 67 PV after 1021 PV.

Acknowledgement

This research was funded by the NSERC, Motorola, and EnviroMetal Industrial Research Chair held by Dr. R.W. Gillham.

References

1. Gillham, R.W.; O'Hannesin, S.F. Ground Water 1994, 32, 958-967.
2. Johnson, T.J.; Scherer, M.M.; Tratnyek, P.G. Environ. Sci. Technol. 1996, 30, 2634-2640.
3. Gillham, R.W.; O'Hannesin, S.F.; Odziemkowski, M.S.;Garcia-Delgado, R.A.; Focht, R.M.; Matulewicz, W.H.; Rhodes, J.E.; 1997 International Containment Technology Conference and Exhibition, St. Pertersburgh, FL., February 9-12, 1997.
4. Odziemkowski, M.S.; Gillham, R.W.; Focht, R.; in Electrochemical Society Proceedings, 1998, Vol. 98-5, 91-102.
5. Gui, L.; Gillham, R.W.; Odziemkowski, M.S.; Environ. Sci. Technol. 2000, 34, 3489-3494.
6. Muftikian, R.; Fernando, Q.; Korte, N.E.; Water Research 1995, 29, 2434-2439.
7. Deng, B.; Campbell, T.J.; Burris, D.R.; Environ. Sci. Technol. 1997, 31, 1185-1190.
8. Ritter, K.; M.Sc. Thesis, University of Waterloo, Waterloo, Ontario, Canada, 2000.
9. Orth, S.W.; Gillham, R.W.; Environ. Sci. Technol. 1996, 30, 66-71.
10. Henry, J. In Metal Finishing – 60th Guidebook and Directory Issue, Vol. 90, No.1 A, 1992; p. 353.

Chapter 15

In Situ Fluidization for Permeable Reactive Barrier Installation and Maintenance

Robert K. Niven

School of Civil Engineering, The University of New South Wales at the Australian Defence Force Academy, Northcott Drive, Canberra, ACT 2600, Australia

In this study, the application of *in situ fluidization* (ISF) for the installation of permeable reactive barriers (PRBs) - specifically zero-valent iron (ZVI) barriers - in sandy soils is investigated by laboratory experiments in columns and tanks. In this method, the soil is fluidized using a water jet, and ZVI is added to the *in situ* fluidized zone to create a PRB. Initially, efforts were focused on "hydraulic matching" of the ZVI particles and recipient sand, to disperse the ZVI vertically throughout the fluidized zone. However, experiments indicated serious difficulties with this approach. Instead, an "extraction method" was developed, in which hydraulically heavier ZVI is added as the jet is withdrawn. The result is a sequence of laminations of ZVI and original sand, extending throughout the former fluidized zone. Substantial cost savings are expected compared to existing PRB installation methods. The implications for PRB installation and maintenance are examined in detail.

Introduction

Since the 1980s, there has been considerable interest worldwide in the use of *permeable reactive barriers* (PRBs) - permeable zones containing reactive or catalytic solids or other fixed reagents - for the *in situ* treatment of contaminated groundwater. Such barriers are normally installed some distance downgradient from the contaminant source zone, enabling degradation or fixation of dissolved contaminants and protection of the downgradient groundwater quality. Initially, PRBs were envisaged as fixed installations, consisting of vertical trenches filled with solid active material. Over time, the concept has evolved to include discrete or discontinuous zones of any geometry, containing some kind of solid-bound or long-life reagent installed by any means (*1*). The main group of contaminants currently being addressed by PRBs are chlorinated solvents, many of which have been found to be catalytically degraded by particles of zero-valent iron (ZVI) (*1,2*). A wide variety of other applications of PRBs have been developed, including lime or organic-filled barriers for the treatment of heavy metals and acid mine drainage, phosphate barriers for various metals, and oxidizing or activated carbon barriers for volatile organic compounds (*1*).

At present, most PRBs are constructed by standard engineering techniques, generally involving isolation of the treatment zone (by sheet piling or other means), excavation of natural soils, backfilling with the reactive material or soil-reagent mixture, and reconnection to the aquifer (*3*). This requires heavy engineering equipment, and thus can only be conducted at high cost. The ZVI used for chlorinated solvent barriers is also quite expensive, and in thick barriers, is applied inefficiently. In many applications, to maintain stability of the excavation, the ZVI must be installed with a biodegradable polymer, which is later broken down using an enzyme (*4*). In the 1990s, efforts were directed to reduce installation costs by the development of "funnel-and-gate" systems, in which relatively impermeable zones (the "gate") were installed to direct groundwater flow through narrower PRB zones (the "funnel") (*5*). Cost savings are achieved by the standard engineering methods used for construction of the impermeable gates, as well as by savings in reagent costs. More recently, a variety of new installation procedures and injection methods have been developed, focussing on the PRB as a loosely-defined and often discontinuous treatment zone (*6,7*). An important driving force behind these methods is the desire to avoid excavation - and its associated high costs and handling difficulties - altogether.

Recently, a new method, *in situ fluidization* (ISF), has been developed for the remediation of subsurface non-aqueous phase liquid (NAPL) and/or metal contamination (*8-13*). The method is applicable principally to sandy soils. In this method, illustrated schematically in Figure 1, a water or water / air jet is lowered from the surface into a sandy soil, producing a sharply defined *in situ*

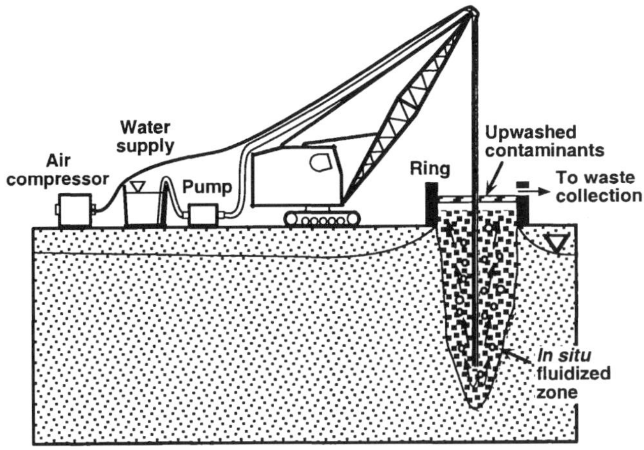

Figure 1. Schematic diagram of ISF for contaminant remediation and/or PRB installation (Reproduced with permission from reference 10. Copyright 1998 National Research Council, Canada.)

fluidized zone (liquified soil) surrounding the jet. Fluidization of the soil - as distinct from *in situ* flushing - has the advantage that it destroys the soil skeleton, releasing both NAPL contaminants and those present on fine particles in the soil (such as heavy metals), which would otherwise remain trapped within the soil. The released contaminants are carried to the ground surface, where they can be recovered for treatment. Laboratory experiments on diesel-contaminated soils indicate that significant reductions in diesel levels (96-99.9%) may be achieved, for a wide range of initial diesel concentrations (10,000 to 150,000 mg/kg), and for fines contents of at least 10% (*8,10-12*). Significant reductions in lead (up to 88%) from soils containing 1500-2000 mg/kg of precipitated lead have also been achieved (*8,11*).

The hydraulics of ISF have been examined in detail by tank experiments and pilot trials in the field (*9,11,13*). ISF can be applied to soils within the sand size range, and occurs quite rapidly, with development of the "asymptotic" *in situ* fluidized zone geometry within minutes. Systematic experiments on sandy soils and derived correlations indicate that a 50 mm (2") ID jet operated at 4.4 L/s (70 gpm) can achieve a fluidized depth of 15 m (45 feet), greater than normal remediation depths (*9,10*). At higher flow rates, greater depths are possible. The method also penetrates laminations of peat or clay, washing such material to the surface (*13*). Finally, ISF has the advantage that small-scale pump and jetting equipment can be used, reducing site mobilization requirements (*10,13*).

The aim of the present study is to examine the possible application of the ISF method for the installation of PRBs, with specific attention to ZVI barriers. In this approach, the solid reagent (such as ZVI) is added to the *in situ* fluidized zone, producing a body of reagent-enriched sand. The desired mass loading of solid is installed by appropriate control of the fluidization and reagent addition processes. Whilst some excess soil will be generated by the process, wholesale excavation - and associated handling difficulties - is avoided. For this reason, PRB installation by ISF is expected to produce substantial cost savings over existing installation methods. The present study follows two preliminary investigations (*14-15*), which are further developed and expanded in detail.

Background Knowledge

Fluidization

Fluidization is the suspension of a granular solid in an upwardly flowing fluid (*16*). Since the drag force applied to each solid particle must overcome its buoyant weight, fluidization does not occur until the upward velocity exceeds the so-called *minimum fluidization velocity*, given by Wen & Yu (*17*):

$$U_{mf} = \frac{2\left\{-150\mu(1-\varepsilon_{mf}) + \sqrt{[150\mu(1-\varepsilon_{mf})]^2 + 7\rho_f \varepsilon_{mf}{}^3 \varphi_p{}^3 d_p{}^3 (\rho_s - \rho_f) g}\right\}}{7\,\rho_f \varphi_p d_p}$$

where μ = fluid dynamic viscosity, ε_{mf} = porosity at minimum fluidization, ρ_f = fluid density; ρ_s = solid density, φ_p = particle shape factor (sphericity); d_p = particle diameter and g = acceleration due to gravity. This velocity is a superficial velocity (*i.e.* averaged over the cross-section of the bed). The above equation is normally simplified using the experimentally determined relations $(1-\varepsilon_{mf})/\varphi_p{}^2\varepsilon_{mf}{}^3 \approx 11$ and $1/\varphi_p\varepsilon_{cmf}{}^3 \approx 14$ at minimum fluidization (*17*):

$$U_{mf} = \frac{1}{\rho_f d_p}\left\{-33.7\mu + \sqrt{[33.7\mu]^2 + 0.0408\rho_f d_p{}^3 (\rho_s - \rho_f) g}\right\}$$

For quartz sands below approximately 0.6 mm diameter, a simpler equation (*16,18*) may be used:

$$U_{mf} \simeq \frac{d_p^{\ 2} g(\rho_s - \rho_f)}{150\ \mu} \cdot \frac{\varphi_p^{\ 2} \varepsilon_{mf}^{\ 3}}{(1-\varepsilon_{mf})} \approx \frac{d_p^{\ 2} g(\rho_s - \rho_f)}{1650\ \mu}$$

There is some debate about the constant 150 in the above three equations. A value of 180 or 200 is preferred by some workers (*16,18-19*).

Two phenomena take place during fluidization, of interest here. The first, *elutriation*, is the washing out of fine particles from a fluidized bed, which occurs when the applied velocity exceeds the terminal velocity (within the fluidized bed) of the fines. In an infinite fluid, the terminal velocity, U_t, at low particle Reynolds numbers ($\mathrm{Re}_t = d_p U_t \rho_f / \mu < 0.01$) is given by Stokes' law:

$$U_t = \frac{d_p^{\ 2} g(\rho_s - \rho_f)}{18\ \mu}$$

At higher Reynolds numbers, the terminal velocity deviates from Stokes law, the effect of which may be determined using correlations based on the drag coefficient (*20*). Within a fluidized bed, the terminal velocity is further affected by the presence of other particles. A procedure to determine terminal velocities within a fluidized bed, involving correction of the infinite-fluid terminal velocity, has been developed (*21*). The effect of particle shape is allowed for using two variables: the sphericity and a "dynamic shape factor" (DSF) (see *21*).

The second phenomenon, *segregation*, is the vertical grading of a fluidized bed by size and/or density. As a general rule, finer and/or less dense particles tend to segregate towards the top of the bed, whilst coarser and/or denser particles normally segregate to the base. Some exceptions to this rule have been observed in variable-density systems, due to variations in bulk density (*22*). As with elutriation, segregation is the result of differences between the in-bed terminal velocities of individual particles. Experiments on binary mixtures of constant density particles indicate partial segregation occurs at a particle size ratio of about 1.3, with complete segregation above 1.6 (*17,23-24*). Whilst variable-density segregation is more complex, in this study it is assumed that the above ratios can be extrapolated to variable density mixtures, in accordance with buoyancy considerations.

In Situ Fluidization

In situ fluidization (ISF) is defined herein as self-bounded fluidization within a granular solid using a jet, whether conducted in the field or in a three-dimensional tank setting. To create the *in situ* fluidized zone, the jet must be operated and lowered from the soil surface. If inserted within a borehole and pressurized at depth, hydraulic fracturing will result. ISF is possible within sandy soils in both the saturated and vadose zones (*10*). The present study, however, will only concern applications in the saturated zone.

Tank experiments (*9,11*) indicate that as the jet is lowered from the surface, an initial "stable" *in situ* fluidized zone is produced, of open ellipsoidal form. This is represented in Figure 2. With increasing jet depth, *H*, the profile remains

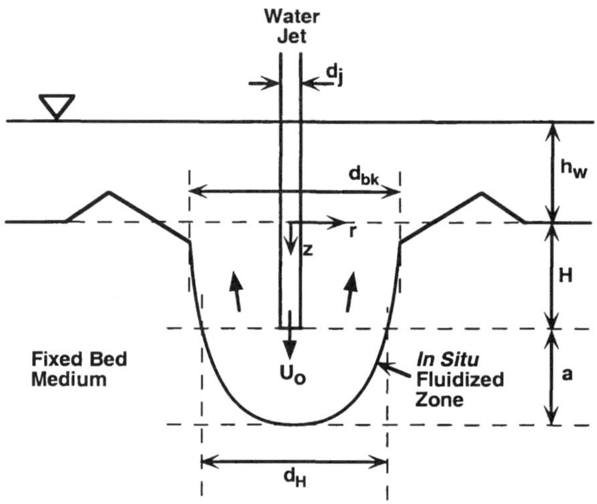

Figure 2. In situ fluidized zone geometry. (Adapted with permission from reference 9. Copyright 1998 International Association for Hydraulic Research.)

unchanged, but at some critical jet depth, H_c, it undergoes a transition to an asymmetric, spouted zone, and thence to a submerged fluidized cavity (*9*). Eventually a point of refusal to fluidization is reached. The stable fluidized zone geometry is reproducible, and not subject to hysteresis. The jet penetration distance, *a*, for $H<H_c$ may be correlated in terms of the densimetric particle Froude number = $\mathrm{Fr}_p = U_o \big/ \sqrt{d_p(\rho_s - \rho_f)g/\rho_f}$ as (*9,11*):

$$a/d_j = 2.94\ \mathrm{Fr_p}^{0.995}\ (d_p/d_j)^{0.493}$$

based on 214 experimental points, where U_o = jet velocity and d_j = internal jet diameter. The critical depth, H_c, is a function of U_{mf} (*9*):

$$\begin{cases} H_c / d_j = 0.098 \left(U_o / U_{mf}\right)^{0.67} \left(d_p / d_j\right)^{-0.49} & \text{, for } d_p / d_j < 0.08 \\ H_c / d_j \leq 0.22 \left(U_o / U_{mf}\right)^{0.89} & \text{, for } d_p / d_j \geq 0.08 \end{cases}$$

The volume of the fluidized zone can be calculated by integration of the ellipsoidal shape, $r(z)$, about the z axis, as approximated by:

$$r(z) \approx \tfrac{1}{2}\, d_{bk} \sqrt{1 - z^2/(H+a)^2}$$

$$\therefore \qquad V_{zone} = \int_0^L \pi\, r(z)^2 dr \approx \frac{\pi}{6} d_{bk}^{\ 2} (H+a) \qquad (1)$$

where d_{bk} is the diameter of the fluidized zone at its break in slope (Figure 2). More accurate calculations are obtained by fitting a variable-depth ellipse (*9*).

The Hydraulic Matching Method

Rationale

To employ ISF for PRB installation, there must be some mechanism to enable dispersal of the solid reagent throughout the fluidized, recipient sand. Initial efforts were therefore directed towards *hydraulic matching* of the ZVI and sand particle sizes, so that the terminal velocity of the ZVI, during fluidization, would match that of the supporting sand. This is represented in Figure 3a. The ZVI would then behave in a similar fashion to the sand, and be dispersed throughout it during ISF. Experiments were conducted to test this approach.

Experimental Method

The experiment was conducted using uniform processed "Dry Anna Bay" quartz sand (d_{50} ~ 230 μm), from ACI Industrial Minerals of Newcastle, NSW. Based on the fluidized-bed terminal velocity (*20,21*), this sand was found to be hydraulically equivalent to ZVI particles of diameter 100 to 113 μm, depending on the porosity. (The parameter values used were, for sand: ρ_s = 2650 kg m^{-3}, φ_p = 0.8 and DSF = 0.6; for iron: ρ_s = 7870 kg m^{-3}, φ_p = 0.6 and DSF = 0.5.) Experiments were conducted using various sizes of ZVI, selected from a range extending from below to above this diameter. For cost reasons, it was decided to

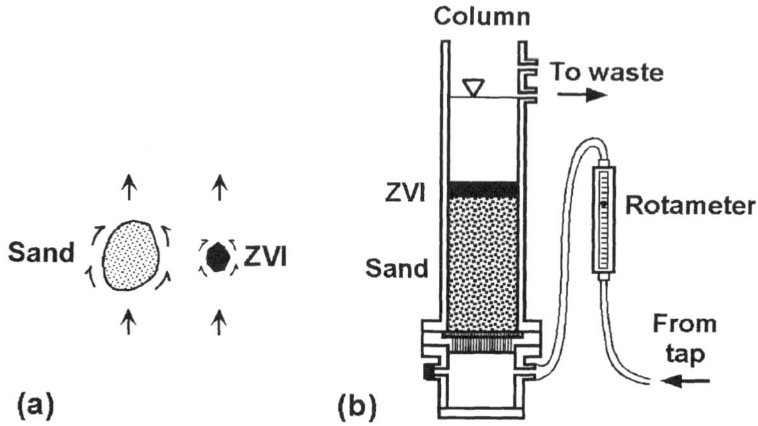

Figure 3. (a) Conceptual diagram of the hydraulic matching method, and (b) column experimental apparatus (Figure 3a adapted with permission from reference 15. Copyright 2001.)

use standard ZVI rather than the catalytic form normally used in PRBs, as the same fluidization principles apply to both (so long as allowance is made for the particle shape). Iron filings, of 99.9% purity and d_{50} ~ 500 μm, were obtained from Chem-Supply Pty Ltd of Gillman, SA. The iron was ground in a steel Humboldt Wedag ring mill to produce six 0.2 to 1 kg batches (Batches 1 to 6) for experimental use, spanning particle sizes d_{50} ~ 85 to 310 μm. Sieve analysis results for the original and ground ZVI batches and the Anna Bay sand are included in Figure 4. A thin dotted line at 110 μm is also shown, indicating the desired ZVI diameter. Calculated d_{50} values are also listed in the figure.

Experiments were conducted in which sand and ZVI, at various mass loadings, were fluidized together in a 100 mm ID perspex column, as represented in Figure 3b. The column consisted of a cylindrical tube with a detachable base, in which the soil is supported on a 200 μm stainless steel mesh overlying a perforated perspex plate. Overflow was permitted through several openings near the top of the column. Potable water was used, for which flow rates were measured by a rotameter. The experimental procedure involved insertion of a known mass of sand (1.3 to 1.6 kg) to the column, which was saturated slowly and then fluidized for 5 to 10 minutes at a flow rate of 2.8 L/min, to wash out the fines from the soil (approx. 1% by mass). The flow was then shut off and the water above the soil drained, producing a sand of porosity 0.51 to 0.54. A known mass (50 to 200 g) of ground ZVI was then delivered to the top of the sand using a long-neck funnel. The sand-ZVI system was then refluidized at 1.6

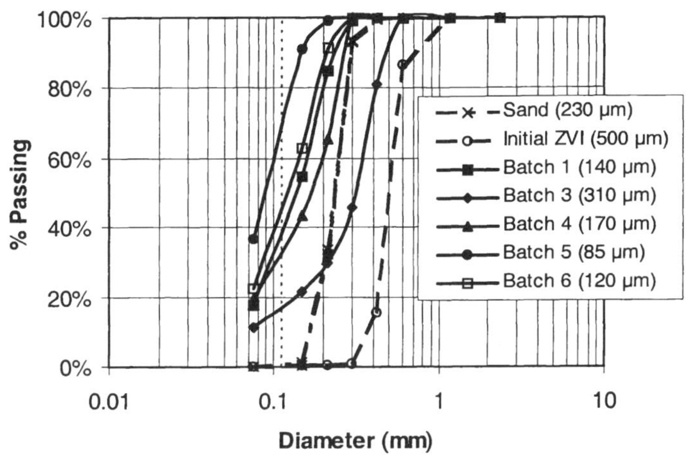

Figure 4: Sieve analysis results for the sand and ZVI batches (the median diameter, d_{50}, for each material is shown in brackets).

to 2.8 L/min., and the resulting effects observed. The experiment was conducted until the turbidity of the effluent was substantially reduced, after which the flow was shut off and the column drained.

To determine the vertical distribution of ZVI in each experiment, the soil was extruded from the column and sliced into separate vertical intervals, which were separately sampled. These were oven-dried at 105°C for 24 hours, split into halves or quarters (depending on quantity) using a riffle box, bagged, and analysed for iron. The analyses were conducted by Ecowise Environmental Pty Ltd of Fyshwick, ACT, using acid digestion and atomic absorption spectrometry. The results were reported as % Fe by mass. For QA/QC purposes, one in ten samples analyzed was a blind duplicate; the results were comparable to their equivalent samples.

To examine the effect of ZVI addition during ISF, a separate experiment was conducted involving jet fluidization of Anna Bay sand within a glass tank (0.75 m long x 0.34 m wide x 0.46 m deep). The experimental set-up was similar to that used in other ISF experiments (*9*). An 11 mm ID half-jet was used, consisting of a vertically sliced brass pipe, sealed along its cut vertical edge. This enabled fluidization against the wall of the tank, allowing observation of the ISF process. An experiment was conducted in which this jet, operated at a velocity of up to 1.75 m s^{-1} (flow rate 5.0 L/min), was lowered 90 mm into the sand to produce a fluidized zone. The fluidized zone profiles were traced on an overhead transparency taped to the tank. A sample of Batch 4 (50 g) was then added to the fluidized zone using a funnel, and the resulting effects observed.

Experimental Observations, Results and Discussion

As expected, fluidization of the sand-ZVI mixture in the column experiments produced a rapid impulse of extremely dark turbidity, due to elutriation of the finer component of each batch of ZVI used. Dramatic mixing of the sand and ZVI in the bed was also observed, which in most experiments settled to produce a significant layer of ZVI (5 to 10 mm thick) at the base. A fair amount of ZVI was observed in motion throughout the bed. After cessation of the experiment and water drainage, a thin (<1 mm) layer of ZVI was left on the bed surface due to filtration of the water. This is an artifact of the experimental procedure, and does not reflect the fluidized ZVI distribution.

The resulting vertical distributions of iron in each column experiment, along with their corresponding mass loadings, are illustrated in Figure 5. Ignoring the Fe results at the top of the column, it is apparent that, as expected, addition of Batch 3 (d_{50} ~ 310 μm) performed poorly, with only 0.3 to 0.4% by mass Fe distributed throughout the column, and 69% Fe at the base. Evidently, this batch was predominantly segregated to the base. In contrast, addition of Batch 4 (170 μm) or 5 (85 μm) performed better, with 1% Fe loading throughout most of

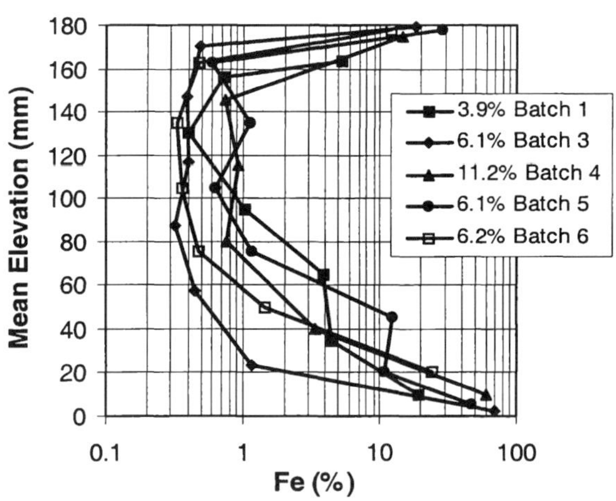

Figure 5. Vertical iron distributions obtained in the column experiments (the initial mass loading of ZVI is shown in brackets)

the column. Batch 1 (140 μm) exhibited a stepped grading, with higher Fe (≥4%) throughout the bottom third of the column. Curiously, Batch 6 (120 μm), which might have been expected to perform well, exhibited a similar distribution to Batch 3.

In the tank experiment, addition of ZVI to the fluidized zone produced an initially opaque turbidity. A fair proportion of ZVI sank through the fluidized zone, and formed a layer of ZVI a short distance above the maximum depth of fluidization, preventing fluidization beneath it. This can be referred to as an "armored" bed (*i.e.*, a layer of heavier material which protects that beneath it from erosion). After about five minutes, scour fluctuations had relocated this armored horizon downwards, almost to the former maximum extent of fluidization. On cessation of flow, some ZVI was carried upwards into the bed during withdrawal of the jet. A photograph taken after withdrawal of the jet is given in Figure 6. The final fluidized profile was measured to give $H+a$ = 197 mm and d_{bk} = 155 mm, from which the fluidized zone volume, calculated as half the volume of eq. (1), is approximately 1.24 L. Assuming a porosity of 50% and insignificant loss of ZVI due to elutriation, the mean loading of ZVI was 3.1% by mass and 1.0% by volume. This was however predominantly present as a segregated layer.

Thus whilst some results of the column and tank experiments were encouraging, none of the experiments produced a useful vertical distribution of ZVI throughout the fluidized zone. It is possible that better results could have been obtained with more uniform batches of ZVI. Whatever the cause, it is clear that hydraulic matching of the sand and ZVI diameters is difficult. In the field, this difficulty will only be exacerbated by natural variations (heterogeneities) in the soil particle size. In consequence, hydraulic matching during ISF must be ruled out as a method for the installation of PRBs in the field.

The Extraction Method

Owing to the lack of success of the hydraulic matching method, an alternative approach was sought to enable PRB construction using ISF. It was decided to exploit the phenomenon of segregation. By adding hydraulically heavier ZVI particles, one can reliably produce a segregated layer at the base of the fluidized zone; thus by adding successive increments of ZVI, with concurrent withdrawal of the jet, one should be able to install a sequence of segregated ZVI layers interbedded with the sand. This could function as a PRB. An experiment was conducted to examine this method.

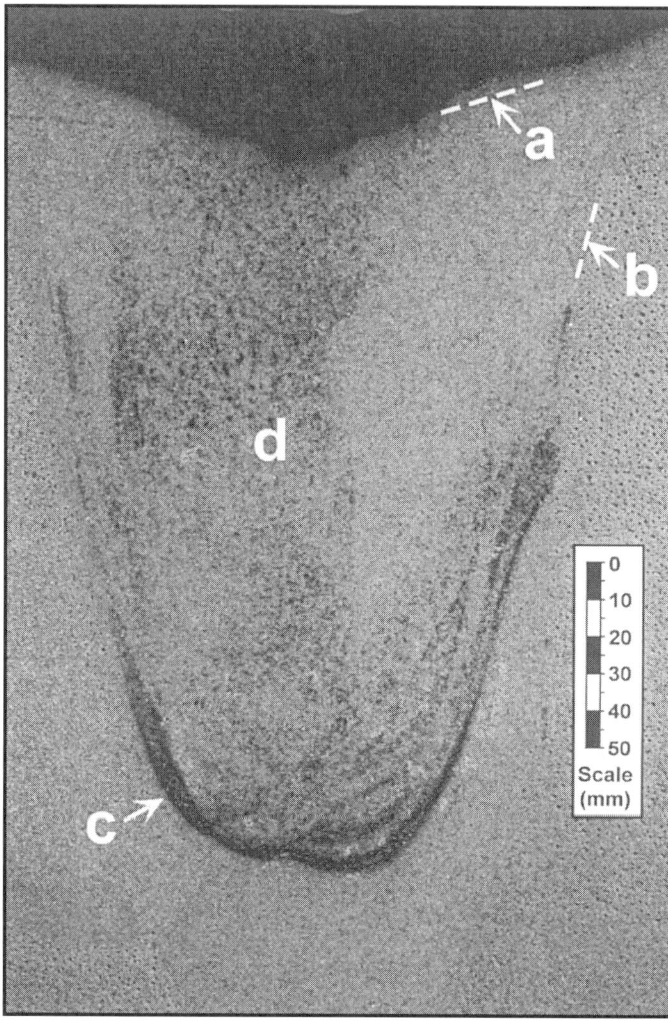

Figure 6. Photograph of ISF "hydraulic matching" experiment in tank, after completion, showing (a) surface of former fluidized zone; (b) former fluidization boundary; (c) segregated ZVI layer of ZVI at base of zone, and (d) scattered ZVI within the fluidized zone.

Experimental Method

An experiment was conducted in an unfluidized part of the same tank used previously, using the equipment and procedure described earlier. In this case the half-jet was lowered to 150 mm depth in the sand, at a jet velocity of 2.10 m s^{-1} (flow rate 6.0 L/min), producing a 300 mm deep *in situ* fluidized zone. A 50 g sample of Batch 4 ZVI was delivered to the fluidized zone using a funnel, following which the jet was withdrawn in increments of 30 mm. At each increment, an additional 50 g ZVI was added. In total, 600 g of ZVI was added in 12 increments over a period of 22.5 minutes. Observations were recorded of the fluidization process.

Experimental Observations and Discussion

As in the previous tank experiment, each addition of ZVI produced a highly turbid effluent, and a segregated ZVI layer, <1 to 5 mm thick, at or near the base of the fluidized zone. As shown in the photograph in Figure 7, the result was a sequence of continuous to near-continuous layers of ZVI and sand. A fair amount of ZVI was also scattered through the recipient sand, illustrated in the inset Figure 7d. The elliptical profile of the ZVI layers is a feature of the ISF process. Measurements of the overall zone indicated $H+a$ = 262 mm and d_{bk} = 162 mm, from which the total fluidized volume is approximately 1.8 L. Again assuming an insignificant loss of ZVI as fines, and a porosity of 50%, the mean loading of ZVI was 25.0% by mass and 8.5% by volume.

From this experiment, it is evident that the ISF extraction method has considerable potential for the *in situ* construction of PRBs, enabling ZVI loadings comparable to those required in field applications. Additional ZVI loadings, even up to 100%, could be achieved by increasing the amount of ZVI added at each stage. By applying the same fluidization principles, the method is equally applicable to the installation of PRBs containing other solid reagents. As ISF (as applied here) requires relatively simple equipment, and avoids the need for excavation, substantial cost savings should be possible compared to existing PRB construction methods.

Other Implications

It is worth commenting on other implications of the ISF process. One possibility currently being examined is the use of ISF for barrier maintenance. Many existing PRBs are reaching the end of their design life, or in some instances, were installed in a sub-optimum setting (*1*). The maintenance

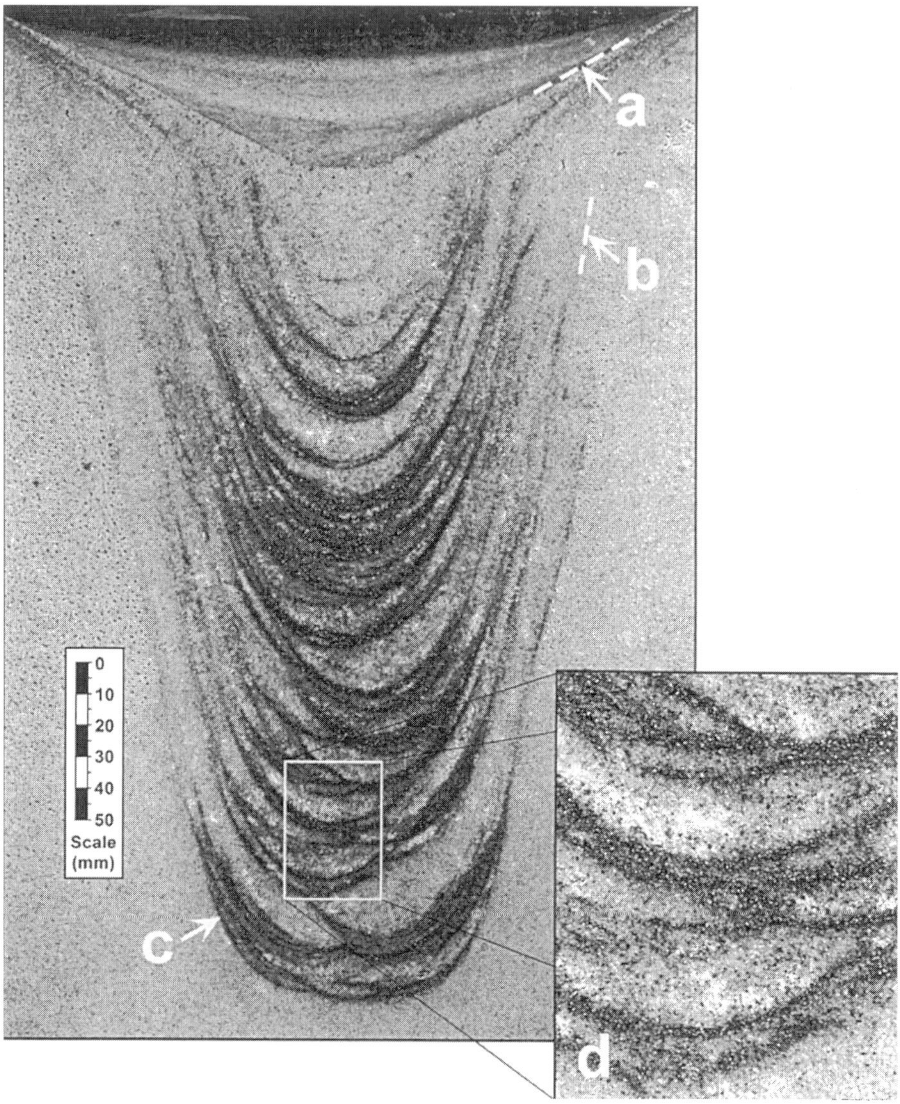

Figure 7. Photograph of ISF "extraction method" experiment in tank, after completion, showing (a) surface of former fluidized zone; (b) one of several former fluidization boundaries; (c) segregated ZVI layers, and (d) inset showing detail of ZVI layering. (Adapted with permission from reference 15. Copyright 2001.)

requirements for these barriers may include physical reconfiguration of the PRB, relocation or removal of the spent reagent, and/or chemical or microbiological modification or reactivation of the reagent surface. For the first two physical processes, in particular, ISF may allow ongoing barrier maintenance without significant capital cost. ISF will also be suitable for those chemical or biological processes best conducted in a fluidized bed, such as those which require particle size separations or multiphase mass transfer.

To extract spent solid reagent, it will in most cases be necessary to remobilize solid particles which are hydraulically heavier than the surrounding sand. Two scenarios are possible: (*i*) relocation of the material downwards, by segregation, and (*ii*) extraction of the solid by some means other than simple fluidization. In the first scenario, ISF could be conducted to *undermine* a PRB, causing the ZVI to collapse downwards, effectively enabling on-site disposal of the material at depth. Alternatively, the segregated ZVI could be recovered as a slurry. Here the method may involve jet pumping or related technologies (*25*).

In the second scenario, two possibilities suggest themselves: (*i*) magnetic separation and (*ii*) *in situ* froth flotation. To test the first option, a column experiment was conducted in which an electromagnet, consisting of a 120 mm long, 15 mm diameter steel tube, wrapped with four windings of thin latex-coated copper wire, was lowered into a fluidized bed containing 9.0% by mass Batch 4 ZVI in 2.2 kg of Anna Bay sand. The electromagnet was operated at 3 amps and 5 volts DC. The procedure was otherwise identical to that described previously. By operating the magnet within the fluidized bed and raising it above the bed, a significant loading of ZVI was observed around each pole of the magnet. A photograph is included in Figure 8. It was found that a significant proportion of the visible iron present could be removed by repeated insertion and withdrawal of the magnet. For *in situ* froth flotation, it is well known that various combinations of sorbents and surfactants promote the attachment of air bubbles to metal sulfide minerals, enabling their separation from "gangue" minerals (*26*). With adaptation, this approach may also lend itself to the recovery of ZVI and other solids.

A further implication of ISF is the possibility of PRB construction within a *submerged fluidized cavity* (see *9*). This is a submerged cavity of fluidized material overlain by unfluidized soil, which forms at depths greater than the critical depth, *i.e.* beyond the depth at which ISF becomes unstable. If the dimensions of the fluidized cavity can be controlled, this may provide a useful technique for the installation of a PRB at depth, without significant disruption of the overlying soil. Application beneath built structures may even be possible. The method bears many similarities to the "borehole mining" method previously used for the recovery of uranium and other granular ores (*27*).

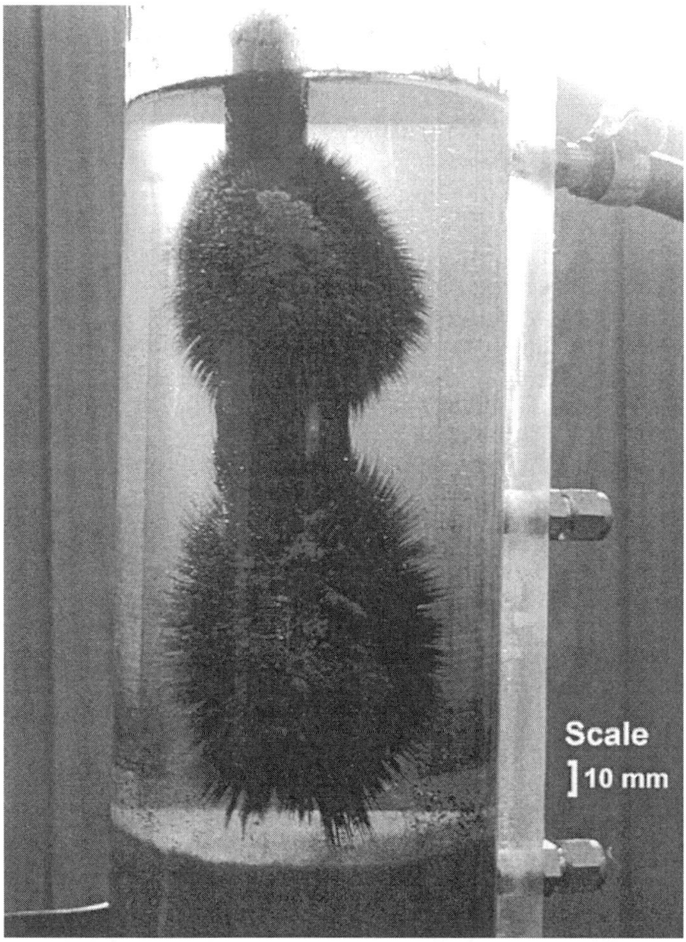

Figure 8. Photograph of magnetic ZVI removal during fluidization, showing ZVI particles clinging to the poles of an electromagnet. (The magnet is shown raised from the fluidized bed, in the overlying water column.)

Conclusions

In conclusion, it is demonstrated by column and tank experiments that ISF has considerable potential for the *in situ* construction of PRBs. As it avoids excavation of the soil, and associated mobilization and handling difficulties, substantial cost savings may be possible compared to existing PRB installation methods. Initial efforts were directed towards the "hydraulic matching" method, in which the ZVI and recipient sand particle sizes were matched hydraulically; however, even in an experimental setting, accurate hydraulic matching proved difficult. Owing to the heterogeneous and non-uniform nature of most real soils, even greater difficulties could be expected in the field. In contrast, an experiment concerning the "extraction method", involving the successive addition of hydraulically heavier ZVI as the jet was withdrawn, produced a sequence of segregated layers of ZVI within the recipient sand. This could function as a PRB. The final ZVI loading (25% by mass) is of a similar order to that required in field applications, and can be varied by appropriate control of the ZVI addition process.

A number of implications of ISF are examined in detail. These include PRB maintenance using ISF, and the possibility of PRB installation within a submerged fluidized cavity.

Further work is required on the use of ISF for PRB installation. This includes examination of the appropriate means of addition of the ZVI or other solid reagent (directly into the fluidized zone or within the water flow), pilot scale field trials, and investigations into the effect of the ellipsoidal ZVI laminations on groundwater flow paths and the dehalogenation reaction process.

Acknowledgments

This project was supported financially by the School of Civil Engineering, The University of New South Wales at the Australian Defence Force Academy, Canberra, Australia, and by an American Chemical Society Corporation Associates travel grant. The assistance of Mike Wilson for the sieve analyses and grinding, Mary Dalton for the electromagnet, and David Paterson and David Sharp for the photography, are gratefully acknowledged.

References

1. Warner, S.D.; Sorel, D. In *Innovative Strategies for the Remediation of Chlorinated Solvents and DNAPLs in the Subsurface, 221st ACS National Meeting, San Diego, CA, April 1-5, 2001*; Henry, S.M.; Bennett, C.R., Eds.; Division of Environmental Chemistry, American Chemical Society: Washington, DC, 2001; Vol. 41(1), pp 1104-1112.

2. Gillham, R.W.; O'Hannesin, S. *Ground Water* **1994,** *32(6)*, 958-967.
3. EnviroMetal Technologies Inc., URL http://www.eti.ca/, accessed 20 July 2001.
4. Focht, R.M., Vogan, J.L. & Krug, T.A. In *Innovative Strategies for the Remediation of Chlorinated Solvents and DNAPLs in the Subsurface, 221st ACS National Meeting, San Diego, CA, April 1-5, 2001*; Henry, S.M.; Bennett, C.R., Eds.; Division of Environmental Chemistry, American Chemical Society: Washington, DC, 2001; Vol. 41(1), pp 1120-1125.
5. Starr, R.C.; Cherry, J.A. *Ground Water* **1994,** *32(3)*, 465-476.
6. Geiger, C.L., Clausen, C.A., Brooks, K. & Major, D. In *Innovative Strategies for the Remediation of Chlorinated Solvents and DNAPLs in the Subsurface, 221st ACS National Meeting, San Diego, CA, April 1-5, 2001*; Henry, S.M.; Bennett, C.R., Eds.; Division of Environmental Chemistry, American Chemical Society: Washington, DC, 2001; Vol. 41(1), pp 1028-1033.
7. Schnell, D.L. In *Innovative Strategies for the Remediation of Chlorinated Solvents and DNAPLs in the Subsurface, 221st ACS National Meeting, San Diego, CA, April 1-5, 2001*; Henry, S.M.; Bennett, C.R., Eds.; Division of Environmental Chemistry, American Chemical Society: Washington, DC, 2001; Vol. 41(1), pp 1114-1119.
8. Khalili, N.; Niven, R.K. In *Environmental Geotechnology, Proceedings of the 3rd International Symposium, June 10-12 1996, San Diego, CA*; Fang, H.-Y.; Inyang, H.I. Eds.; Technomic Publishing Co.: Lancaster, PN, 1996; Vol. 1, pp 745-754.
9. Niven, R.K.; Khalili, N. *J. Hydraulic Research* **1998,** *36(2),* 199-228.
10. Niven, R.K.; Khalili, N. *Can. Geotech. J.* **1998,** *35(6),* 938-960.
11. Niven, R.K. Ph.D. thesis, The University of New South Wales, Sydney, NSW, Australia, 1998.
12. Niven, R.K.; Khalili, N.; Hibbert, D.B. *Chem. Eng. Sci.* **2000,** *55,* 3013-3032 and 3033-3051.
13. Niven, R.K.; Khalili, N. In *Proceedings of the 2000 Contaminated Site Remediation Conference, Melbourne, Victoria, 4-8 December 2000*; Johnston, C.D. Ed.; Centre for Groundwater Studies, CSIRO Land and Water: Wembley, WA, Australia, 2000; Vol. 2, pp 679-686.
14. Niven, R.K. In *Innovative Strategies for the Remediation of Chlorinated Solvents and DNAPLs in the Subsurface, 221st ACS National Meeting, San Diego, CA, April 1-5, 2001*; Henry, S.M.; Bennett, C.R., Eds.; Division of Environmental Chemistry, American Chemical Society: Washington, DC, 2001; Vol. 41(1), p 1113.
15. Niven, R.K. *International Containment & Remediation Technology Conference and Exhibition, Orlando, Florida, 10-13 June 2001*; Florida State University: Tallahassee, FL, 2001, accepted for publication.
16. Leva, M. *Fluidization*; McGraw-Hill Book Co.: NY, 1959.
17. Wen, C.Y.; Yu, Y.H. *Chem. Eng. Progr. Symp. Series* **1966,** *62(62),* 100-111.

18. Ergun, S. *Chem. Eng. Progr.* **1952,** *48*, 89-94.
19. Macdonald, T.F.; El-Sayer, M.S.; Mow, K.; Dullien, F.A.L. *Ind. Eng. Chem. Fundam.* **1979,** *18,* 199-208.
20. Clift, R.; Grace, J.R.; Weber, M.E. *Bubbles, Drops and Particles;* Academic Press, Inc., NY, 1978.
21. Cleasby, J.L.; Fan, K. *J. Envir. Engrg. Div. ASCE* **1981,** *107(EE3)*, 455-471.
22. Gibilaro, L.G.; di Felice, R.; Waldram, S.P.; Foscolo, P.U. *Chem. Eng. Sci.* **1986**, *41(2)*, 379-387.
23. Richardson, J.F.; Zaki, W.N. *Trans. Inst. Chem. Engrs.* **1954,** *32,* 35-53.
24. Al-Dibouni, M.; Garside, J. *Trans. Inst. Chem. Engrs.* **1979,** *57,* 94-103.
25. Richardson, T.W.; McNair, E.C., Jr. *A Guide to the Planning and Hydraulic Design of Jet Pump Remedial Sand Bypassing Systems*; Instruction Report HL-81-1; Office, Chief of Engineers, U.S. Army: Washington, DC, 1981.
26. Sutherland, K.L.; Wark, I.W. *Principles of Flotation;* Australasian Institute of Mining & Metallurgy: Melbourne, Vic., Australia, 1955.
27. Savanick, G.A. *Borehole (Slurry) Mining of Coal, Uraniferous Sandstone, Oil Sands and Phosphate Ore*; Report of Investigations 9101; U.S. Department of the Interior: Washington, DC, 1987.

Chapter 16

Installation of Dispersed Iron Permeable Reactive Treatment Zones Using Pneumatic Injection

Deborah L. Schnell[1] and James Mack[2]

[1]Pneumatic Fracturing, Inc., 1718 Springtown Road, Alpha, NJ 08865–4634
[2]Pneumatic Fracturing. Inc., 25 Starview Drive, Neshanic Station, NJ 08853

The use of iron as a reductant for groundwater containing chlorinated solvents is gaining acceptance in the remediation field. However, the limitations of the technology lie in the depths to which they can be easily installed. One method developed to address this issue uses the patented pneumatic fracturing process to emplace reactive media such as iron filings. *Pneumatic injection*, the technology, creates a dispersed iron permeable reactive treatment zone (PRTZ) where the iron filings are intermixed with the formation in a calculated ratio of iron:soil. The dispersed iron PRTZ eliminates certain concerns associated with solid iron walls, such as biofouling and mineral precipitation, and the treatment zone is installed over a greater lateral distance, allowing for longer residence times for reduction. This significantly reduces the potential for breakthrough. A controlled process, the pneumatic injection method uses high pressures and flows to emplace the reactive media at targeted depths, reducing the amount of iron needed to achieve reduction, and hence, costs.

Significant research has been conducted on the reaction kinetics involved in the reduction of chlorinated solvents by the use of iron. Laboratory batch and column studies have consistently demonstrated successful reduction, and the process has been expanded to field scale at over 70 sites in North America, Europe, and Australia. Most of the *in situ* installations have been shallow at depths less than 50 ft (15.2 m). These shallow installations incorporate trenching or the funnel-and-gate method, a patented technology, which involves driving sheet piling to form a rectangular treatment cell. The native material in the cell is replaced by a mixture of iron filings and silica sand, producing a zone that is more permeable than the surrounding subsurface. The limitation of these methods is the depth to which a reactive cell can be installed, since trench excavations become difficult below a depth of approximately 30 ft (9.1 m).

In order to extend *in situ* reactive dechlorination to greater depths, the Hazardous Substance Management Research Center (HSMRC) located at New Jersey Institute of Technology (NJIT) has developed a new method for injecting iron and other reductants directly into the subsurface. The injection system is a modification of the patented, field-proven pneumatic fracturing system (U.S. Patent No. 5,032,042, 1991), which has been applied to *in situ* clean-ups over the last several years. The new emplacement method (U.S. Patent No. 5,908,267, 1996), a dry media injection, allows for direct injection of iron filings into unconsolidated geologic formations. The process uses nitrogen gas as a medium to carry the iron into the formation and creates a zone 10 to 15 ft (3 to 4.5 m) thick in which the iron is dispersed within the formation. A schematic of the pneumatic injection method is depicted in Figure 1. This injection process causes minimal site disturbance, can be applied at targeted depths with precision, can be applied near or beneath structures and/or utilities, and is cost-comparative with other installation methods.

Case Problem

This section discusses a case study to aid the reader in understanding the design parameters required prior to installation of a dispersed PRTZ using pneumatic injection.

Plume Description

The site is located in an industrial area of California. The facility manufactured a variety of silica-based and carbon-based products for aerospace,

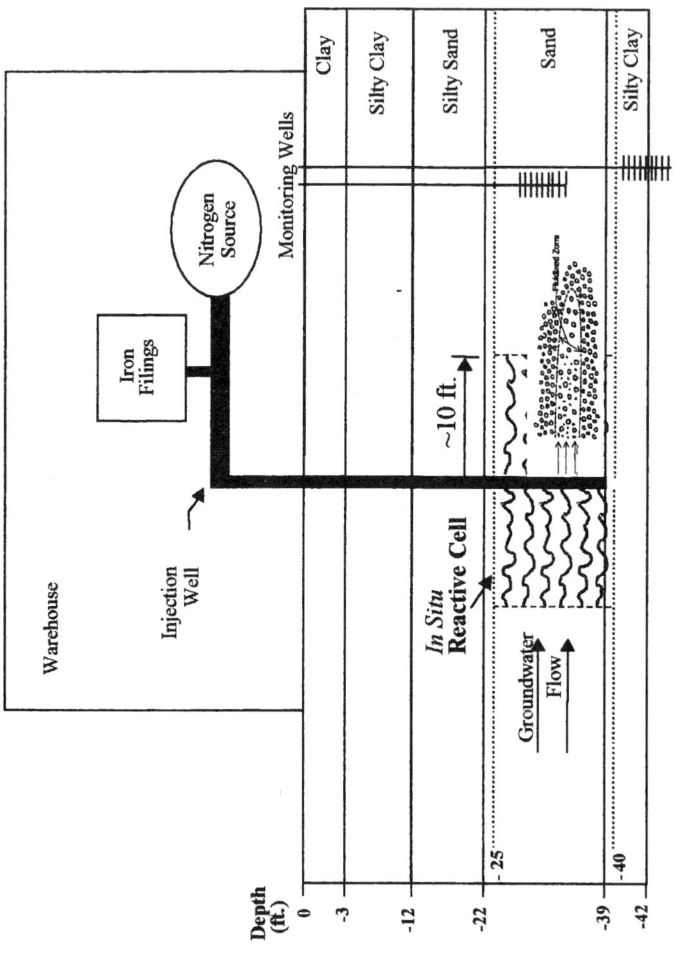

Figure 1. Schematic of Pneumatic Injection Method

defense, marine, and commercial applications. Solvents and resins used at the facility have included 1,1,1-trichloroethane (TCA), toluene, acetone, epoxy resins, phenolic resins, trichloroethene (TCE), perchloroethene (PCE), and isopropyl alcohol.

The source of the chlorinated compounds contamination is a solvent/cleaning area. A cone penetrometer test (CPT) investigation was conducted along the southern boundary of the site, demonstrating that the chlorinated compound concentrations were more prevalent in two separate transmissive zones, one at approximately 25 ft (7.6 m) below grade, and the other at approximately 50 ft (15.2 m) below grade. The highest concentration of TCE (270,000 μg/L) was detected at 57 ft (17.4 m) below grade. According to a recent report, the PCE concentrations across the facility range from 3 μg/L to 25,000 μg/L, and TCE concentrations range from 5 μg/L to 140,000 μg/L.

Regulatory Driver

The site is located in an urbanized area, directly upgradient from residential homes. Several remedial alternatives were evaluated for treatment: iron PRTZ, pump and treat, source removal by dual phase extraction, and monitored natural attenuation (MNA). Based on a cost analysis of these technologies and the need to control off-site migration of the dissolved plume, an iron PRTZ was selected for installation at the site boundary, coupled with MNA.

Hydrogeology

The site is located in coastal plain. Site investigations confirmed that predominantly fine-grained sediments are interbedded with sand layers from ground surface to a depth of at least 102 ft (31.1 m). A continuously sampled boring determined that there are three localized transmissive zones overlain and underlain by poorly permeable silty clay and clay units within the first 100 ft (30.5 m). These transmissive zones have been verified by geotechnical analyses and are described below:

- The interval between 26 ft (7.9 m) and 31 ft (9.5 m) below grade is silty sand, described as well graded, fine to coarse, and saturated. This interval appears to be the first water-bearing zone, which corresponds to the identification of a similar unit in other borings in the vicinity.

- The second interval is between 42 ft (12.8 m) and 60 ft (18.3 m) below grade with various lithologies within this interval. Since there are several material changes within short vertical distances, they are grouped together to form one zone.

- The third interval is the zone between 81 ft (24.7 m) and 92 ft (28.0 m) below grade. This interval contains saturated, silty sand between 85 ft (25.9 m) and 92 ft (28.0 m) below grade. Dry silt is above this interval, and damp sandy silt is above the dry silt.

It appears that most of the impacted groundwater is flowing through these three zones, particularly the upper two. Isolating the groundwater flow to selected transmissive intervals has important consequences for the PRTZ design and even more importantly to the installation costs, particularly when using the pneumatic injection method.

Dispersed Iron PRTZ Design

Steps required for the design of a dispersed iron PRTZ are discussed in this section. These components include site-specific information, preliminary design, treatability study, and final design. A general discussion of each topic is presented, as well as comparative information as it relates to the case problem from the previous section.

Site Information Required

Four key components are required for design of any PRTZ:

- Stratigraphy
- Geotechnical information
- Hydrogeology
- Contaminant concentrations.

This information aids the engineer in evaluating the design of the PRTZ, as well as the selection of the installation method.

Stratigraphy

Detailed analysis of the stratigraphy is important, especially if the pneumatic injection process is selected as the installation method. This can be accomplished with continuously sampled soil borings, CPT equipment, or conductivity probes. One approach would be to first use various probe logging equipment to define the stratigraphy, and then perform continuous borings at the most representative locations to collect appropriate soil samples.

Geotechnical Analysis

Geotechnical analyses should be conducted on soil samples collected from the continuous core, especially if there are identifiable transmissive zones. Specific samples from the soil boring are analyzed for geotechnical parameters, including grain size distribution, density, permeability, and porosity. The sample intervals chosen should be based on the stratigraphic analysis, boring log descriptions, and what appear to be the most transmissive zones by visual observation.

Hydrogeology

To accurately calculate residence times and treatment zone thickness, an understanding of the site hydrogeology is important to the PRTZ design. The primary information required for these calculations is groundwater velocity obtained from the permeability, porosity, and hydraulic gradient. Permeability can be obtained from pumping tests, falling head tests, and geotechnical analyses of specific soil samples. It is important to understand these measurements for each targeted transmissive interval when dealing with a multi-layer system. The pneumatic injection process allows for adjustments during installation to address specific design criteria for a particular interval.

Contaminant Concentrations

The PRTZ is designed for worst-case contaminant loadings. Calculations are performed to determine half-lives, residence times and ultimately, treatment zone thickness. Breakdown or daughter products are also considered to assure complete degradation. The pneumatic injection process can be adjusted to modify the iron/soil mixture and radius of influence to account for sections of the PRTZ where higher contaminant loadings are present.

Preliminary Design

Items to be included in the preliminary design are location, depth, iron:soil ratio, soil grain size distribution and permeability, thickness of the treatment zone, and residence time needed to achieve required reduction of primary and daughter contaminant concentrations.

Some of the factors considered in choosing the most appropriate iron for the PRTZ are size, shape and density. Ideally, iron should have the same or larger particle size as the soil to minimize interstitial packing of the particles. If the iron used is smaller than the soil it is applied to, the hydraulic conductivity of the treatment zone may decrease and groundwater will flow around or below the PRTZ. In addition a clean, "degreased" iron particle with a high specific surface area is necessary. Laboratory studies at NJIT found that gray cast iron filings were the most effective in degrading TCE. The particles are porous, have a large surface area, and are inexpensive *(1)*.

The iron size is determined by the use of grain size distribution curves of targeted transmissive zones. Sieve size information for various sizes of iron filings obtained from an iron distributor (e.g., Connelly-GPM, Inc.) are plotted against the grain size information obtained from geotechnical analyses of the transmissive zones. For example, Figure 2 shows the grain size curve for the iron filings selected for this zone plotted against the site soil from the case problem. For this zone, the –18/+50 (1 mm to 0.3 mm) sieve opening of iron was chosen as a best fit to the site soil. Conversely, a combination of 50% of –18/+50 (1 mm to 0.3 mm) sieve opening of iron and 50% of –40/+60 (0.425 mm to 0.250 mm) sieve opening of iron filings were chosen as a best fit for the mid-zone.

The iron content is a function of the calculations required for degradation based on a solid iron PRTZ. These calculations consider half-lives, maximum contaminant loadings, and groundwater flow rates. Based on the first-order decay equation and literature values of the rate constant for degradation of PCE and its daughter products, half-lives for each of the products were calculated. Then residence time was estimated using the calculated half-lives while considering the maximum contaminant loading that the PRTZ will experience and the effluent goals. Finally, the PRTZ thickness is calculated using the site-specific groundwater flow velocity. For example, a 0.5 ft (0.15 m) thick solid iron wall was calculated to be sufficient in achieving the desired reduction rates from the case problem. Since the pneumatic injection process produces a dispersed mixture of iron and soil over a broader thickness (i.e., 10 to 15 ft or 3.0 to 4.5 m), the solid iron calculation must be adjusted to allow for an iron:soil ratio over a greater lateral distance. Dispersing the required 0.5 ft (0.15 m) thick iron over a wider thickness results in 5% iron content by volume. A nominal factor of safety was added to the design, resulting in an 8% by volume iron:soil ratio.

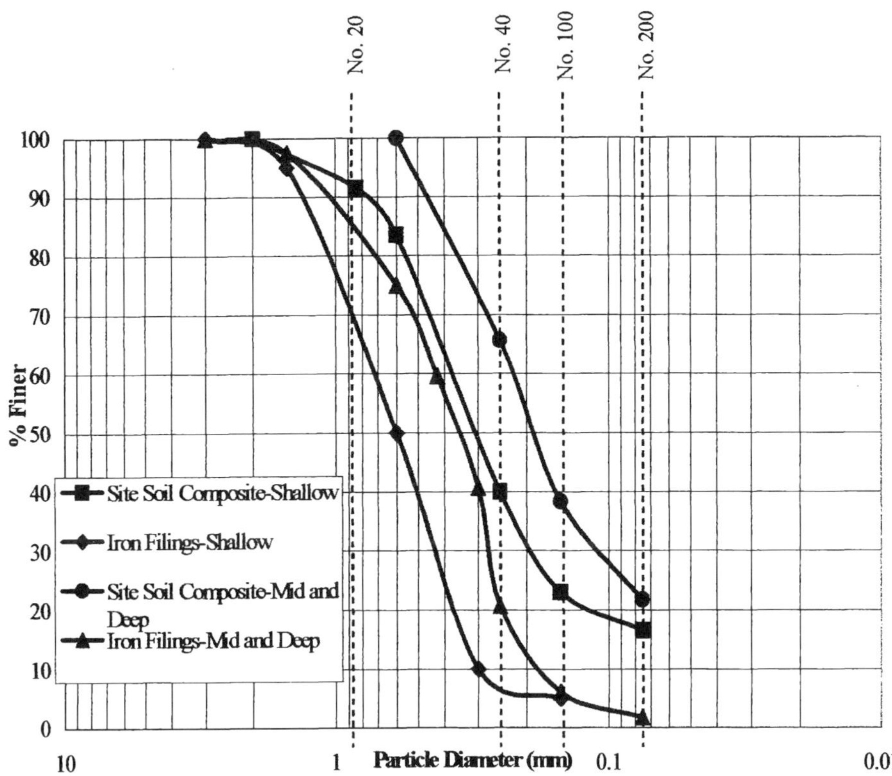

Note: Site soil composite was obtained from geotechnical analyses of site soil. -18+50 iron filings were used for shallow zone. A 50/50 mixture of -18+50 and -40+60 iron filings was used for the mid and deep zones.

Figure 2. Iron Size Modeling of the Shallow Transmissive Zone from Case Problem

To evaluate the reaction kinetics and establish a basis for final design, column studies should be performed using site-specific soil and groundwater data. The column studies are designed to mimic site conditions as closely as possible in order to provide an accurate basis for the full-scale PRTZ layout.

For the case problem, a total of five columns were constructed, three of which were designed to evaluate iron degradation rates in the three transmissive zones and two were assigned as control columns. An illustration of a typical column design is depicted in Figure 3. Control columns were used to provide Quality Assurance/Quality Control of the treatability study. Groundwater from the site was used for contaminant loading into the columns, and water flow through the columns was adjusted to simulate site groundwater flow rates.

Three of the columns contained soil mixed with iron filings. The iron particle sizes were fitted to the grain size distribution curves from the geotechnical analyses of the three transmissive zones as discussed in the previous section. The iron:soil ratio used in the treatability study columns was 8% by volume. Sample ports were placed within the columns, and samples were collected at multiple sample ports at periodic intervals during the study. Samples were analyzed for PCE, TCE, cis-1,2-dichloroethene (cis-1,2-DCE), and vinyl chloride.

Results of Treatability Study

The objectives of the treatability study are to determine contaminant degradation rates, half-lives and the required residence times in the reactive zone in order to achieve the desired end reductions. The information is used for design of the full-scale PRTZ.

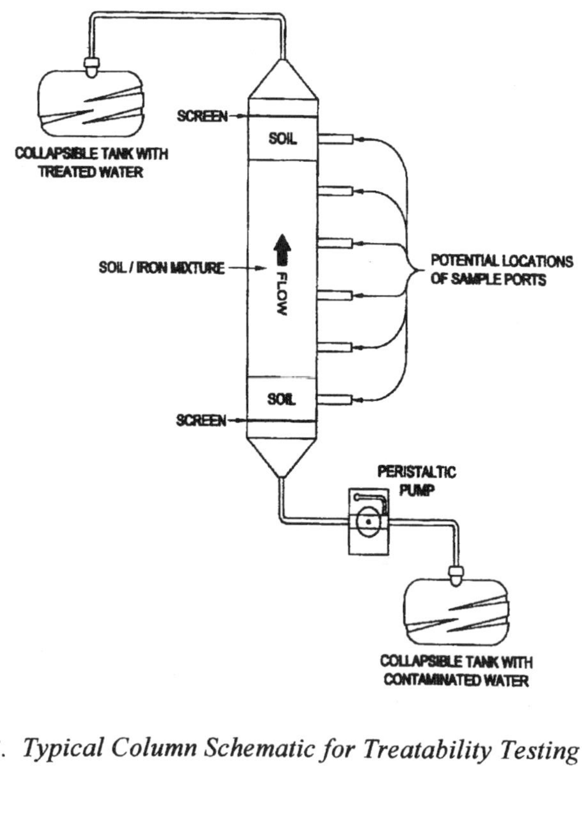

Figure 3. Typical Column Schematic for Treatability Testing

Permeability

An important component of the design process is to match the soil grain size with iron particle size so that there is not a decrease in permeability but rather a slight increase. Grain size distribution curves are developed for soils and matched with iron particle grain size distribution curves to identify the compatible iron, as described below.

From the case problem, the addition of iron increased the permeability of the iron/soil mixture for each zone. Mixing a ratio of 8% of the coarse iron (i.e., particle size range from –18/+50 or 1 mm to 0.3 mm) within the soil in the shallow zone resulted in a permeability of 7.28 E-03 cm/sec (20.6 ft/day), an increase of 64% in permeability. Permeability was calculated based on a method provided for construction dewatering *(2)*. The mid-depth zone had a mixture of 50% fine (i.e., particle size range from –40/+60 or 0.425 mm to 0.250 mm) and 50% coarse iron. An 8% addition of this blended iron within the soil of this zone increases the permeability by 115% to 1.48 E-03 cm/sec (4.2 ft/day). Thus, the iron particle size had the desired effect, which was to increase the soil permeability within the treatment zone.

Degradation Rates (Half-Lives)

Another important component of design is to determine site-specific degradation rates, which are converted to compound-specific half-lives. This information, coupled with groundwater flow velocity, defines residence time requirements which, in turn, determine the treatment zone thickness. Because the pneumatic fracturing dry media injection process creates a treatment zone that is a mixture of iron and soil, the treatment zone is not pure iron. Since literature degradation rates are based on pure iron and do not represent the iron distribution pattern achieved by the pneumatic fracturing dry media injection method that will be used to actually construct the PRTZ, site-specific degradation rates need to be developed based upon a mixture of iron and soil (e.g., in the case problem, 8% iron:soil ratio). This section discusses the basis for calculating these degradation rates and half-lives for PCE, TCE, cis-1-2-DCE and vinyl chloride for the case problem.

Samples were extracted from each column based on a time interval from T_0 or from the start of the test. Utilizing a flow rate of 0.22 mL/min, which was based on site-specific hydrogeologic information, the data was evaluated as if the water traveling through the column was a slug so that at T_x the slug of water would have moved from one point in the column (e.g., bottom) to another point in the column. Although the entire data sets for each column were evaluated, these data were reduced to calculate specific degradation rates for each zone based on this interpretation.

The degradation of chlorinated volatile organic compounds (VOCs) in the presence of iron is consistent with first-order decay kinetics. The equation used to determine the degradation time to achieve site goals is $t = (\ln (C/C_o)) / -K$; where t is residence time, C is desired concentration or effluent, C_o is original concentration or influent, and K is a rate constant. Half-lives can be calculated using the following equation: $t_{1/2} = (\ln 2) / K$.

These equations were used in conjunction with the degradation rates calculated from the treatability study to obtain the site-specific half-lives for the VOCs in the three transmissive zones. Half-life calculations for the various VOC compounds of the case problem are contained in Table I. As discussed, these half-lives are representative of a mixture of iron and soil, not of pure iron typically presented in literature. The degradation rates are a function of the amount of iron in the soil mixture and the type of iron (coarse *vs.* fine). Fine iron has greater surface areas and therefore more aggressive reaction rates.

Table I. Degradation Rates Calculated for an 8% Iron: Soil Dispersed Iron PRTZ

Compound	$t_{1/2}$ *(hours)*		
	Shallow	*Mid-Depth*	*Deep*
PCE	26.7	22	14.3
TCE	13.8	8.6	9.7
Cis-1,2-DCE	47	47	47

Residence Time

The residence time required in the PRTZ to achieve the desired end-point concentrations is one of the most important pieces of design data. For the case problem, the desired end-point concentrations are Federal Groundwater MCLs. The maximum dissolved phase concentration from available data was used in order to assure that a conservative approach is applied to the residence time calculation for each transmissive zone.

The half-life degradation rates shown in Table I were used to calculate residence time in the PRTZ to achieve the desired end point concentrations. It is assumed that there is concurrent production and degradation of each VOC in solution. Therefore, each daughter product was calculated individually based on degradation of each contaminant and summed as a worst case scenario.

Residence time is then coupled with groundwater flow rate calculations to determine the thickness required for the PRTZ. Residence time calculations for the three transmissive zones are contained in Table II.

Table II. Residence Time Requirements to Achieve Federal MCLs

Compound	*Starting Concentration (µg/L)*	*Ending Concentration (µg/L)*	*Residence Time (hrs)*	*(days)*
ShallowTransmissive Zone				
PCE	3,700	5	254	10.6
TCE	62,000	5	188	7.8
Cis-1,2-DCE	19,410	6	539	22.5
Mid-Depth Transmissive Zone				
PCE	9,900	5	237	9.9
TCE	270,000	5	134	5.6
Cis-1,2-DCE	75,370	6	629	26.2
Deep Transmissive Zone				
PCE	34	5	40	1.7
TCE	4,800	5	95	4.0
Cis-1,2-DCE	1,210	6	353	14.7

The thickness of the PRTZ is dependent upon the residence time required to achieve complete degradation of all contaminants to the required MCLs and the groundwater flow velocity. Treatment zone thickness for each transmissive zone was calculated using hydraulic conductivities that represent iron/soil mixtures contained in the case problem. Based upon the maximum residence time necessary in each transmissive zone, the following thicknesses were determined:

- Shallow transmissive zone: 6 ft (1.8 m) thick
- Mid-depth transmissive zone: 1.3 ft (0.4 m) thick
- Deep transmissive zone: 0.9 ft (0.3 m) thick.

Since the pneumatic injection process is able to attain a radius of influence of 10 ft to 15 ft (3.0 m to 4.6 m), this installation method would be appropriate in achieving these calculated thicknesses.

Final Design

The information discussed above is combined with additional requirements such as shape of the PRTZ, nozzle installation methods, injection pattern, injection points (i.e., number and spacing), thickness of injection lifts, radius of influence, performance monitoring requirements and injection operational monitoring to develop the full-scale design.

The surgical precision in directing the iron placement is one of the unique aspects of the pneumatic injection process. This allows for significant flexibility in placing iron within the various transmissive zones. For example, in areas where groundwater VOC concentrations are higher (i.e., the plume core), iron injection can be performed in 6 in (0.15 m) lifts, rather than 12 in (0.30 m) lifts, to assure that adequate iron is available for degradation. Additionally, rather than building one continuous PRTZ throughout the vertical thickness of the zone, the iron can be emplaced in the impacted transmissive zones. This approach can reduce iron requirements substantially and is more efficient and cost-effective.

For the case problem, approximately 42 injection points, spaced 20 ft (6.1 m) apart were determined for design. At each injection point, injections would be performed in four directions per lift to form a cylinder of iron mixed with soil. The radius of influence for the iron injections is estimated at 12 ft to15 ft (3.7 m to 4.6 m), so the PRTZ would consist of a series of overlapping cylinders of iron and soil forming a 10 ft (3.0 m) thick continuous treatment zone. A conceptual diagram of a treatment zone cell is depicted in Figure 4.

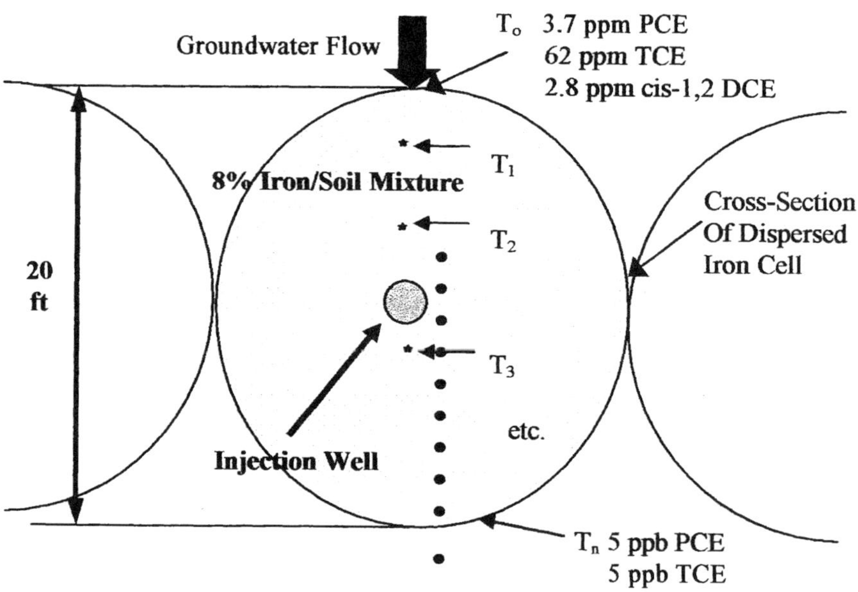

Figure 4. Conceptual Diagram of Dispersed Iron PRTZ Pneumatic Injection Process (Plan View)

Pneumatic Injection Method

Pneumatic injection is an extension of the patented, field proven pneumatic fracturing technology. Pneumatic fracturing was originally developed as a permeability enhancement technique for contaminated geologic formations such as fine-grained soils and bedrock. The basic fracturing technology has been available commercially for several years and has been successfully integrated with vapor extraction, pump and treat, and dual phase extraction in a number of waste clean-ups.

Pneumatic injection uses a unique injection nozzle and other equipment to emplace the iron into the soil. Other variants of the pneumatic injection process include the installation of extended radius wells for free product recovery and the injection of nutrients for bioaugmentation and biostimulation.

The following section discusses the mechanisms, equipment and construction methods of installing an iron PRTZ.

Transport Mechanism

Two basic transport mechanisms are responsible for migration and distribution of iron into the formation. These are transport through a fluidized aggregate lens and transport through a discrete, open fracture *(3)*. Transport through fluidized lenses occurs when the treated formation is cohesiveless and the injected gas velocities are sufficient to keep the geologic particles in suspension. The pore space dilation and oscillatory particle motion allow passage of the injected solid media through the fluidized geologic matrix. An important consideration with this transport mechanism is the maintenance of a dilute condition of the injected dry media with respect to the carrier gas in order to prevent clogging and achieve maximum dry media penetration of the soil formation. The fluidization velocity is a function of soil porosity, shape and size range of the soil particles and viscosity of the injection fluid.

Particle movement through discrete, open fractures, the second transport mechanism, is encountered when the dry media injection process is applied to a fine-textured cohesive formation. In this transport mechanism, the dry media particles remain suspended as long as the critical suspension velocity is maintained. As velocities attenuate radically away from the injection point, particle transport changes to saltation and banking. The field of sediment transport provides models for this transport mechanism.

One significant factor must be considered when using the pneumatic injection method for PRTZ installation - geologic material is not homogeneous, which will have an impact upon the distribution and uniformity of the injected iron particles. As mentioned previously, it is very important to have a thorough understanding of site-specific stratigraphy because the injection process can be adjusted to account for variations in soil characteristics. The fluidization process overcomes much of the heterogeneity along the actual injection pathway, which ranges from 0.25 to 0.5 ft (0.08 to 0.15 m) in thickness, and by varying the gas injection pressure and duration, the fluidization condition can be sustained and expanded along the flow pathway, homogenizing the soil within the injection zone. Yet, care should be taken to avoid injecting at contact boundaries, particularly in cases of significant variations in soil particle size.

Equipment Description

Generally, the dry media injection equipment consists of four basic components:

- Injection nozzle and associated piping
- The compressed gas source
- Dry media feed system
- System operation controls.

The injection nozzle is advanced to the maximum depth required for the PRTZ. The nozzle is designed to inject media in a horizontal or planar format in a 360° circumference. The nozzle itself is directional, so that each injection covers an area of 90°. To achieve a 360° circumference, the nozzle is rotated between injections four or five times. The nozzle is designed to produce a high velocity so that the maximum acceleration of gases and media is achieved immediately before entry into the soil formation. A protective closure is placed over the exit port to prevent clogging. After the 360° injection has been completed, the nozzle is retracted an appropriate distance and the process is repeated. In this manner, 360° planar cylinders of iron/soil mixtures are "stacked" to build a zone of iron/soil spanning the interval desired. Each injection interval is six inches to one foot thick (0.15 m to 0.3 m).

PRTZ Construction

When constructing a PRTZ utilizing the pneumatic injection process, important considerations are verification of treatment zone thickness, continuity and density. These items are achieved by controlling key installation parameters such as injection direction and depth, injection duration, weight of iron injected, pressure and gas flow rate, and gas volume used.

During PRTZ construction, technicians and the injection system operator continuously monitor these parameters. Pneumatic ball valves on the system permit the operator to control the gas flow rate, pressure, and iron feed. Gauges and regulators throughout the system provide data on gas flow rate, pressure, back pressure, and weight of media injected. After each injection, performance data are compared to the calculated schedule of values to evaluate the radius of influence of each injection. As long as the injection parameters fall within ±10% of the appropriate values, the necessary radius of influence is achieved. However, the injection will be repeated, if the actual values fall outside of the predetermined margin of safety.

The distribution of the iron is a function of gas flow rate, the rate of media fed into the gas stream, and the ratio of pressurized gas to dry media. Pneumatically operated ball valves are used to control the media feed. These valves are preset to turn to a percent open, which has been calculated based upon formation soil type and expected radius of influence. Typically for iron particles, the ratio of gas to iron should be between 500:1 to 2,000:1 *(3)*. The iron is introduced into the gas stream at the predetermined feed rate until the scale indicates the desired weight has been achieved. At that point, the system is turned off, the nozzle is rotated, and the process is repeated.

To ensure consistent iron density throughout the injection zone, the operator controls the gas flow rate and volume during the injection event. The iron is introduced into the gas stream once the fluidized path within the subsurface has been established. At this initial state, iron particles are carried to the extremes of the radius of influence. The operator then begins to reduce the gas flow rate and volume at a controlled pace, so that the outer portions of the fluidized zone begin to collapse, embedding the iron within the soil matrix. This process results in iron being evenly distributed throughout the injection radius.

Dispersed vs. Solid PRTZ

The concept that mixed iron/soil reactive zones are equally as effective as pure iron walls in degrading chlorinated compounds is supported by both the literature and project-specific treatability studies *(4)*. Numerous column studies of mixed iron and soil have indicated that a significant portion of the contaminant degradation takes place in the first third of the column; most likely this same process occurs in the field. An iron/soil treatment zone allows gradual movement of groundwater through the reactive media and results in gradual changes in groundwater chemistry, compared to the more radical changes a pure iron wall creates.

The driving forces for precipitate formation are changes in pH and Eh. A treatment zone of pure iron provides a radical variation in formation geochemistry, producing significant shifts in pH and Eh. The literature indicates that reactive walls constructed of pure iron can cause pH increase as high as 10 and Eh reductions to –400 to –500 mV; however, iron/soil mixtures have pH changes of only 1 to 1.5 units and Eh values of –200 to –350 mV *(5)*. The less aggressive alterations in groundwater geochemistry, coupled with a longer reaction distance, result in the formation of less precipitates over a larger volume of the treatment zone. Thus, iron/soil mixed treatment zones will form precipitates at a slower rate and have a longer performance period.

Additionally, the iron grains in pure iron walls all touch the other, forming a continuous iron environment. As the iron corrodes, it first forms gelatinous rinds around the individual grains which then begin to harden as the minerals precipitate. The hardening precipitate coalesces, eventually forming a sheet of precipitate on the upgradient side of the wall. With an iron/soil treatment zone, individual grains will corrode and rust, but these grains are interspersed within the soil matrix. Some precipitate will still form around the grains and on the grain surface but coalescence with other precipitates will be significantly decreased. In one study *(6)*, over 40 pore volumes were passed through a column containing 15% of 100-mesh (0.15 mm) iron with no measurable decrease in the reactivity of the iron material. Therefore, in general, a dispersed iron wall (i.e., iron mixed with soil) is more resistant to fouling/scaling than a solid iron wall.

There are other potential advantages of using a dispersed reactive treatment zone versus trying to inject a discrete, vertical reactive wall. First, there is less likelihood a dispersed treatment zone will contain holes (i.e., zones with no iron) since the iron is spread over a lateral distance of up to 15 ft (4.6 m) or more. Thus, a dispersed treatment zone uses transverse groundwater dispersion as an advantage since it increases the probability that the solvent will contact the reductant. Residence time in a dispersed PRTZ is also increased relative to a thin, discrete wall. In a thin, discrete wall, continuity is essential since it is too narrow to benefit from dispersion and residence time is relatively short (e.g., 6 hours).

Second, the media is injected horizontally, corresponding to more nearly the natural planes of stratification or bedding present in many geologic formations. Conversely, vertically injected walls must cross such natural planes, with the result that layered heterogeneities become an obstacle rather than an aid.

Finally, it permits the creation of preferential flow zones by choosing a reductant grain size which is larger that the grain size of the native aquifer.

These coarse reactant stringers will transmit a large proportion of the groundwater, while simultaneously reducing it.

Case History of PRTZ Installed with Pneumatic Injection

The following section discusses a site-specific application of a dispersed iron PRTZ installed using pneumatic injection.

Background

The site for the PRTZ is a former electrical transformer reclamation facility located in the Mid-West. As a result of past industrial operations, chlorinated solvents leaked into the subsurface entering both the unsaturated and saturated zones.

The facility is located within the alluvial floodplain and has surficial soils composed of recent alluvium consisting of interbedded silt, clay, and fine sand. This fine-grained topstratum transitions rather abruptly into a coarse-grained substratum at a depth of approximately 20 ft (6.1), and bedrock is located at an average depth of approximately 100 feet across the site.

The groundwater table at the site varies from 18 to 22 ft (5.5 to 6.7 m) below grade. Aquifer conditions are predominantly unconfined although during periods of high water table, a semi-confining condition temporarily occurs due to the low permeability of the topstratum. Hydrogeologic studies identified a source area in the unsaturated zone (subsequently removed), which caused a gradual but continuous release of solvent into the saturated coarse substratum below. The hydraulic conductivity of the substratum is estimated to range between 1.0 E-01 cm/sec (283 ft/day) and 1.0 E-03 cm/sec (2.83 ft/day) and like the soil grain size, generally increases with depth. Maximum TCE concentrations within the plume generally fluctuate between 70 mg/L and 370 mg/L.

Concept Design

The *in situ* PRTZ was created by injecting iron through two wells located 10 feet apart. Injections were made at multiple depths and directions to create a distributed network of reactive media within the sand substratum. The network of reactive media was designed to resemble a cylinder in cross section and was shown to dechlorinate the solvent abiotically as long as sufficient residence time is provided. The PRTZ was positioned downgradient of the source area towards the south edge of the delineated plume. Vertically, the treatment zone measured 12 ft (3.6 m), extending from a depth of 25 ft (7.6 m) to a depth of 37 ft (11.3 m).

Based upon available permeability and gradient data, the average groundwater velocity of the medium to fine sand stratum across the test area was estimated at 1.1 E-04 cm/sec (0.3 ft/day). Thus, during the operational period, an estimated 6 pore volumes of groundwater passed through the 10 ft (3.0 m) wide treatment zone. Since the iron is spread over a lateral distance of up to 15 ft (4.6 m) or more, residence time for the 10 ft (3.0 m) wide cell was estimated to range between 30 to 60 days.

Field Test

Pneumatic fracturing was used to inject the iron into the subsurface at various depths, forming a reactive media treatment zone or PRTZ. The treatment zone was formed using two injection points, positioned 10 ft (3.0 m) apart, each with a radius of influence of 5 ft (1.5 m). The length of the reactive media cell totaled 20 ft (6.1 m). The PRTZ was positioned perpendicular to the groundwater flow such that the groundwater will travel through the treatment zone. The injections were conducted at depths from 25 to 37 ft (7.6 to 11.3 m) below grade, resulting in a cell volume of approximately 2400 cubic feet (68.0 cubic meters).

Initially, steel casings were installed from the surface to a depth of 20 ft (6.1 m). The injection nozzle was inserted to the bottom of the casing and then was jetted to the maximum depth of injection into the subsurface using water to displace the formation materials (running sands). Once the injection nozzle was placed at depth, pressurized nitrogen was injected into the formation forming a fluidized bed of sand. Following the nitrogen injection, the iron particles were added to the nitrogen flow stream and injected into the formation. After the injection was completed, the nozzle was rotated 90° at the same depth and the process was continued. Four injections in four separate directions at one depth were completed, forming a radial path of iron. The nozzle was raised one foot (0.3 m), and the process was repeated.

A total of 1600 lbs (725.8 kg) of iron was injected into the reactive zone: 1,000 lbs (453.6 kg) in IW-#1 and 600 lbs (272 kg) in IW-#2. This equated to a concentration of 0.04% (by mass) if the iron is assumed to be uniformly distributed over the entire treatment zone. To compare with other reactive wall configurations, an equivalent wall width may be calculated. Assuming the injected iron was mixed with the soil to form a wall 1-foot thick, the equivalent concentration would be 6.7%. Kinetic calculations performed during design indicated that this mass of iron would cause a measurable reduction, which was the goal of this PRTZ. In order to accomplish more complete reduction, higher concentrations of iron would be required.

Results and Conclusions

In order to evaluate the field performance of the reactive zone, groundwater samples were extracted from the site wells at regular intervals before injection to establish a baseline, and again following injection to check for reductions.

An overall review of the sampling results suggests that the PRTZ was effective in reducing TCE concentrations in the groundwater. The first evidence was in the downgradient wells, which consistently showed a distinct and sustained reduction in TCE concentration following injection of the iron. As depicted in Figure 5, the data for the wells closest to the injection point exhibited a large initial concentration dip immediately after injection, which was attributed to solvent stripping by the carrier gas. Approximately 60 days following injection, the downgradient wells continued to exhibit reductions in the range of 30 to 60% compared with baseline TCE concentrations.

Conclusions

Integration of the pneumatic injection process and reactive media was designed specifically for this PRTZ. To determine the feasibility of the integration, laboratory studies were performed on both the site soils and the iron to be injected. Field demonstrations following bench scale tank testing were also performed to determine the prospect of injection into a saturated sand formation.

Extensive testing performed on the iron, including efficiency of degradation of TCE and enhancement of permeability, was successful in determining the type of iron to be used to form the reactive media cell. Field demonstrations confirmed the plausibility of injecting into this type of geologic formation and assisted in modifying the nozzle design.

Groundwater samples were collected prior to and following injection of the reactive media zone. Results of the sampling indicate a substantial reduction in TCE concentrations in each well. These results confirm the effectiveness of the emplacement of the iron filings using pneumatic injection.

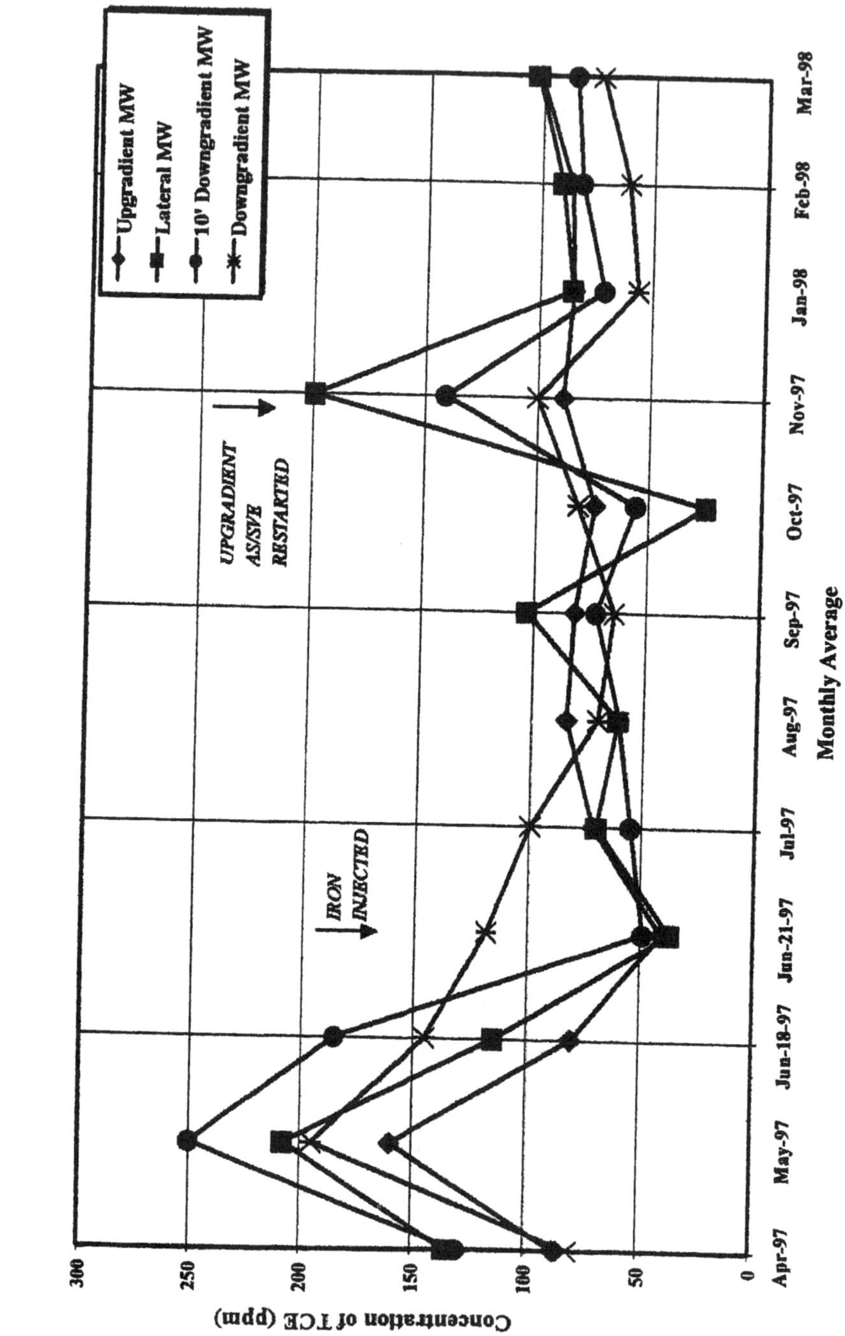
Upgradient MW
Lateral MW
10' Downgradient MW
Downgradient MW
UPGRADIENT AS/SVE RESTARTED
IRON INJECTED
Concentration of TCE (ppm)
300
250
200
150
100
50
0
Apr-97
May-97
Jun-18-97
Jun-21-97
Jul-97
Aug-97
Sep-97
Oct-97
Nov-97
Jan-98
Feb-98
Mar-98
Monthly Average

Discussion

Pneumatic injection of iron filings creates a dispersed iron/soil mixture PRTZ. The process is effective in building PRTZs to treat deep VOC contaminated groundwater in a cost-effective manner. Site-specific hydrogeologic and geotechnical information is needed to design the iron/soil mixture and the basic PRTZ components such as thickness and shape. The pneumatic injection process is patented technology that uses high pressure and flow of a carrier gas to first fluidize the formation particles and then emplace the iron filings within the soil matrix. The process is controlled at the subsurface by experienced system operators to verify wall continuity, thickness and iron density.

Dispersed iron PRTZs appear to have certain performance advantages over solid iron PRTZs. The distribution of iron within the soil matrix appears to minimize precipitation and biofouling, as opposed to solid iron walls. This indicates that dispersed iron PRTZ can have longer operational lives with less maintenance requirements.

References

1. *Integration of Pneumatic Fracturing with Bioremediation and Reactive Media;* Schnell, D.L.; M.S. Project: New Jersey Institute of Technology, Newark, NJ; 1997.

2. Powers, J. Patrick; *Construction Dewatering: New Methods and Applications*, John Wiley & Sons, 1992.

3. Schuring, J.R.; Boland, T.M.; King, T.C.; McGonigal, S.T.; *Treating Non-Naturally Occurring Subsurface Soil Contaminants with Pneumatic Injection of Dry Media*, U.S. Patent No. 5,908,267, 1999.

4. EnviroMetal Technologies, Inc.; *Anticipated Iron Lifetime in a Permeable Reactive Barrier – TM2.13*; Memo; June 1998.

5. Gillham, R.W.; O'Hannesin, S.F.; *Enhanced Degradation of Halogenated Aliphatics by Zero-Valent Iron*; Ground Water; **1994**, 32 (6), 958-967.

6. Orth, W.S.; Gillham, R.W.; *Dechlorination of Trichloroethene in Aqueous Solution using Fe^0;* Environmental Science & Technology; **1996**, *30*, 66-71.

7. Schnell, D.L.; Boland, T.M.; Schuring, J.R.; Parks, C.B.; *Pneumatic Injection of Iron to Treat Chlorinated Solvents*; presented at the 4th International Symposium on Environmental Geotechnology and Global Sustainable Development (ASCE), Boston, MA, 1998.

Chapter 17

Performance Monitoring of a Permeable Reactive Barrier at the Somersworth, New Hampshire Landfill Superfund Site

Timothy Sivavec[1], Thomas Krug[2], Karen Berry-Spark[2], and Robert Focht[3]

[1]GE Corporate Research and Development Center, One Research Circle, Niskayuna, NY 12309
[2]GeoSyntec Consultants, 60 Research Lane, Suite 206, Guelph, Ontario N1G 5B2, Canada
[3]EnviroMetal Technologies, Inc., Waterloo, Ontario N2V 2G6, Canada

Over a six-month period, the performance of a pilot-scale, zero-valent iron permeable reactive barrier (PRB) at the Somersworth, NH Sanitary Landfill Superfund Site was evaluated. The 21-ft long x 34-ft deep x 28-inch wide PRB was installed in November 1999 to test a construction technique that used a biodegradable polymer slurry to support an open excavation while a granular iron/sand mixture was placed into the subsurface. Criteria used to assess the PRB's performance included (1) hydraulic testing to evaluate potential fluid viscosity effects related to the use of the biopolymer, (2) monitoring of VOCs, groundwater parameters (TOC, TDS, pH, DO, ORP, specific conductance, viscosity) and inorganic parameters (including metals, major ions and nutrients), (3) microbial characterization of groundwater, (4) reactivity testing of cored iron material, and (5) advanced surface analysis of cored iron material.

As permeable reactive barriers (PRBs) containing zero-valent iron become more widely used to remediate contaminated groundwaters, there remains uncertainty in the prediction of their long-term performance. While a number of accelerated aging laboratory and pilot-scale tests have not indicated any significant performance issues caused by the build-up of surface precipitates or bio-fouling, there has been relatively little performance data collected in the field at pilot- or full-scale installations (1-6).

An increasingly popular and potentially cost-effective construction method for PRBs is the biopolymer slurry technique (7-9). Installation of an iron-based PRB using biopolymer slurry trenching is similar to constructing a conventional impermeable bentonite slurry wall. As a trench is excavated, biopolymer (*e.g.*, guar gum) is added as liquid shoring to provide stability to the trench walls. Excavation can continue through the biopolymer without the need for dewatering. Granular iron is then placed into the trench through the slurry using a tremie tube.

While the biopolymer trench construction method had been employed at a limited number of sites prior to the start of the constructability test performed at the Somersworth Landfill Site, some uncertainties existed regarding the use of this construction method. Specific objectives of the constructability test at the site, therefore, were to evaluate whether:

1. the iron/sand mixture could be placed as specified without separation of the iron and sand;
2. the guar gum could be broken down and/or flushed from the PRB within a reasonable timeframe;
3. the use and removal of the guar gum would have an adverse effect on the permeability of the sand/iron mixture or aquifer in the vicinity of the PRB;
4. the use and removal of the guar gum would have an adverse effect on the reactivity of the granular iron; and
5. the use and removal of the guar gum and associated materials would have an adverse effect on the geochemical or microbial conditions in the vicinity of the PRB.

Site Description and Characteristics

The Somersworth Sanitary Landfill Superfund Site is a 26-acre landfill that was constructed in the early 1930s on the site of a former sand and gravel quarry. The landfill was used to dispose of household trash, business refuse, and industrial wastes. Waste was burned at the landfill until 1958. From 1958 to 1981, the waste material was placed in excavated areas, compacted, and covered with soil. In 1981, use of the landfill stopped when the City of Somersworth began disposing of its municipal waste at a regional incinerator. In 1981, the City of Somersworth implemented a closure plan for the landfill that involved the covering of a portion of the landfill with clean fill. Volatile organic compounds (VOCs), principally tetrachloroethene (PCE), trichloroethene (TCE), 1,2-dichloroethene (1,2-DCE) and vinyl chloride (VC), were found to be present in the groundwater.

The site is characterized by sands and gravels having a hydraulic conductivity of approximately 0.02 cm/second. The hydraulic gradient varies from 0.01-0.004 ft/ft near the edge of the waste. The top of the water table ranges from less than 2 ft to about 20 ft below ground surface. As much as 10% of the waste is located below the water table. The aquifer is 30 to 40 ft thick.

Since the PRB technology had not been implemented at the scale of the Somersworth site or in a landfill setting, an initial pilot-scale wall was installed in 1996 using a vibratory-hammer/caisson method. This pilot-scale PRB system consisted of an 8-ft-diameter "gate" of iron between layers of pea gravel. Slurry walls measuring 4.5 ft in length funneled groundwater through the gate. Due to problems with the installation of the funnel-and-gate pilot, the full-scale design called for a continuous wall. A second pilot installation that used a biopolymer slurry-supported, open trench method was therefore undertaken and is described below.

PRB Installation and Development

In late November of 1999, a 25-ft long trench was excavated using a CAT 330 backhoe with an extended boom and a 24-inch wide bucket. The depth to bedrock was approximately 34 ft. Iron beams (30-inch wide) were placed 21 feet apart at either end of the trench to control the placement of the sand-iron mixture and to demarcate the ends of the test section. The narrowest width of the trench was measured to be 28 inches and the appropriate percentage of iron required was calculated to be 90% by weight. The portions of the test section outside of the iron beam ends were backfilled with soil that had been previously excavated.

Sand and granular iron (source: Connelly-GPM, -8+50 mesh) were mixed using concrete mixing trucks. The trucks arrived with a pre-measured quantitiy of sand (3464 pounds). Granular iron (33,000 pounds) was added to the concrete mixing truck at the site, along with city water to wet the iron/sand mixture prior to placement in the trench. Between 450 and 600 gallons of water were added per truck to wet the mixture.

A plastic-lined slurry containment area was constructed beside the working platform at a lower elevation to hold the excess slurry displaced from the trench as the iron/sand mixture was added to the trench. A shallow diversion trench was dug from one end of the test section to the slurry containment area allowing the slurry to flow by gravity from the trench to the containment area.

The wetted iron/sand mixture (90/10% by weight) was placed into the trench using a tremie pipe. The tremie pipe consisted of a 12-inch diameter metal pipe that was initially lowered to the bottom of the trench. Initially, some care was taken while placing the iron/sand mixture in the trench through the tremie pipe to minimize the potential for segregation and contact between the iron/sand mixture and the biopolymer slurry by placing a paper swab in the tremie pipe before starting to place the iron/sand mixture. However, this approach was considered ineffective and no further attempt was made to prevent the biopolymer slurry from filling the bottom of the tremie tube ahead of the iron/sand mixture. Two temporary development wells (6-inch diameter slotted PVC) were placed in the trench prior to the addition of the iron/sand mixture to allow for later removal of biopolymer and groundwater from the trench.

The trench was then developed to remove and encourage the degradation of the residual biopolymer. Development was initiated by pumping fluid from the two development wells at a rate of approximately 10 gallons per minute. The initial 300 gallons of fluid pumped from the wells was transferred to the slurry containment area. After this, the extracted fluid was discharged to the top of the exposed iron/sand mixture. One gallon of high pH enzyme breaker (Rantec LEB-H) was then added to the fluid being discharged to the top of the exposed iron/sand mixture. Water was extracted from the well and discharged to the top of the iron for approximately 6 hours until the viscosity of the fluid was reduced such that one quart of fluid passed through a Marsh funnel in 28 seconds. The two development wells were then pulled out.

Following the development of the trench, the backhoe was used to remove material to expose a clean surface of the iron/sand mixture and geotextile was placed on this surface. The geotextile was extended vertically along the sides of the trench and five feet of clay was placed with the geotextile envelope in 12-inch lifts and compacted by tamping using the backhoe bucket. After the other site work was complete, the biopolymer in the containment area was broken, solidified with the spoils and hauled to the top of the landfill for disposal.

Cored Material Testing

Approximately four weeks after the trench development was completed, cored samples of the iron/sand mixture in the test section were collected at seven locations (Figure 1) using either a split-spoon or direct-push (Geoprobe) sampling technique. Where possible, samples were collected through the entire vertical profile of the test section. The iron/sand mixtures were visually

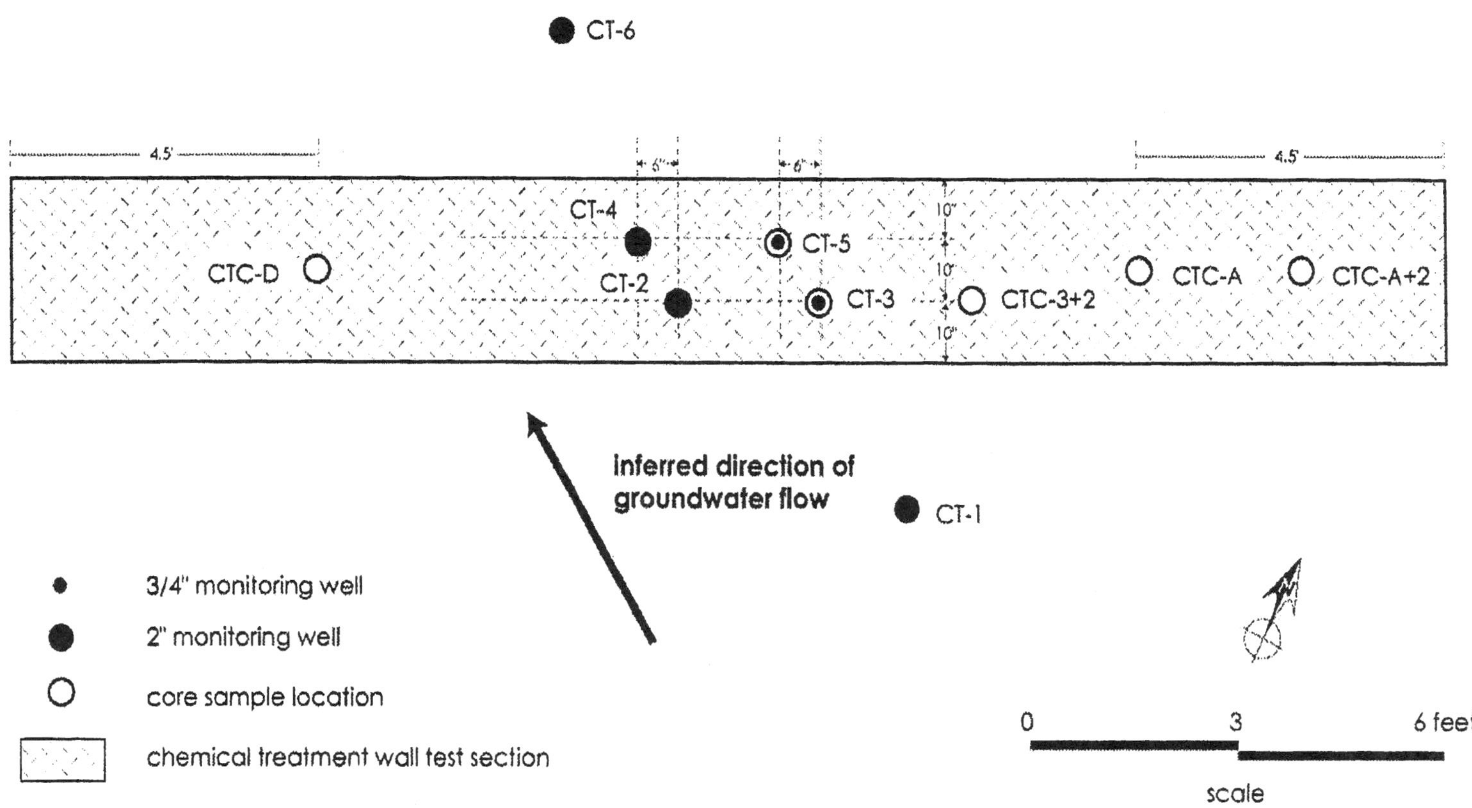

Figure 1. Locations of core samples and monitoring wells in the pilot-scale PRB test section at the Somersworth, NH Landfill Superfund Site.

inspected and photographed immediately after being brought to the surface. The recovery of the iron/sand mixture by the samplers was low (generally 20 to 75%). The samples collected were considered to be highly disturbed.

A total of five borings were drilled within the test section down to the bedrock surface. The borings confirmed the presence of bedrock at a depth of about 34 feet below the top of the working platform. The iron/sand mixture was observed in all five borings; however, approximately 2 feet of sandy material on top of the bedrock surface was observed in CT-5. The samples from boring CT-5 showed a consistent iron/sand mixture throughout the boring down to within 1.8 feet of the bedrock surface. Samples from the bottom 1.8 feet of the boring indicated the presence of sand with very little or no iron. It is suspected that this sand moved in to the bottom of the trench after the last pass with the excavator bucket was made and before the iron/sand mixture was added to the trench. This finding highlights the need to minimize the time between the excavation of the trench and the placement of the iron/sand mixture and to minimize the disturbance of the sides of the trench. The remaining borings suggested that at the iron/sand mixture was continuous from the top of the test section to the bedrock surface. The samples from boring CT-3, located less than one foot away from CT-5, indicated the presence of the iron/sand mixture throughout the entire vertical profile of the test section to the bedrock surface. This finding indicated that the sand above the bedrock surface observed in boring CT-5 was a localized phenemena.

Nine separate samples collected from the test section and a composite of samples from borings CTC-A and CTC-D were analyzed using magnetic separation. The iron content ranged from 72-85%, indicating that the composition of the PRB test section was relatively homogeneous. The mean iron content of the nine samples was 81%, which was equal to the iron content of the composite sample. The iron content of the iron/sand mixture was less than expected based on the proportions of iron and sand specified for construction. The reason for this discrepancy between the measures and expected iron contents is believed to be due to the bias in the results introduced in the sample collection procedure. It is believed that as a result of its higher density, more iron was lost during sample collection than sand.

Grain size distributions were also determined for unmixed sand and iron, a composite sample of six boring at various depths, and samples from borings CTC-D and CTC-3+2. The grain size distribution for the iron portions of the samples collected from the test section was very similar to that of the unmixed iron. The grain size distribution for the sand portion of samples collected from the test section was, however, consistently finer grained than the unmixed sand. This decrease in distribution in the sand may be the result of abrasion of the sand particles during mixing with the iron in the cement trucks or the result of bias during sample collection. The close agreement of the grain size distributions for iron and sand from the different borings suggested that no significant particle segregation occurred within the test section.

Bench-scale reactivity testing was completed using the composite sample from six borings. The degradation rates for the chlorinated ethenes calculated for the results of this testing were higher than those measured in bench-scale studies conducted previously using site groundwater. Table 1 presents a summary of these half-lives. The degradation rates observed in the composite sample suggested that the degradation rates within the test section were not impaired by the biopolymer construction method. The improvement in the chlorinated ethene degradation rates observed may have resulted from biological processes. The samples submitted for reactivity testing were collected from the test section several weeks after construction and may have been innoculated with the microbiological consortia from the site, which has been previously demonstrated to biodegrade chlorinated ethenes.

Table I. Summary of Chlorinated Ethene Half-Lives from Bench-Scale Reactivity Tests Performed at 10°C

	unmixed iron		iron from test section[1]	
	half-life, hr	r^2	half-life, hr	r^2
tetrachloroethene	1.9	0.94	1.6	0.99
trichloroethene	1.1	0.96	0.6	0.97
cis-1,2-dichloroethene	2.0	0.96	1.1	0.99
trans-1,2-dichloroethene	1.0	0.88	0.4	1.00
vinyl chloride	1.6	0.97	0.9	0.99

[1] composite of six boring samples

A number of samples were subjected to surface analysis using x-ray photoelectron spectroscopy (XPS) with depth profiling and scanning electron microscopy (SEM). Traces of guar gum residue or its degradation products on the iron surfaces were detected by XPS (C1s peaks at 286.2 eV). Surface concentrations of carbonate were no higher in the test section iron samples than in the control iron (approximately 2%), as would be expected given the short time the iron was in contact with site groundwater. Similarly, calcium concentrations were very low (0.2-0.3%), and not statistically much higher than that measured in the control iron (0.1%). SEM showed no significant differences between iron grain boundaries of cored and as-received (virgin) iron.

Groundwater Wells and Monitoring

Six PVC monitoring wells were installed in and around the test section along a transect in the direction of site groundwater flow as shown in Figure 1. Two 2-inch diameter monitoring wells and two ¾-inch diameter monitoring wells were installed in the granular iron test section. The 2-inch wells were used for hydraulic testing and in-well monitoring probes and were installed in borings used for the collection of samples of the iron/sand mixture. The ¾-inch wells were used for collection of groundwater samples and were located approximately 24 inches away from the 2-inch wells. In addition, 2-inch diameter monitoring wells were installed in the overburden upgradient and downgradient of the test section. These wells were used for hydraulic testing, in-well monitoring probes and groundwater sampling. The monitoring wells have a screened interval of 15 feet in length and the bottoms of the well screens are approximately 6 feet above the bedrock surface. The wells installed in the test section were constructed without filter packs, while the wells installed in the overburden have filter packs. All wells have bentonite seals and lockable protective casings.

Groundwater samples were collected from four monitoring wells in and around the PRB test section (CT-1, CT-3, CT-5, and CT-6) on three occasions for: (1) measurement or analysis of TOC, TDS, pH, ORP, DO, specific conductance, temperature, dissolved iron, viscosity, and biomass (by phospholipid fatty acid [PFLA] analysis; and (2) VOCs to monitor VOC destruction in the sand/iron mixture. The sampling was conducted at approximately three-week intervals beginning six weeks after construction of the test section. Measurements of pH, ORP DO, specific conductance, temperature, and viscosity were conducted in the field using low-flow purge sampling, while other analyses were conducted on samples submitted to an analytical laboratory.

In addition to the monitoring events described above, dedicated, in-well data-logging probes (YSI-600 XLM) were installed in the four 2-inch wells set along a transect through the PRB test section. Groundwater elevation, ORP, pH, temperature, specific conductance, and DO were monitored at 4-hour intervals over a 6-month period. The in-well probes were used to provide greater data density than could be obtained using low-flow purge methods. This data would allow for correlation of groundwater field parameters with changes in barrier performance (*e.g.*, biopolymer breakdown, potential groundwater mounding). In addition, the in-well probes are preferred over low-flow purge methods in that they minimize hydraulic perturbations near the PRB test section.

Field measurement of pH, ORP and four other groundwater quality parameters collected along the flow path through the test section are displayed in Figures 2 and 3. Groundwater samples from within the test section, as expected, had a higher pH and lower specific conductance than those of the surrounding aquifer. The higher pH in the test section is due to the anaerobic reduction of

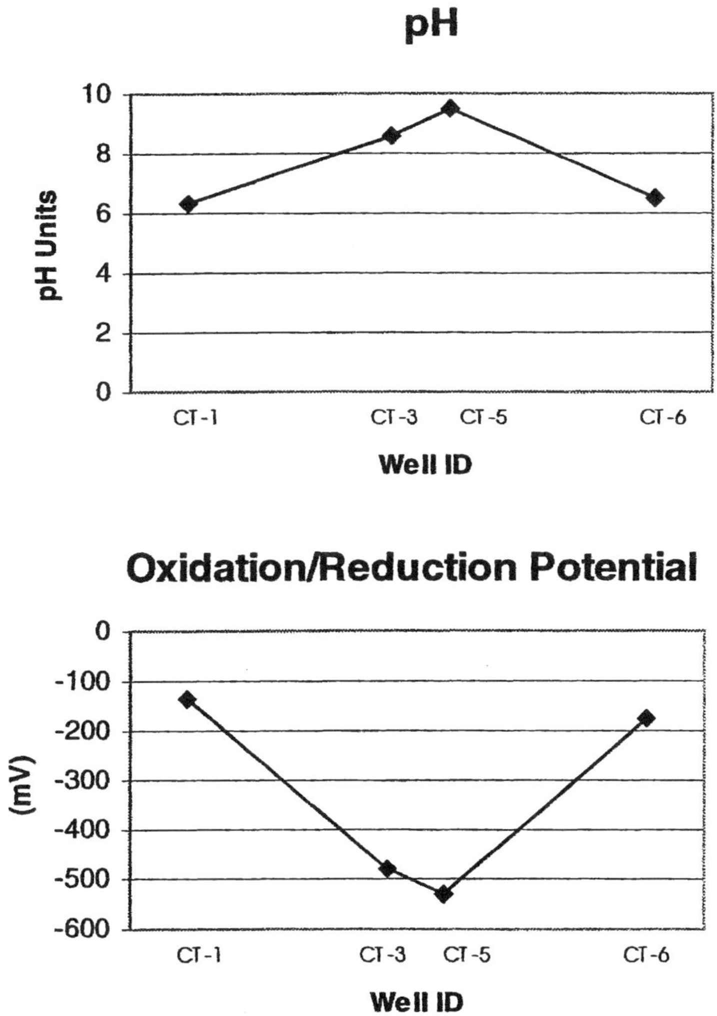

Figure 2. Field parameters measured along flow path through the PRB test section (low-flow purge/flow-through cell method).

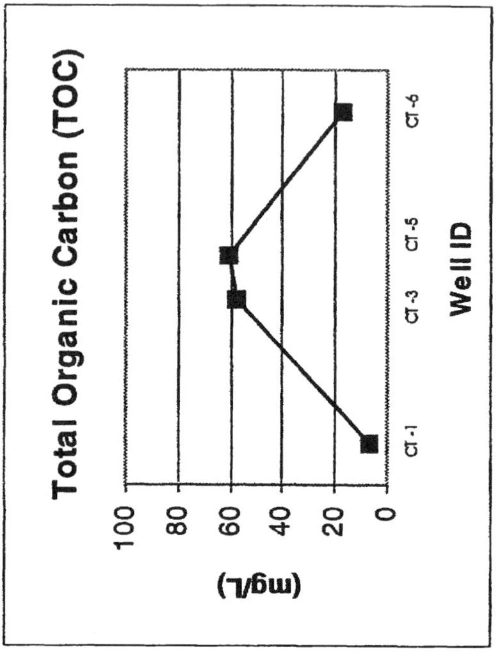
Total Organic Carbon (TOC)
(mg/L)
100
80
60
40
20
0
CT-1
CT-3
CT-5
CT-6
Well ID

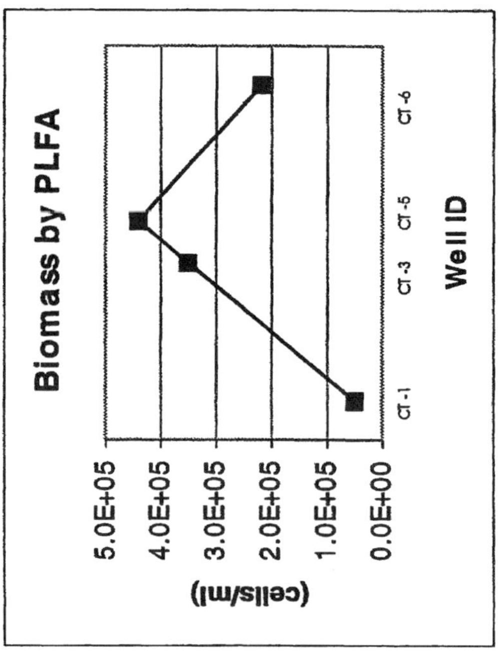
Biomass by PLFA
(cells/ml)
5.0E+05
4.0E+05
3.0E+05
2.0E+05
1.0E+05
0.0E+00
CT-1
CT-3
CT-5
CT-6
Well ID

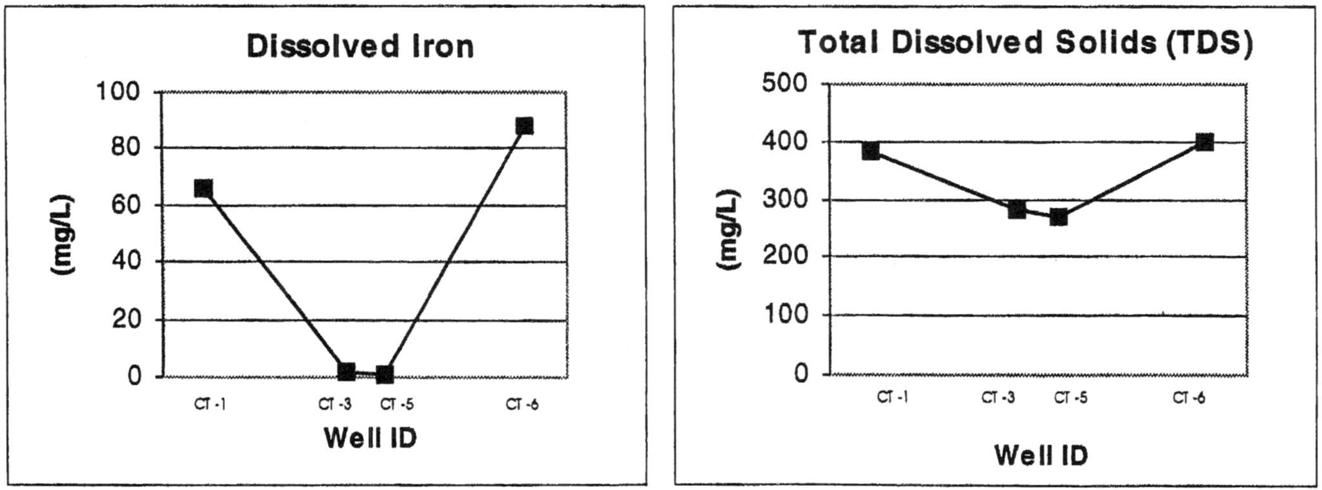

Figure 3. General groundwater quality data measured in wells along flow path through the PRB test section.

water by zero-valent iron (equation 1) which produces hydroxyl ions, resulting in an elevated pH within the test section. The pH change shifts the carbonate equilibrium (equation 2) and precipitates carbonate minerals (equations 3, 4),

$$Fe^{0} + 2H_2O \rightarrow Fe^{2+} + H_2(g) + 2OH^{-} \quad (1)$$

$$HCO_3^{-} + OH^{-} \leftrightarrow CO_3^{2-} + H_2O \qquad (pK_a = 10.3) \quad (2)$$

$$Fe^{2+} + CO_3^{2-} \rightarrow FeCO_3(s) \qquad (K_{FeCO3} = 3.2 \times 10^{-11}) \quad (3)$$

$$Ca^{2+} + CO_3^{2-} \rightarrow CaCO_3(s) \qquad (K_{CaCO3} = 2.8 \times 10^{-9}) \quad (4)$$

which decreases the specific conductance of the groundwater by removing Fe^{2+}, Ca^{2+}, and CO_3^{2-} from solution in the iron test section. Downgradient of the test section, groundwater rapidly re-equilibrated with carbonate minerals and resulted in a downgradient specific conductance and pH comparable to that measured upgradient.

Groundwater samples from the PRB test section also had lower dissolved iron and TDS concentrations than the samples from the surrounding aquifer, as would be predicted with the precipitation of carbonate minerals in the PRB test section. Groundwater samples from the test section had higher TOC and biomass concentrations than the samples from the upgradient aquifer, most likely indicative of the presence of some residual biopolymer in the test section.

The DO field measurements made using a flow-through cell under low-flow purge sampling conditions were positively biased because of the sampling technique introduces oxygen while the groundwater sample is brought to the surface. The DO concentrations measured with the in-well probes were consistently lower than field measurements (Figure 4), and were almost all less than the manufacturer's specified instrument resolution of 0.2 mg/L. The in-well probe data are considered to be more reliable than field measurements and indicate that the aquifer and the test section were anaerobic, as expected.

The ORP of samples collected from the aquifer using a flow-through cell/low-flow purge sampling method was slightly reducing and ranged from –195 to -112 mV over the three monitoring events. The ORP of samples collected from the PRB test section was more highly reducing than the aquifer and ranged from –676 to –402 mV. The ORP measurements made using the in-well probes confirmed the reducing conditions within the test section and aquifer; however, the in-well probes measured even more highly reducing conditions than indicated by field sampling (Figure 5). This difference appears to be the result of bias in the introduction of oxygen in the purged samples, as similarly low ORP measurements were measured in the laboratory when the same YSI probes were placed in a mock well set-up containing 90/10 granular

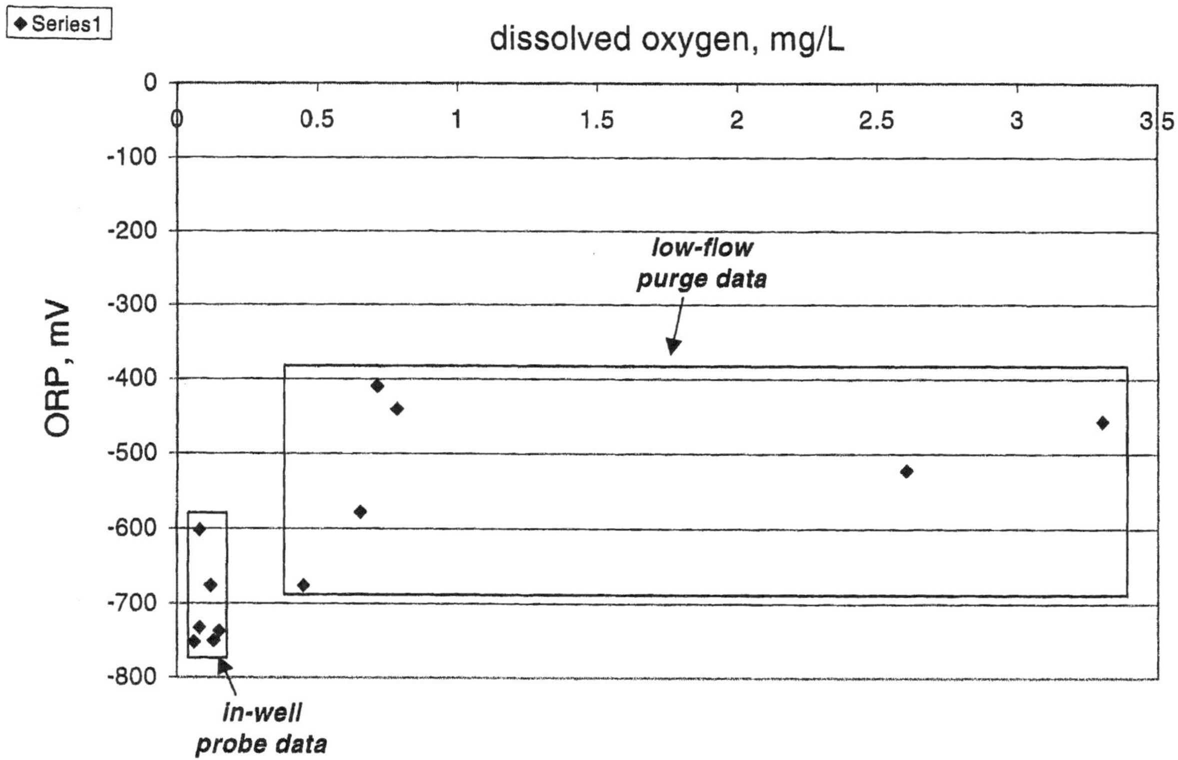

Figure 4. Comparison of ORP vs. DO data sets collected using in-well, YSI probes and low-flow purge/flow-through cell methods.

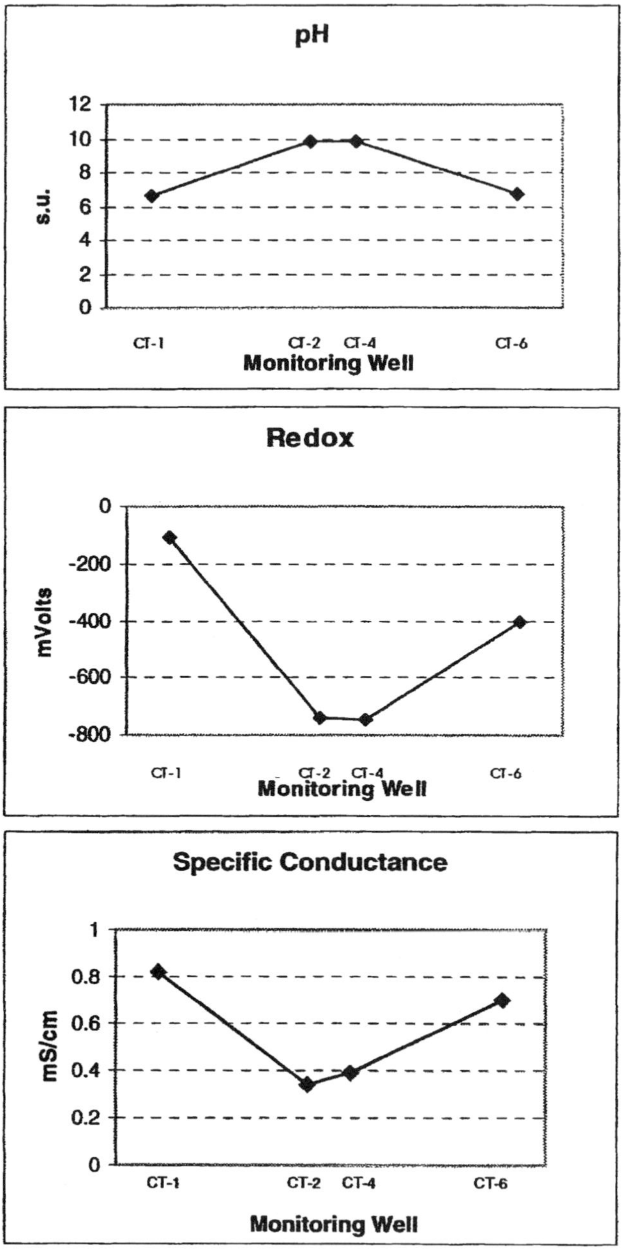

Figure 5. Representative pH, oxidation-reduction potential and specific conductance measurements obtained using devoted, in-well, groundwater parameter probes (YSI) in wells along a transect through the pilot PRB.

iron/sand. A comparison plot showing the complete data set of ORP and DO measurements in the PRB test section collected by field sampling and by the in-well, YSI probes is shown in Figure 4.

No measurable differences in the viscosity of fluid samples from wells installed in the aquifer or the PRB test section were observed. These data indicate that any amount of biopolymer still present in the test section was not significant enough to affect fluid viscosity within levels discernable by this testing method. However, groundwater samples from the test section had higher TOC and biomass concentrations than samples from the upgradient aquifer, indicating the presence of some residual biopolymer in the test section (Figure 3).

Figure 6 presents the PCE, TCE, cis-1,2-DCE and VC concentration data in wells along the flow path through the test section for the third monitoring event. The distribution of the chlorinated ethenes suggests that these compounds were degraded within the test section and that the groundwater flowing from the test section (as indicated by monitoring well CT-5) meets the ICLs. Samples of groundwater from the upgradient well (CT-1) contained elevated concentrations of PCE, TCE, cis-1,2-DCE and VC (up to 200 micrograms per liter (μg/L), 230 μg/L, 140 μg/L, and 33 μg/L, respectively). No detectable concentration of any chlorinated ethene was measured within the test section (CT-3 and CT-5). Samples of groundwater from the downgradient monitoring well (CT-6) contained some detectable concentrations of chlorinated ethenes, even though no detectable concentrations are flowing through the test section. Chlorinated ethenes that became sorbed onto the aquifer solids downgradient of the test section prior to the PRB installation are believed to be desorbing into the groundwater downgradient as groundwater flushes through these contaminated solids. The less chlorinated ethenes (cis-1,2-DCE and VC) are present at higher concentrations because biodegradation processes are degrading the more highly chlorinated ethenes.

Hydraulic Testing

Figure 7 summarizes the estimated hydraulic conductivity values calculated from slug tests performed in CT-1, CT-2, CT-4, and CT-6. In general, water level responses to slug addition or withdrawal were very rapid. Oscillatory responses were observed for the wells screened within the PRB test section (CT-2 and CT-4). Log-linear response data were evaluated using the method of Bouwer and Rice (10), while analysis of oscillatory response data were evaluated using the method of Kipp (11). The software program ADEPT was used to calculate hydraulic conductivity using these methods (12).

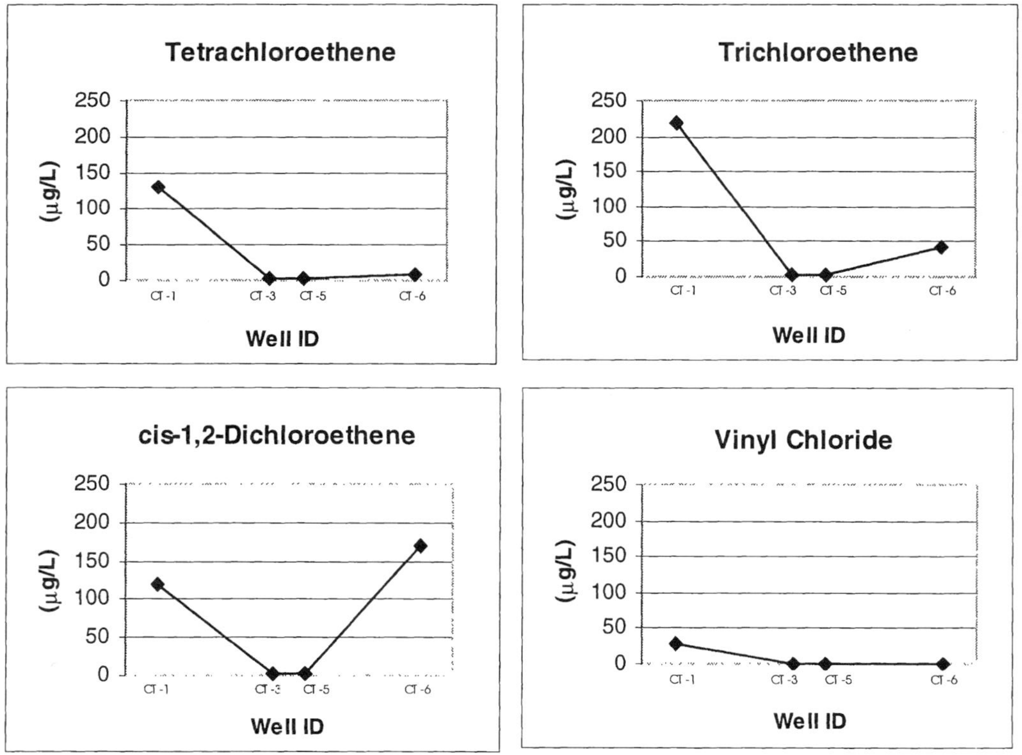

Figure 6. Chlorinated ethenes measured in wells along flow path through PRB test section (February 2000 monitoring event).

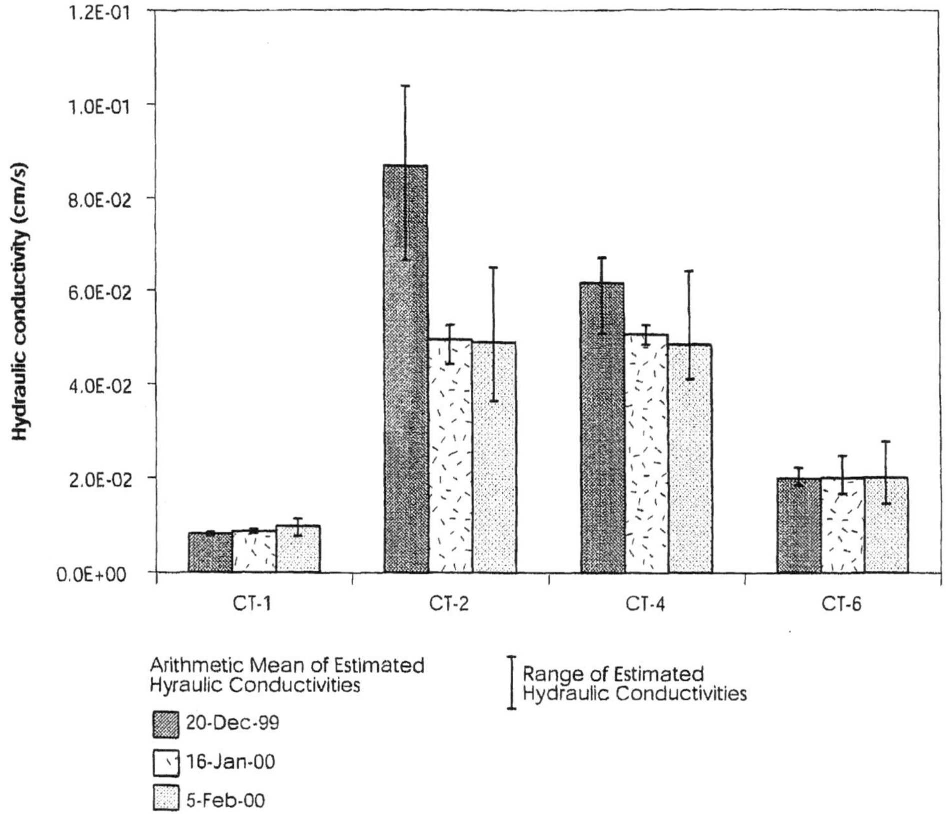

Figure 7. Summary of hydraulic conductivity testing.

The estimated hydraulic conductivity of the aquifer (CT-1 and CT-6) shows measurement variability but no apparent temporal trends. This variability is attributable to variability inherent with the measurement method. The estimated hydraulic conductivity of the PRB test section, however, decreased between the first and second monitoring event but was stable between the second and third monitoring event. This decrease may be attributable to settling of the iron/sand mixture around the screens of the test section monitoring wells as the wells were developed following installation.

The arithmetic mean of the estimated hydraulic conductivity of the iron/sand mixture in the test section is $5.6x10^{-2}$ cm/sec. The estimated hydraulic conductivity of the aquifer ranged from $9.2x10^{-3}$ cm/sec at CT-1 to $2.0x10^{-2}$ cm/sec at CT-6 and was lower than that of the test section by at least a factor of two.

The slug test data, along with the measurements of the viscosity of the fluid from the wells within the test section, demonstrate that the amount of residual biopolymer in the test section is not significant enough to have a measurable effect on the hydraulics of the PRB test section.

Acknowledgements

The authors acknowledge Dr. Robert Puls and the National Risk Management Research Laboratory, U.S. EPA, Ada, OK for use of four YSI-600 XLM sondes in this pilot-scale PRB evaluation. This paper was presented at the Spring National Meeting of the American Chemical Society held in San Diego, CA on April 1-5, 2001.

References

1. Vogan, J.L.; Butler, B.J.; Odziemkowski, M.S.; Friday, G.; Gillham, R.W. *Designing and Applying Treatment Technologies: Remediation of Chlorinated and Recalcitrant Compounds*; Battelle Press, Columbus, OH, 1998, pp. 163-168.
2. O'Hannesin, S.F.; Gillham, R.W. *Groundwater*, **1998**, *36(1),* 164-170.
3. Puls, R.W; Blowes, D.W.; Gillham, R.W. *J. Haz. Mater.* **1999**, *68*, 109-124.
4. Vogan, J.L.; Focht, R.M.; Clark, D.K.; Graham, D.L. *J. Haz. Mater.* **1999**, *68*, 97-108.
5. Mackenzie, P.D.; Horney, D. P.; Sivavec. T.M. *J. Haz. Mater.* **1999**, *68*, 1-17.

6. Powell, R.M.; Blowes, D.W.; Gillham, R.W.; Schultz, D.; Sivavec, T.M.; Puls, R.W.; Vogan, J.L.; Powell, P.D.; Landis, R. *Permeable Reactive Barrier Technologies for Contaminant Remediation*, U.S. EPA, Sept. 1998, EPA/600/R-98/125.
7. Focht, R.M.; Vogan, J.L. *Prepr. Paper ACS Natl. Meeting, Am. Chem. Soc., Div. Environ. Chem.* **2001**, *41(1)*, 1120-1125.
8. Day, S.R.; O'Hannesin, S.F.; Marsden, L. *J. Haz. Mater.* **1999**, *B67*, 285-297.
9. Gavaskar, A.R.; Gupta, N.; Sass, B.M.; Janosy, R.J.; O'Sullivan, D. *Permeable Barriers for Groundwater Remediation: Design, Construction and Monitoring;* Battelle Press, Columbus, OH, 1998.
10. Bouwer, H.; Rice, R.C. *Water Resources Research*, **1976**, *12(3)*, 423.
11. Kipp, K.L. *Water Resources Research*, **1985**, *21(9)*, 1397.
12. Levy, B.S. ADEPT Software for Aquifer Data Evaluation. CHESS, Inc., 1999.

Chapter 18

Performance Monitoring and Dissolved Hydrogen Measurements at a Permeable Zero Valent Iron Reactive Barrier

Dominique Sorel[1,3], Scott D. Warner[1], Bettina L. Longino[1], Jim H. Honniball[1], and Lisa A. Hamilton[2]

[1]Geomatrix Consultants, 2101 Webster Street, 12th Floor, Oakland, CA 94612
[2]General Electric Company, 640 Freedom Business Center, King of Prussia, PA 19406
[3]Current address: S.S. Papadopulos & Associates, 217 Church Street, San Francisco, CA 94114–1310

The first commercial full-scale permeable zero-valent iron reactive barrier (PRB) was installed in 1994 in Sunnyvale, California to treat dissolved chlorinated hydrocarbons present in groundwater. Performance monitoring results, including recently measured dissolved hydrogen concentrations are presented. Research conducted by others indicates that hydrogen evolution may be an indicator of the performance of the dehalogenation reaction. Chlorinated hydrocarbons concentrations are consistently below groundwater cleanup standards within the treatment zone. Treatment appears not to decrease from precipitation of inorganic species in the upgradient gravel zone. The high levels of dissolved hydrogen within the PRB indicate that the corrosion process remains strong more than six years after installation.

Introduction

The Sunnyvale Permeable Reactive Barrier (PRB) provides performance data from the longest-running commercial installation of an in situ iron wall. The purpose of this paper is to provide an update of the monitoring results, and summarize the long-term performance of the remedy. Of particular interest, hydrogen gas concentrations were recently measured for the first time since installation. Monitoring of hydrogen concentrations over time may be useful in evaluating the change in zero-valent iron reactivity. It could potentially serve as a treatment longevity indicator.

Site History

VOCs, mainly trichloroethylene (TCE), cis-1,2-dichloroethylene (cis-1,2-DCE), and vinyl chloride (VC), were detected in soil and groundwater beneath the site in 1983. Source excavation was performed in 1986. A groundwater extraction system, installed in 1987, provided hydraulic capture, but was expected to require many decades to centuries to lower groundwater VOC concentrations to cleanup standards (5 micrograms per liter [μg/L] for TCE, 6 μg/L for cis-1,2-DCE, and 0.5 μg/L for VC).

In late 1991, a PRB composed of granular iron was identified as a possible remediation method to replace the groundwater extraction and treatment system. Laboratory batch and column studies were performed by the University of Waterloo and a nine-month on-site pilot test was performed in 1992. The pilot test successfully demonstrated the effectiveness of the treatment process and provided information to design a full-scale system. A design for a full-scale application of the technology was prepared in 1993. Regulatory approval for the system, which included removing the groundwater extraction and treatment system, was granted by the CRWQCB in November 1993. Construction of the PRB and slurry wall system occurred between November 1994 and February 1995. Groundwater monitoring is ongoing, and in 1999 the "Five-Year Status and Effectiveness Evaluation Report" was submitted to and approved by the CRWQCB. This final remedy is being maintained and assessment of the progress is ongoing.

Hydrogeologic Conditions

The VOC-affected aquifer lies between approximately 10 and 20 feet below ground surface. The thickness of the aquifer is variable and ranges between approximately 2 and 10 feet. The aquifer is composed of interbedded silt, sand, and clay, generally is semi-confined, and is separated from the next deepest aquifer by 55 to 65 feet of silty clay. The magnitude and direction of the shallow, horizontal hydraulic gradient is approximately 0.004 foot-per-foot northward, and shifts of up to 50 degrees from northwest to northeast have been observed. Groundwater velocity appears to range between 0.5 and 1 ft/day and the magnitude of the vertical gradient is approximately 0.005 foot-per-foot upward. Results of aquifer hydraulic testing suggest an average transmissivity of approximately 375 ft^2/day.

Remedy Construction Details

A discussion of the design and construction of the "Sunnyvale PRB" was first presented at the 1995 American Chemical Society National Meeting. It has been described in Yamane et al. (*1*) and Szerdy et al. (*2*) and consists of the design elements described below (Figure 1):

- an approximately 40-ft-long by 15-ft-deep by 8-ft-wide treatment cell consisting of a 4-ft-wide flow-through zone of 100 percent granular iron placed between 2-ft-wide upgradient and downgradient pea gravel sections. The cell was keyed vertically approximately 2 feet into the top of the underlying aquitard. Based on the results of bench-scale and pilot testing, the 4-ft flow-through thickness of the treatment zone provides the required minimum 2-day groundwater residence time to achieve cleanup standards;
- lateral low-permeability barriers constructed on the east and west sides of the treatment cell extending approximately 300 feet and 225 feet upgradient of the treatment zone respectively. The slurry walls are approximately 22 feet deep, and are keyed about 2 feet into the underlying aquitard material; and
- a lateral hydraulic barrier constructed of steel sheet piles extends approximately 20 feet downgradient (north) of the eastern end of the treatment cell.

The treatment cell was constructed by driving sheet piles, excavating the soil within the sheet piled area, backfilling the trench with granular iron and pea gravel, and extracting the sheet piles. During sheet pile extraction, it is likely that some granular iron mixed with the pea gravel in the upgradient sections of the PRB. As the treatment wall was backfilled, 4 monitoring wells were placed in the granular iron within approximately 1 foot of the downgradient edge (PRB wells). In 1997, 2 additional PRB wells were installed in the center of the granular iron section (2 feet from the downgradient edge). Two piezometers were placed in the upgradient pea gravel section of the PRB (upgradient performance wells), and 3 piezometers were installed approximately 5 to 10 feet downgradient of the PRB (downgradient performance wells).

Performance Monitoring Results

Site groundwater levels are monitored quarterly. More than 20 sampling events have been conducted since 1995 with groundwater samples analyzed for VOCs by gas chromatography. Additional samples have been periodically collected and analyzed for inorganic species, dissolved hydrocarbon gases, and volatile fatty acids. Field water quality parameters (pH, temperature, dissolved oxygen, oxidation-reduction potential [redox], and specific conductance) are measured using a flow-through cell. Groundwater samples are collected from PRB wells using low-flow purging methods. The general results of more than five years of monitoring are summarized in the following key points (Figure 2):

- Site water levels generally follow seasonal regional fluctuations. Local hydraulic mounding is intermittently present within the treatment zone in the rainy winter months and dissipates after the rainy season.
- Concentrations of target VOCs, including VC, are reduced to cleanup standards within the PRB.
- Dissolved oxygen concentrations are generally lower within the treatment zone (less than 0.8 mg/L). The average redox potential of groundwater (Eh ranging from -300 to -500 mV) within the treatment zone is in agreement with expected conditions for this type of treatment process (*3*). pH values within the PRB (pH $\cong$ 11) are higher than in samples from the native aquifer (pH $\cong$ 7.5), while the downgradient pH returns to near neutral (pH $\cong$ 7.8) within approximately 5 feet of the PRB due to buffering by the native aquifer. pH within the treatment zone appears to have increased from 10 to slightly over 11 since installation of the remedy.
- Depressed values of cation and bicarbonate ratios in samples collected from PRB wells are consistent with markedly lower specific conductance values measured in PRB well samples than in upgradient performance well samples. This indicates that mineral precipitation is occurring within the PRB. Sodium and potassium are conserved as loss of other species occurs. These trends are expected, and are consistent with an anticipated upward shift in pH within the treatment zone for groundwater that is saturated with respect to bicarbonate species under ambient conditions. Evidence of mineral precipitation is first observed in the upgradient pea gravel section, thus protecting the 100 percent iron zone from significant loss of porosity.
- Concentrations of dissolved hydrocarbon gases suggest that dechlorination is occurring both within and outside the PRB. Biological activity within the PRB is low relative to upgradient and downgradient groundwater. The dechlorination process within the treatment zone is predominantly abiotic

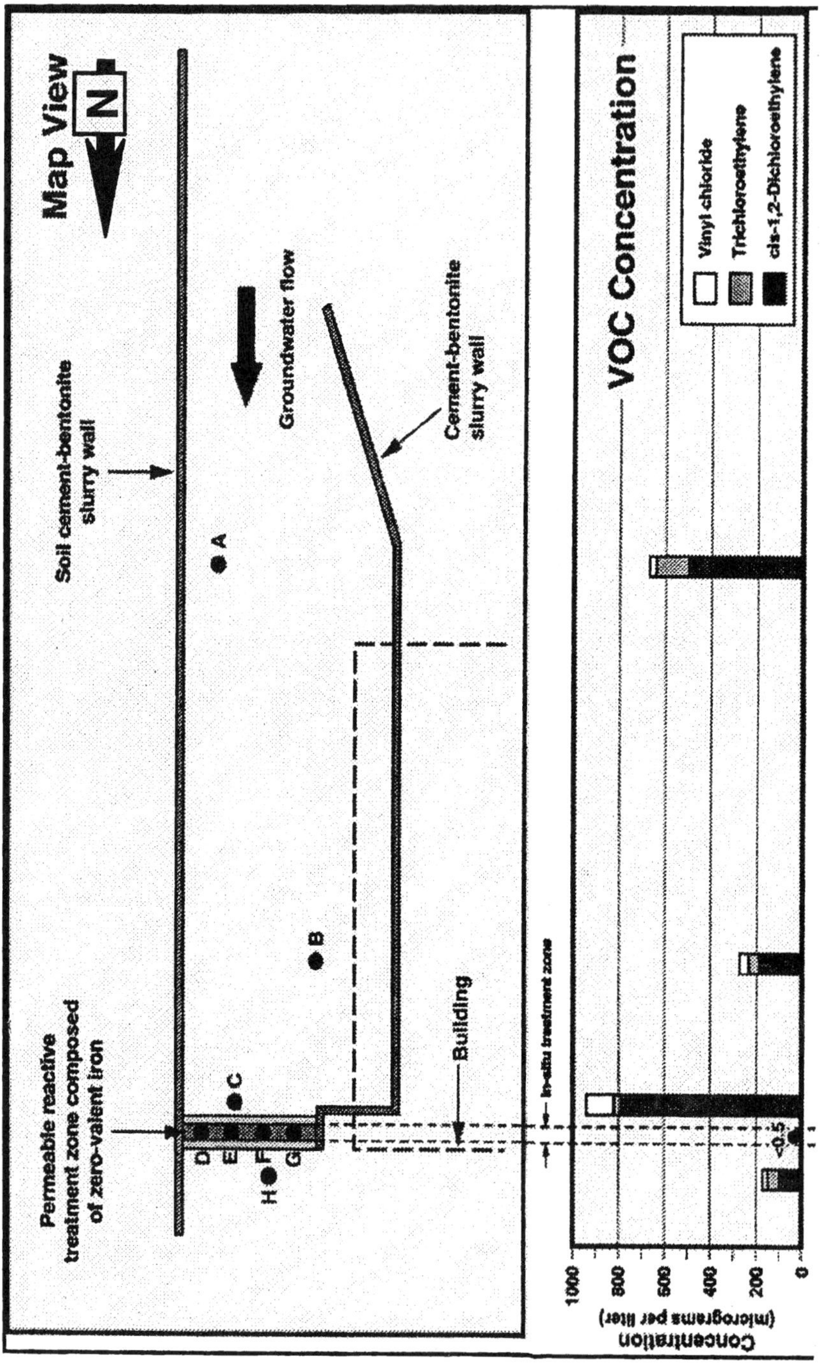
Map View
N
Soil cement-bentonite slurry wall
Groundwater flow
Cement-bentonite slurry wall
A
B
C
D
E
F
G
H
Permeable reactive treatment zone composed of zero-valent iron
Building
In-situ treatment zone
VOC Concentration
Vinyl chloride
Trichloroethylene
cis-1,2-Dichloroethylene
<0.5
Concentration (micrograms per liter)
1000
800
600
400
200
0

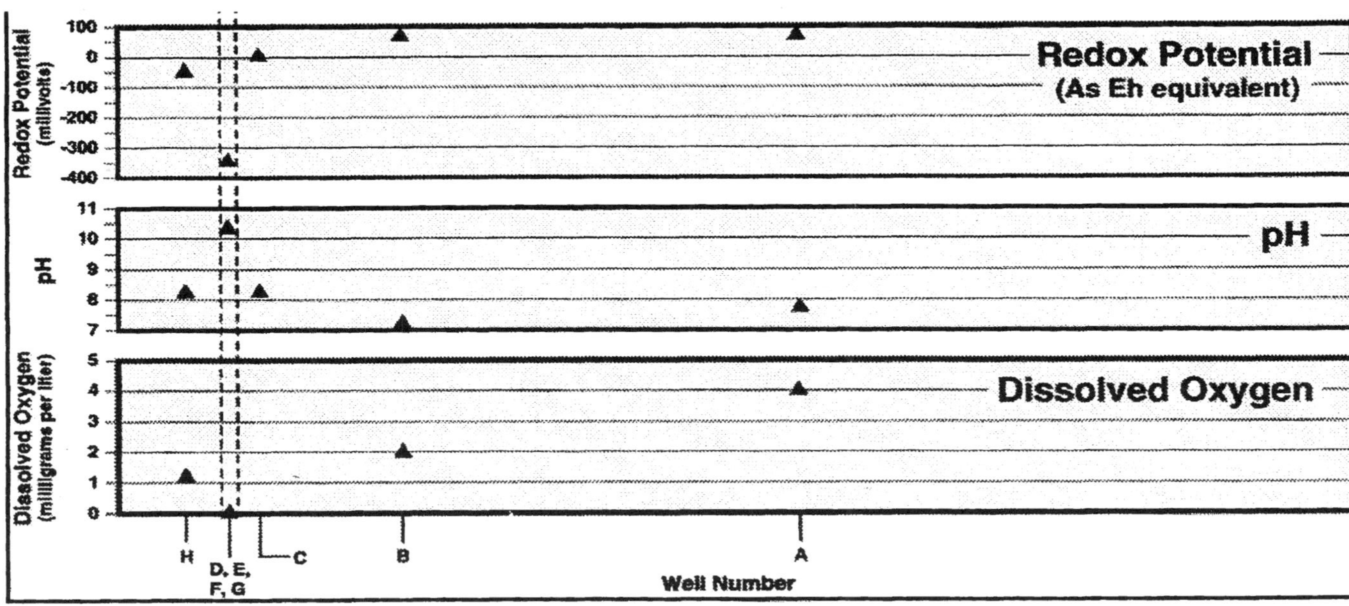

Figure 1. Diagram showing a schematic representation of the PRB and selected monitoring results. (Reproduced with permission from *International Conference on Remediation of Chlorinated and Recalcitrant Compounds;* Battelle Memorial Institute; 1998, p 147. Copyright 1998 Battelle Memorial Institute.)

with no indication of biofouling or excess biological activity that may influence reaction chemistry.

Dissolved Hydrogen Monitoring Results

Hydrogen gas is generated by the process of zero-valent iron corrosion when water is used as the oxidant (*4,5*). Also, others have suggested that hydrogen evolution can impact the hydraulic performance of permeable reactive barriers (PRBs) because of the generation of a gas phase or bacterial growth (*6*). Rather than a completely abiotic system, the dehalogenation observed in a zero-valent iron medium can be sustained by microbial populations using hydrogen as an electron donor (*7*).

Dissolved hydrogen concentrations were measured in groundwater samples collected from monitoring wells installed within and adjacent to the PRB. Measurements of dissolved hydrogen were taken as part of a larger data collection event to provide performance assessment data of the PRB system. For this monitoring event, the bubble-strip method was used to collect the samples, which were analyzed by Microseeps, Inc. of Pittsburgh, Pennsylvania.

Concentrations of dissolved hydrogen in background groundwater samples (i.e., located more than 150 feet from the PRB) ranged from approximately 2 to 15 nanomolar (nM), with higher values measured within an area where natural biodegradation of chemical constituents is suspected to occur. Within the 100 percent zero-valent iron zone of the PRB, dissolved hydrogen concentrations approached the solubility limit of hydrogen (greater than 600,000 nM). Dissolved hydrogen concentrations were approximately 100 nM in samples collected 10 feet upgradient of the PRB; ambient concentrations significantly greater than 30 to 40 nM are unusual in typical groundwater systems. The occurrence of dissolved hydrogen in the near upgradient samples at concentrations 10 times greater than background indicates the presence of a hydrogen "halo" around the granular iron zone.

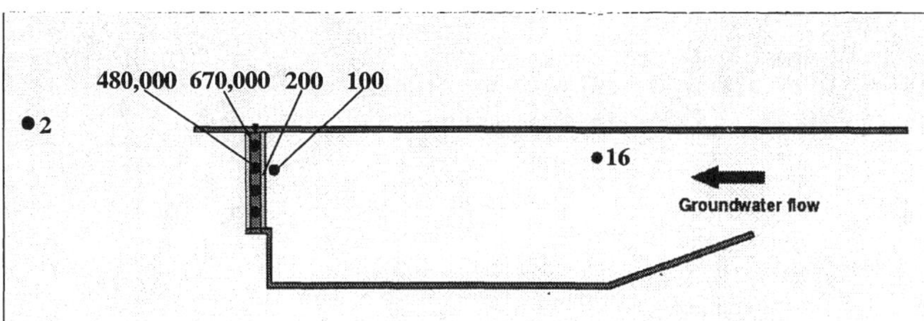

H_2 Solubility = 1.6ppm = 0.02 L/L= 800,000 nM

Figure 2. Dissolved hydrogen gas concentrations in nanomolar (nM)

Conclusion

VOCs are passively removed from site groundwater before flowing downgradient and offsite as indicated by concentrations below groundwater cleanup standards in wells within the treatment zone. Treatment of VOCs is not affected by seasonal water level fluctuations. The effectiveness of the treatment process appears not to have decreased as a result of precipitation of inorganic species in the upgradient pea gravel zone. The distribution of hydrogen concentrations is consistent with the distribution of other general water quality parameters monitored at the site (e.g. redox, oxygen, inorganic parameters).

The high concentrations of dissolved hydrogen within the PRB indicate that the corrosion process remains strong more than six years after PRB installation. Additional data need to be collected to evaluate the usefulness of hydrogen concentration measurements within a PRB system with respect to the evaluation of long-term performance.

References

1. Yamane, C. L.; Warner, S. D.;Gallinatti, J. D.; Szerdy, F. S.; Delfino, T. A. Hankins, D. A.;Vogan, J. L. *209th National Meeting, American Chemical Society*, Anaheim, California, **1995** ,Vol. 35, No. 1, 792-795.
2. Szerdy, F. S.; Gallinatti, J. D.; Warner, S. D.; Yamane, C. L.; Hankins, D. A.; Vogan, J. L. *Non-Aqueous Phase Liquids (NAPLs) in the Subsurface Environment: Envionmental Engineering Division*, ASCE National Convention, American Society for Civil Engineers, Washington, D.C., **1996**; 245-256.
3. Gillham, R.W.; O'Hannesin, S.F. *Ground Water* **1994,** *32* (6), 958-967.
4. Matheson, L. J.; Tratnyek, P. G. *Environ. Sci. Technol.* **1994,** *28*, 2045-2053.
5. Reardon, E. J.; *Environ. Sci. Technol.* **1995**, *29,* 2936-2945.
6. Bokermann C.; Dahmke, A.; Steiof, M. *Proceedings of the Second International Conference on Remediation of Chlorinated and Recalcitrant Compounds*, Monterey, CA, Battelle Press, Columbus, OH, **2000,** 433-439.
7. Weathers, L. J.; Parkin, G. F.; Alvarez, P.J. *Environ. Sci. Technol.* **1997,** 31, 880-885.

Chapter 19

Using Ultrasound for Restoring Iron Activity in Permeable Reactive Barriers

Cherie L. Geiger[1], Christian A. Clausen[1], Debra R. Reinhart[2], Christina M. Clausen[1], Nancy Ruiz[3], and Jacqueline Quinn[4]

Departments of [1]Chemistry and [2]Civil and Environmental Engineering, University of Central Florida, 4000 Central Florida Boulevard, Orlando, FL 32816–2366 (fax: 407–823–2252; cgeiger@mail.ucf.edu)
[3]GeoSyntec Consultants, 2100 Main Street, Suite 150, Huntington Beach, CA 92648
[4]NASA, Kennedy Space Center, Kennedy Space Center, FL 32899

Ultrasound application has been shown to be an effective treatment for increasing reactivity of iron used in permeable reactive barriers (PRBs). Since iron goes through corrosion reactions during the remediation process, precipitates form on the iron surface thus leaving the permeable reactive barrier less effective and can even alter groundwater flow around the wall. After several years of laboratory study, ultrasound was used at two different PRB field sites. Results show that ultrasound application improves half-lives from 21-67% (dependent on transducer power and duration of exposure) over untreated iron samples from the same PRB treatment site.

Introduction

Chlorinated organic compounds, and trichloroethene (TCE) in particular, may be the most difficult and costly challenge to groundwater remediation due to low drinking water regulatory limits and widespread occurrence. Zero-valent transition metals, particularly iron, present a unique opportunity to provide passive, in situ treatment to degrade chlorinated organics rather than simply transfer them from the subsurface to another medium (1,2). In light of the potential for passive remediation afforded by permeable iron barriers (3), further study is imperative to explore their merits and minimize their deficiencies. Degradation rates increase with the degree of chlorination, increasing iron mass to solution volume ratio, and increasing iron surface area to mass ratio (4,5). Besides iron surface area, factors influencing rates include the treatment that the iron receives prior to use, and the oxygen level and buffering capacity of the sample solution (6).

Reduction in dechlorination rates and flow problems have been linked to deposition of material, mainly carbonates in highly alkaline water, on the iron surface (7-9). Bridging between iron particles can result in plugged pore spaces at the upstream face of a barrier and interfere with the hydraulics of a remediation system (10). In an aquifer with significant alkalinity, and especially under even moderately aerobic conditions, eventual blinding of the iron surface and reduced reactivity will occur at the upstream portion of the permeable barrier, where the greatest contaminant mass transfer takes place. A technology that can maintain or restore wall performance is needed and the use of ultrasound for this purpose may be applicable.

Ultrasonics refers to periodic stress waves that occur at frequencies in excess of 20 kHz. At their upper extreme, ultrasonic frequencies are so high that their extremely short wavelengths are comparable to the agitation of molecules caused by heat (11). Effects produced by ultrasonic waves arise from cavitation, which occur in those regions of a liquid which are subjected to rapidly alternating pressures of high amplitude. These high pressures are relieved by the radiation of intense shock waves (12). High energy ultrasound has been observed to degrade organic compounds (13,14), suggesting a potential remedial approach for contaminated waters. However, its more common application as a cleaning process may provide an opportunity to enhance a zero-valent iron wall through surface maintenance. Acid-washed nickel particles have been observed to be several hundred-fold times more active in hydrogenation reactions after being sonicated (15). Improved efficiency and increased activity due to surface changes created by sonication have also been noted in lithium (16), zinc (17), and copper catalysts (18).

Although ultrasound is normally used in liquid mediums, it has been found that ultrasonic energy can pass through saturated particulate. Experiments have been reported that show ultrasonic energy can pass through water saturated soil in energy levels adequate to clean an iron surface to a distance of 54-cm (19). These findings and the results of the work presented in this paper show that there is reason to pursue ultrasound as a methodology for restoring the efficiency of iron containing PRBs.

The objective of this research was to investigate the application of ultrasonic energy to rejuvenate iron surfaces in a PRB with the goal of enhancing/restoring the rate of trichloroethylene (TCE) degradation.

Extensive batch studies were conducted to examine the impact of ultrasound on iron under various conditions. The effect of ultrasound on TCE degradation (no iron) was initially studied to determine if there was any destruction from the ultrasound treatment itself. Variables that were examined include the following: the type of iron (100-mesh or coarse iron filings), ultrasound treatment and length of use, and the condition of the iron (acid-washed, unwashed). Samples were monitored for concentrations of TCE, dichloroethylene (DCE) isomers, vinyl chloride and non-chlorinated, two-carbon compounds using a purge-and-trap concentrator and GC/FID. Rate constants were determined both with and without ultrasound application.

In addition to laboratory studies, a research PRB was installed at Cape Canaveral Air Station , Florida to test the effectiveness of ultrasound in field use. The field site is located at Launch Complex 34 (LC34), Building 21900H, and is commonly referred to as the Engineering Support Building (ESB), was used to clean rocket parts (primarily with TCE) during the 1960s. Recent estimates indicate that over 20,000 kg of TCE remain beneath this building as DNAPL and continue to function as a contaminant source, fueling a greater than 1-km^2 plume. The PRB placed at this site was not intended to treat the entire plume but was instead intended for a research project testing the construction technique and the use of ultrasound. Several observation and monitoring wells (10-cm diameter) were placed in front of, inside, and behind the PRB. Two different ultrasound transducers have been used at this site: 4.4-cm diameter, 0.61-m length, 1000-W, 40 kHz and 9.50-cm diameter, 1.22-m length, 3000 W, 25 kHz.

A second field site was treated with ultrasound in July, 2000. The same transducers were used. Two 10-cm wells were installed to accommodate the 25-kHz transducer and the 40-kHz transducer was used in an existing 5-cm well. Experiments to determine the effectiveness of treatment were carried out in an identical manner as at the Canaveral Air Station PRB and yielded similar results.

Methodology

Laboratory Experiments

Iron was used in either an unwashed (as received state) or acid-washed condition. The procedure for acid washing the iron is as follows: approximately 50 mL of 5% sulfuric acid solution were added to five grams of dry, 100-mesh or Peerless iron. The mixture was allowed to stand for 10 minutes under periodic shaking and the acid solution was decanted. The resulting "wet" iron was washed repeatedly with de-ionized water to ensure that no acid remained. In the experiments using unwashed iron, the iron was simply used as received from the manufacturer. The iron was transferred to a 1.0 L Tedlar® bag and sealed. Five hundred mL of natural ground water was added to the bag through the septum/port. The bag was then purged for 30 minutes with nitrogen gas. All remaining headspace in the bag was evacuated.

For experiments that included ultrasound exposure, bags were then placed in a 600-W, 20-kHz Branson ultrasonic bath with an ultrasonic intensity of 0.32 W/cm^2. Ultrasound was applied for periods of 30 minutes to 3 hours, depending on the experiment. To minimize any temperature effects, water flowed continuously through the bath during use. The ultrasonic energy used was not sufficient to degrade the organic compounds under investigation.

TCE/methanol solution was injected via syringe, through the septum, to obtain approximately 10-mg/L solution of TCE in water. The bag was placed on a shaker table for 15 minutes and a 13-mL sample was removed using a 20.0-mL gas-tight syringe. Subsequent 13-mL samples were taken at random time intervals for the next 14 days in an attempt to establish a TCE disappearance rate. Analytical data was gathered using a Tekmar LSC2 purge and trap concentrator with a 10-position autosampler. A gas chromatograph (Hewlett Packard 5890) with a flame ionization detector was used for identification and quantification. The column used was a DB624 (0.25 i.d. X 60 m) purchased from Supelco.

Groundwater Chemistry

Major physical and chemical properties for the natural groundwater used in this study are summarized in Table I. "Field" indicates parameters which were measured with a Corning Checkmate M90 field instrument with interchangeable probes. The groundwater is distinguished by significant hardness (444mg/L as $CaCO_3$) and alkalinity (289 mg/L as $CaCO_3$).

TABLE I. Major physical and chemical parameters of groundwater from East Coast of Central Florida.

Parameter	*Value*	*Units*	*Method*
pH	7.32	---	Field
Dissolved Oxygen	0.1	mg/L	Field
Conductivity	1923	μS	Field
Ca	81.4	mg/L	EPA 6010
Mg	57.7	mg/L	EPA 6010
Fe	0.059	mg/L	EPA 6010
Hardness	444	mg/L as $CaCO_3$	EPA 130
SO_4^{2-}	71	mg/L	EPA 300
Total Alkalinity	289	mg/L as $CaCO_3$	EPA 310.1
Total Dissolved Solids	1100	mg/L	EPA 160.1

Organic Analysis

Liquid samples were analyzed for TCE and daughter products (dichloroethene isomers, vinyl chloride, and ethene) following EPA Method 624. A five-mL portion was injected with 1.0 μL of internal standard (5000 mg/L bromochloromethane in methanol) then transferred to a purge vial. Helium was bubbled through the vial at 22°C for 11 minutes to transfer the TCE onto a Vocarb 3000 trap. The trap desorb time was four minutes at 250°C and the bake time was seven minutes at 260°C. A Hewlett-Packard gas chromatograph Model 5890, equipped with a flame-ionization detector (FID) and a 0.25-mm i.d., 60-m long Vocol capillary column, was programmed for a three-minute hold at 60°C, a 15°C/min ramp rate to 180°C, and a three-minute hold at this final temperature.

Gas samples were analyzed using a Perkin Elmer 5800 GC equipped with an FID. The column used was a 10% carbowax packed column, 6-ft. length with 1/8-inch diameter, purchased through Supelco. Oven temperature was isothermal at 35°C for 30 minutes and the nitrogen flow rate was 10.0 mL/minute.

Scanning Electron Micrographs (SEM)

The morphology of the iron surface and deposits before and after sonication were examined via SEM. Samples were prepared with a gold-palladium mixture applied using a Hummer VI-A Sputtering System (EM Corporation, Chestnut

Hill, MA) with an argon plasma and then observed with an Amray 1810 Scanning Electron Microscope (Bedford, MA).

Specific Surface Area

Iron powder and filings were analyzed for specific surface area before and after sonication. Samples were rinsed with acetone and dried with nitrogen. To prevent oxidation, sonicated and acid-washed samples were stored under nitrogen in screw-cap glass vials. Measurements were taken using a Porous Materials Inc. (PMI) Brunauer-Emmett Teller (BET) Sorptometer Model 201 (Ithaca, NY) using nitrogen as the adsorbate.

Column Studies

Column studies were conducted to better approximate the effects of ultrasound in a soil matrix before the field demonstrations were attempted. Columns were constructed of 1.0-m length, 10.0-cm inside diameter acrylic tubing with eight sample ports installed along the length of the column. Flow through the columns was vertical up-flow with a residence time of 12-hrs. The column material was 20% Peerless iron flakes and 80% native soil obtained from the East Coast of Central Florida.

A sonication probe was installed at the entrance end of the column and it extended approximately 30.5-cm into the column material. After obvious oxidation of the iron in the column, ultrasound was applied for a period of approximately 30-minutes. Kinetic data was acquired over the life of the column.

Field Demonstrations

Ultrasound was applied by use of a submergible 1,000-W or 3,000-W resonator that could be lowered into the wells (5.0 or 10.0-cm diameter) and held at various depths for the intended treatment period. Core samples were taken (within 30-cm of the well) before and after treatment and the iron was removed from these samples under a nitrogen atmosphere. Measured amounts of iron were then placed in zero-head space vials. Deoxygenated, deionized water was added leaving 5-mL of headspace. The vials were then closed with crimp-top septum caps and spiked with a TCE/methanol solution to yield concentrations of 10 mg/L. The vials were covered and placed on a shaker table. Headspace analysis was performed (using equipment described above) and half-lives for TCE disappearance were calculated for samples that had been exposed to ultrasound and for samples taken prior to ultrasound treatment.

Results and Discussion

Effect of Ultrasound on TCE Concentration

In order to study the effects of ultrasound on the enhancement of iron for the degradation of TCE, the impact of ultrasound on TCE alone also had to be determined. A pair of iron blank bags were prepared and then exposed to ultrasound for up to three hours with samples taken at 30 minutes, one, two, and three hours. The only by-product formation observed was 1,1-DCE from the analysis of the samples taken at three hours. The concentration was only 1.0-ug/L, which is also the detection limit of the analysis. While other researchers have used ultrasound at higher energy levels to degrade chlorinated organics (14, 20), that effect was not otherwise observed in this experiment due to the limited application of ultrasonic energy. The disappearance of TCE during ultrasound was attributed to volatilization and this hypothesis is supported by the formation of a small (approximately 170-200 μL) gas bubble in the bags.

The experiment was continued for two weeks with no further ultrasound treatment to determine if there was any loss of TCE through the bag itself. During this time, the loss of TCE from the Tedlar® bags was minimal with 98% of the TCE accounted for (aqueous concentration and mass in gaseous state) when the experiment was over indicating that the bags were not significantly losing TCE under the conditions used for this work.

First-Order Rate Constant Impact

The optimum ultrasound exposure time for iron was determined by exposing different sets of bags (loaded with de-oxygenated water and iron filings with no headspace) to 0.5 hours, one hour, two hours and three hours of ultrasound treatment prior to spiking with TCE. TCE was then added and the bags were allowed to equilibrate for 24-hours before rate studies were performed. It was found that even 0.5 hours of ultrasound exposure yields some increase in TCE degradation rate constants. A significant increase in rate constants occurred when the duration of ultrasound exposure was increased to one hour. Beyond one hour of exposure the increase in TCE degradation rate constant increased only slightly (Figure 1). Based on this study, it was concluded that the major cleaning effects of ultrasound take place within the first hour (at this energy level). It was decided that ultrasound treatments of two hours would be sufficient and treatment beyond that point could not be justified for the purposes of these studies.

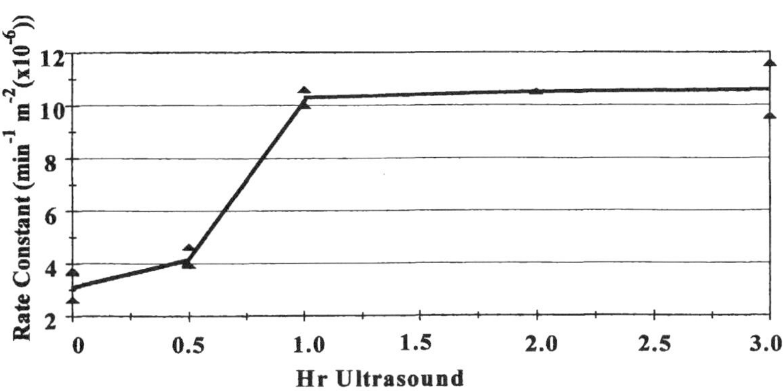

Figure 1. Effect of duration of ultrasound treatment on rate constants for the disappearance of TCE.

The finding that even 30-minutes of ultrasound treatment improves TCE degradation rate constants supports previous work describing possible mechanisms of electron transfer through thin oxide layers. The unwashed iron filings are heavily coated with oxide (this observation based on the orange-brown powdery coating on the particles) which acts to passivate the dehalogenation reaction at the surface. Brief periods of even low energy ultrasound serve to knock away some of the more loosely held debris. This exposes Fe(0) through pitting (causing discreet increases in dehalogenation reactions) and a thin oxide layer which has been shown to allow for electron transfer reactions to occur (21-23).

Table II summarizes average first-order rate constants, specific surface areas and the ratio of sonicated to unsonicated surface areas. Rate constants were calculated after 24 hours of exposure to TCE to allow equilibration with the iron surface. All correlation coefficients for the linear, first-order fit used for the calculation of rate constants exceeded 0.99.

Table II. Comparison of first order rate constants ($hr^{-1} \bullet g\ Fe^{-1}(x\ 10^{-3})$) and surface area ($m^2/g$) before and after US for iron under different conditions.

Iron	*Condition*	*No US*	*US*	*No US*	*US*	*%change*
		k_{obs}, 1st Order Rate Constants		Surface Area (m^2/g)		
100-mesh	Unwashed	3.75	5.00	3.09	4.12	+33.3
	Acid-washed	6.42	6.54	4.28	5.40	+26.2
	Oxidized	8.46	14.40	49.16	22.33	-54.6
Iron Filings	Unwashed	2.70	3.36	0.77	2.07	+169
	Acid-washed	4.74	4.68	1.92	2.91	+51.6
	Oxidized	6.96	12.36	24.29	14.53	-40.2

Exposure to ultrasonic energy generated increased first-order degradation rate constants with few exceptions. Sonication of acid-washed iron, both powder and filings, yielded rate constants within two percent of their unsonicated baseline rate constants, statistically insignificant. Exposure at this energy level and duration may not provide additional cleaning to an acid-washed surface. The first-order rate constant for unwashed filings increased approximately 24 percent after sonication. This post-sonication increase strongly suggests that iron with an occluded surface can benefit from exposure to ultrasonic energy.

The high rate constants for oxidized irons relative to their acid-washed counterparts seemed unreasonable, considering that the oxidized surface should be passivated. Because acid-washed and oxidized iron bags were operated under the same temperature and mixing conditions, there was no reason to expect that volatilization phenomena could lead to any observed differences. The stable TCE concentrations observed for control bags also show that volatilization was not an issue. To gain insight into this situation, the concentration of total degradation products for the acid-washed and oxidized iron were examined over time.

Scherer, et al. (24) summarized three possible modes of reduction reactions at the interface between solid (oxide or metal/oxide) and the solution phase: localized defects, resonance tunneling or complexation. Because the oxidized iron may feature areas where any of these three mechanisms could occur, it was expected that some degradation products would be observed. However, the oxide layer on intentionally oxidized iron was relatively thick yielding a surface with many layers of overlapping oxide, far greater than the approximately 10 Å thickness suggested as a limit for direct tunneling of electrons (24).

Acid-washed filings generated degradation product concentrations as high as 9 mg/L while oxidized filings generally yielded stable concentrations of no more than 1 mg/L. Since the first-order rate constant for the disappearance of TCE with oxidized filings was 46.8 percent higher than its counterpart for acid-washed filings yet the concentration of by-products was much lower for the oxidized iron system, it was hypothesized that another physical phenomenon was taking place. These findings suggest that for oxidized iron, TCE disappearance reflects sorption with little accompanying degradation to less chlorinated by-products.

The role of sorption to unreactive sites is not entirely understood. Burris, et al. (25) performed experiments to determine the population and effects of non-reactive sites on the iron surface. His work suggested that most of the iron surface is unreactive sites and that it is unclear what happens to the TCE that sorbs to them. Allen-King et al. (26) published similar work only with the 1,2-DCE isomers. Both articles infer that the rates of sorption of individual compounds can be used to adjust degradation rate constants to a more accurate representation of the chemical reactions taking place at the surface.

Non-chlorinated Fraction of Products

Because of the capacity of ultrasound to activate catalyst surfaces (16-18) and the observed increases in rate constants for most iron samples, the question arose whether sonication changes the degradation process or simply increases the rate. For acid-washed 100-mesh iron and acid washed filings, sonication increased the non-chlorinated fraction of by-products about five percent relative to unsonicated iron; however, there was only a slight difference found in post-sonication versus pre-sonication rate constants (Table II). A more striking contrast was observed for unwashed filings where the non-chlorinated fraction of by-products increased by about 10 percent. The divergence in two-carbon product composition due to sonication was noted as early as one half-life of TCE loss.

In the case of acid-washed iron, the rate of TCE disappearance remained almost constant regardless of sonication history but the percentage of non-chlorinated by-products increased from 5-7%. If sonication were only creating more active sites then the rate of TCE disappearance would have increased with no effect on the speciation of the by-products. The increase in the fraction of non-chlorinated by-products can be attributed to a change in the interaction between TCE (or its by-products) and the active sites on the solid surface. TCE molecules must come into contact and are more strongly sorbed to ultrasound generated active sites than non-ultrasound generated sites (as opposed to desorbing back into solution), and therefore continue through the mechanism (mechanistic pathway suggested by Roberts, et al. (27)) to complete dechlorination.

If the above hypothesis is true and active sites are held by single molecules through complete dechlorination (with prior ultrasound treatment) then fewer TCE molecules would come in contact with active sites per unit of time. Therefore, these data suggest an increase in both number of active sites and increased affinity to the chlorinated solvent and chlorinated by-products.

Considering that the goal of this technology is to completely de-chlorinate the parent compound, these data represent an important benefit of ultrasound treatment. The use of ultrasound can decrease the time necessary to convert the parent compound to fully de-chlorinated end products. Thus, exposure to ultrasound makes the degradation process more efficient as well as more rapid.

Specific Surface Area

The unwashed and acid-washed filings and powder, both before and after sonication, exhibited specific surface areas within the range reported in the literature (28), 0.05 to 4.0 m^2/g (Table II). The specific surface areas for oxidized and groundwater-soaked irons exceed these values, but information was not found in the technical literature for comparison. Sonicated unwashed samples of both the powder and the filings produced specific surface areas within an average of six percent of those measured for their unsonicated acid-washed counterparts.

In the case of acid-washed iron, increases in specific surface area appear related to increases in first-order rate constants for TCE disappearance. This occurrence was independent of sonication history. For unsonicated filings, acid-washed iron exhibited a specific surface area 2.5 times and a rate constant 1.8 times that of unwashed iron. For sonicated filings, acid-washed iron exhibited a specific surface area 1.4 times and a rate constant 1.4 times that of unwashed iron.

However, increases in specific surface area due to sonication did not always produce similar increases in rate constants. For unwashed filings, the specific surface area after sonication was 2.7 times that of unsonicated iron, but the rate constant was 1.2 times that of unsonicated iron. For acid-washed filings, the sonicated specific surface area was 1.5 times that of unsonicated iron, but the rate constant was essentially equal to that of unsonicated iron. Unsonicated oxidized filings exhibited a specific surface area 12.6 times that of unsonicated acid-washed filings. Sonicated oxidized filings yielded a specific surface area five times that of sonicated acid-washed filings. However, as discussed previously, the rate constants measured for oxidized iron undoubtedly represent sorption rather than degradation. Iron surface quality, and not merely specific surface area, is crucial in determining the suitability of an iron for degrading chlorinated solvents.

Scanning Electron Micrographs

The work presented in this paper correlates reduction in reaction rate constants with increased Fe surface obstruction. Other researchers (8) have also used SEMs for visual inspection of iron samples and to show the extent of corrosion products and debris at the water/Fe surface interface.

In order to determine if ultrasound would be effective when used on material corroded by the groundwater (deoxygenated) environment or simple oxidation from air, iron was exposed to these conditions and then treated with ultrasound. Unwashed iron filings that have been soaked in natural groundwater for several months are shown in Figures 2a and 2b before and after exposure to ultrasound. Before ultrasound treatment (Figure 2a), the surface is almost completely occluded with corrosion products and debris of various sizes. After ultrasound treatment (Figure 2b) the surface is noticeably cleaner with little surface debris. The SEMs shown illustrate the effects that natural groundwater can have on iron in the subsurface.

Intentionally oxidized iron filings (both before and after ultrasound) are shown in Figures 2c-d. The iron shows heavy surface debris before ultrasound treatment (Figures 2c and 2d respectively) with large accumulations of oxidation products apparent on the surface. After exposure to ultrasound, the surface is noticeably less obstructed and some fine cracking of the oxide surface is apparent. The cracks in the surface present a new pathway for electron transfer through diffusion of the aqueous solution to the Fe^0 surface as well as more Fe^{+2} diffusion from the bottom of the pit (24).

SEMs were also taken of 100-mesh iron with similar results. Heavily corroded iron is most likely to have apparent pitting and cracks in the surface after exposure to ultrasound. Since the data shown suggests that iron with an obstructed surface area will have lower rate constants, a technology such as ultrasound that can clean the surface in situ may be a necessary component of treatment walls under certain groundwater conditions.

Column Studies

Kinetics of the TCE degradation in the column gradually decreased as observed by the increasing half-life over time (Figure 3). A significant increase in half-life was observed when oxygen was allowed to enter the column via the solution entrance at the bottom of the column. In addition to the increase in half-life, orange coloring of the iron (throughout the length) was visible through the acrylic column. Ultrasound was applied to the column through a transducer inserted into the bottom of the column that extended 30.5-cm up the column length through the center of the column (axial). After ultrasound application, the half-life returned to near original values.

During the application of ultrasound, the orange color of the column matrix material previously visible disappeared and the effluent from the top of the column turned orange with obvious fine particulate. The pressure drop across the column that had built up with the corrosion process declined through the ultrasound application. No plugging occurred during the ultrasound application process. Although the particulate was not saved and analyzed, it was deduced that it was a form of iron oxide (due to the presence of oxygen in the column). The ultrasound application removed the oxide from the iron surface in small enough particles that it posed no problems with plugging of the column material.

Field Application of Ultrasound

The first field site where ultrasound treatment was demonstrated was the Canaveral Air Station, Launch Complex-34 PRB. The construction of this PRB allowed for vertical core sampling of the barrier material so samples were

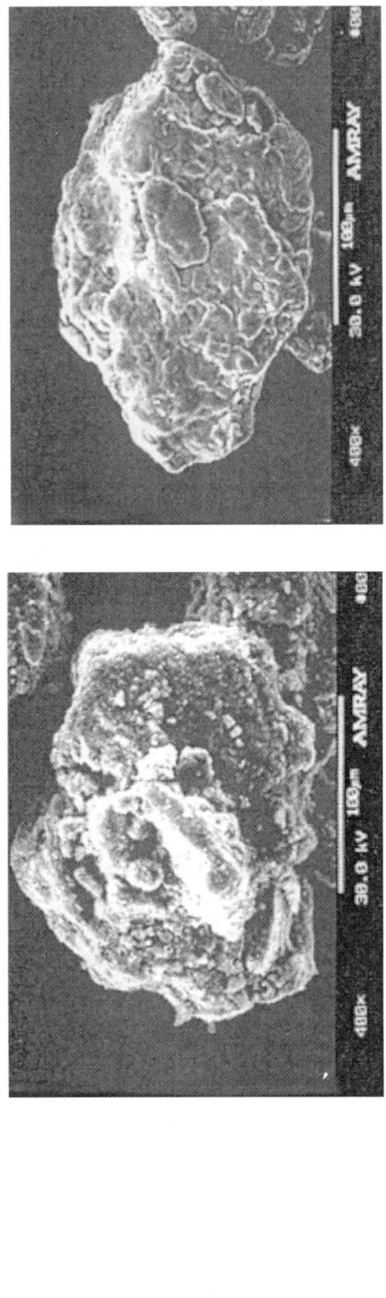

(a) (b)

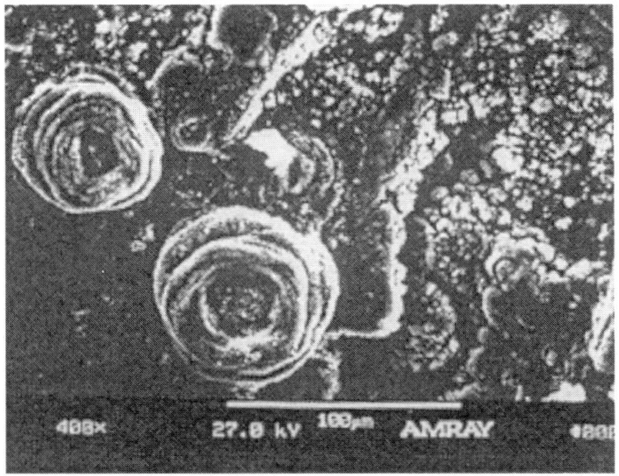

(c)

(d)

Figures 2 (a-d). SEMs iron filings before and after sonication:
(a) iron filings soaked in natural ground water for three months and (b) after sonication for two-hours, (c) air oxidized iron filings and (d) after sonication for two-hours.

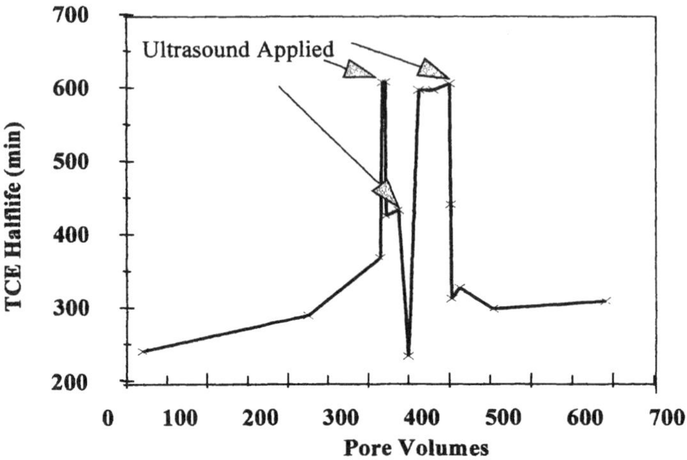

Figure 3. Effects of ultrasound on TCE disappearance half-llife in column experiments. Sharp spikes in half-life indicate points where column material was exposed to oxygen.

obtained from the entire depth of the PRB. As described earlier, core samples were taken prior to ultrasound treatment and after treatment within 30-cm of the monitoring wells where ultrasound was applied. Time of ultrasound exposure and power of the submergible transducer were both examined and compared to determine maximum effectiveness.

Iron taken from all depths showed an improvement after field application of ultrasound. Table III shows percentage improvement for TCE degradation half-lives after ultrasound application (as compared to samples taken pre-ultrasound) at various depths of treatment. Half-lives improved from 21 to 67 percent compared to samples taken before ultrasound treatment. These data show that exposure time and power used are both important aspects of treatment. The lower power unit (1,000-W, 4.40-cm diameter) did not yield as great improvement as did the larger, higher power unit (3,000-W, 9.50-cm diameter). However, many PRBs have 5.0-cm monitoring wells in place and the smaller unit would not require further installation of wells and the improvement shown may be all that is required to bring PRB performance up to the necessary level.

The second PRB (funnel and gate system) used for field demonstration was constructed with a four-foot thick bentonite cap over the reactive media. This was only of concern in that diagonal sample coring had to be used to retrieve samples of iron. Coring was accomplished at 25° from the horizontal so samples taken before and after ultrasound application within 33-cm from the well center were used for analysis of rate constants. The time of application for both units was 90-minutes at each depth.

Table III. Rate constant before and after ultrasound for field experiments.

Field Site	*Relative Depth*	*30-min at 40 kHz*	*90-min at 40 kHz*	*30-min at 25 kHz*	*90-min at 25 kHz*
		Percentage Improvement of Half-Life Compared to Unsonicated Samples			
1	7-12 ft.	24	41	58	67
1	13-18 ft.	11	28	41	66
1	18-26 ft	21	33	45	59
2		N/A	40	N/A	64

The half-lives for iron taken within the site of the 40-kHz (1,000-W, 4.40-cm diameter) transducer application improved an average of 40 % over the samples not exposed to ultrasound. The half-lives for iron taken within the site of the 25-kHz (3,000-W, 9.50-cm diameter) transducer application improved an average of 64 % over the samples not exposed to ultrasound. Again, the higher power transducer yielded higher increases in iron activity but the lower power application yielded significant changes in half-life that may provide enough improvement if the PRB reactive iron is not corroded too extensively.

A project is currently underway at the Launch Complex-34 PRB at Canaveral Air Station, Florida to further study the longevity of the sonication affects. One of the goals of the project is to accelerate the natural aging of the PRB through increased ground water flow (via pumping) and to monitor the duration of the improvements achieved through the use of ultrasound.

CONCLUSIONS

The reaction of TCE with iron has been shown to be dependent upon active surface area. As corrosion products and precipitates build up on the surface, the reaction with TCE is diminished because the active surface area is decreased. Ultrasound has been demonstrated to remove corrosion products, precipitates and other debris from the iron surface and (verified by SEM) causes some pitting and cracking of oxidized iron surfaces. This serves to increase the active iron surface and provide modes for electron transfer to take place and therefore, improves the rate of the dehalogenation reaction. Laboratory studies have also shown that exposure to ultrasound increases the production of non-chlorinated by-products of the dehalogenation reaction.

Field demonstrations have shown that ultrasound also works through saturated subsurface media. Decreases in half-life are significant and are associated with time of exposure to ultrasound and power output of the transducer used. The larger, higher power transducer yields greater improvements of half-life but the lower powered unit can be used in smaller wells and still provides significant changes in iron activity. Using either transducer, the application process is low-cost, requires little time and is easily mobilized.

References

1. Gillham, R. W. and S. F. O'Hannesin. (1994). *Ground Water*, 32, 958-967.
2. Matheson, L. J. and P. G. Tratnyek. (1994). *Environmental Science and Technology*, 28, 2045-2053.
3. Gillham, R. W. and D. R. Burris. (1992). Recent Developments in Permeable In Situ Treatment Walls for Remediation of Contaminated Groundwater. Proceedings from Subsurface Restoration Conference, the Third International Conference on Ground Water Quality, Dallas TX, June 21-24.
4. Gillham, R. W. and S. F. O'Hannesin. (1992). Metal-Catalyzed Abiotic Degradation of Halogenated Organic Compounds. Presentation at Modern Trends in Hydrogeology; the Conference of the Canadian National Chapter, International Association of Hydrogeologists; Hamilton, Ontario, Canada, May 11-13.
5. Vogan, J. L., J. K. Seaberg, B. Gnabasik, and S. O'Hannesin. (1994). Evaluation of In situ Groundwater Remediation by Metal Enhanced Reductive -Dehalogenation - Laboratory Column Studies and Groundwater Flow Modeling. Presented at the 87th Annual Meeting and Exhibition of the Air and Waste Management Association, Cincinnati OH, June 19-24.
6. Sivavec, T. M. and D. P. Horney. (1995). Reductive Dechlorination of Chlorinated Ethenes by Iron Metal. Preprinted extended abstract, presented before the Division of Environmental Chemistry, American Chemical Society, Anaheim, CA, April 2-7.
7. Agrawal, A., P. G. Tratnyek, P. Stoffyn-Egli, and L. Liang. (1995). "Processed Affecting Nitro Reductions by Iron Metal: Mineralogical Consequences of Precipitation in Aqueous Carbonates Environments," preprinted extended abstract, presented before the Division of Environmental Chemistry, American Chemical Society, Anaheim, CA, April 2-7.
8. Johnson, T. L. and P. G. Tratnyek. (1995). Dechlorination of Carbon Tetrachloride by Iron Metal: the Role of Competing Corrosion Reactions. Preprinted extended abstract, presented before the Division of Environmental Chemistry, American Chemical Society, Anaheim, CA, April 2-7.

9. Mackenzie, P. D., D. P. Horney and T. M. Sivavec. (1999). *Journal of Hazardous Materials*, 68, 1-17.
10. Geiger, C. L., C. A. Clausen, D. R. Reinhart, N. Ruiz, A. S. Sonawane and J. Quinn.The Use of Ultrasound to Restore the Dehalogenation Activity of Iron in Permeable Reactive Barriers. Proceedings of the 2001 International Containment & Remediation Technology Conference, Orlando, Fl. June 23-26, 2001.
11. Heuter, T. F. and R. H. Bolt. (1955). Sonics. New York:Wiley.
12. Brown, B. and J. E. Goodman. (1965). High Intensity Ultrasonics. London, Iliffe Books: Princeton, N.J.
13. Bhatnagar, A. and H. M. Cheung. (1994). *Environmental Science and Technology*, 28, 1481-1486.
14. Hoffmann, M. R., I. Hua, and R. Hochemer. (1996). *Ultrasonics Sonochemistry*, 3, S163-S172.
15. Suslick, K.S. and D. J. Casadonle. (1987). *Journal of American Chemical Society*, 109, 3459.
16. Lindley, P. (1990). Sonochemical Aspects of Inorganic and Organometallic Chemistry. In T. J. Mason (Ed.) Chemistry with Ultrasound, Critical Reports on Applied Chemistry Volume 28, Elsevier Science Publishing: New York.
17. Suslick, K. S. and S. J. Doktycz. (1990). Effects of Ultrasound on Surfaces and Solids. In T. J. Mason (Ed.) Advances in Sonochemistry, Vol. 1, (pp. 197-230). JAI Press:London.
18. Lindley, J., T. J. Mason, and J. P. Lorimer. (1987). *Ultrasonics*, 25, 45-48.
19. Cannata, M. A., (1998). Master's Thesis. University of Central Florida.
20. Hung, H. M. and M. R. Hoffmann. (1998). *Environmental Science and Technology*, 32, 3011-3016.
21. Finklea, H. O. (1988). In: Semiconductor electrodes: Finklea, H. O., Ed. Elsheivier, Amsterdam, pp. 1-42.
22. Klausen, J. S. P. Truber, S. B. Haderlein and R. P. Schwarzenbach. (1995). *Environmental Science and Technology,* 29, 2396-2404.
23. Kreigman-King, M. R. and M. Reinhard. (1991) In: Organic Substances and Sediments in Water. R. Baker, Ed. Lewis, MI, Vol. 2, pp. 349-364.
24. Scherer, M. M., B. A. Balko, And P. G. Tratnyek. (1999). The Role of Oxides in Reduction Reactions at the Metal-Water Interface: In Mineral-Water Interfacial Reactions Kinetics and Mechanisms. ACS Symposium Series 715. D. L. Sparks and T. J. Grundle, Eds. American Chemical Society, Washington D. C.
25. Burris, D. R., T. J. Campbell and V. S. Manoranian. (1995). *Environmental Science and Technology,* 29, 2850-2855.
26. Allen-King, R. M., R. M. Halket and D. R. Burris.(1997) *Environmental Toxicology and Chemistry*. Vol. 16, 3, 424-429.
27. Roberts, A. L., L. A. Totten, W. A. Arnold, D. R. Burris, T. J. Campbell. (1996) *Environmental Science and Technology,* 30,2654-2659.
28. Johnson, T. L., M. M. Scherer, and P. G. Tratnyek. (1996). *Environmental Science and Technology*, 30, 2634-2640.

Author Index

Subject Index

B

C

D

P

R

S